T0400221

Climate Change and Plants

Footprints of Climate Variability on Plant Diversity
Series Editor: Shah Fahad

Climate Change and Plants: Biodiversity, Growth and Interactions
Shah Fahad, Osman Sönmez, Shah Saud, Depeng Wang, Chao Wu, Muhammad Adnan and Veysel Turan

Developing Climate Resilient Crops: Improving Global Food Security and Safety
Shah Fahad, Osman Sönmez, Shah Saud, Depeng Wang, Chao Wu, Muhammad Adnan and Veysel Turan

Sustainable Soil and Land Management and Climate Change
Shah Fahad, Osman Sönmez, Veysel Turan, Muhammad Adnan, Shah Saud, Chao Wu and Depeng Wang

Plant Growth Regulators for Climate-Smart Agriculture
Shah Fahad, Osman Sönmez, Veysel Turan, Muhammad Adnan, Shah Saud, Chao Wu and Depeng Wang

Climate Change and Plants

Biodiversity, Growth and Interactions

Edited by
Shah Fahad, Osman Sönmez, Shah Saud, Depeng Wang,
Chao Wu, Muhammad Adnan and Veysel Turan

CRC Press is an imprint of the
Taylor & Francis Group, an **informa** business

First edition published 2021
by CRC Press
6000 Broken Sound Parkway NW, Suite 300, Boca Raton, FL 33487-2742

and by CRC Press
2 Park Square, Milton Park, Abingdon, Oxon, OX14 4RN

CRC Press is an imprint of Taylor & Francis Group, LLC

Library of Congress Cataloging-in-Publication Data
Names: Fahad, Shah (Assistant professor in agriculture), editor.
Title: Climate change and plants : biodiversity, growth and interactions /
edited by Shah Fahad, Osman Sonmez, Shah Saud, Depeng Wang, Chao Wu,
Muhammad Adnan, and Veysel Turan.
Description: First edition. | Boca Raton : CRC Press, 2021. |
Series: Footprints of climate variability on plant diversity | Includes
bibliographical references and index.
Identifiers: LCCN 2020058419 | ISBN 9780367623272 (hardback) | ISBN
9781003108931 (ebook)
Subjects: LCSH: Vegetation and climate. |
Plant-atmosphere relationships. | Climatic changes.
Classification: LCC QK754.5 .C55 2021 | DDC 581.7/22--dc23
LC record available at https://lccn.loc.gov/2020058419

ISBN: 978-0-367-62327-2 (hbk)
ISBN: 978-0-367-62330-2 (pbk)
ISBN: 978-1-003-10893-1 (ebk)

Typeset in Times New Roman
by MPS Limited, Dehradun

Contents

Acknowledgements

Words are bounded and knowledge is limited in praising ALLAH, the Instant and Sustaining Source of all Mercy and Kindness, and the Sustainer of the Worlds. My greatest and ultimate gratitude is due to ALLAH (Subhanahu wa Taqadus). I thank ALLAH with all my humility for everything that I can think of. His generous blessing and exaltation succeeded my thoughts and caused my ambitions to thrive and to bare the cherished fruit of my modest efforts in the form of this piece of literature from the blooming spring of blossoming knowledge. May ALLAH forgive my failings and weaknesses, strengthen and enliven my faith in HIM, and endow me with knowledge and wisdom. All praise and respect are for Holy Prophet Muhammad *Salle Allah Alleh Wassalam,* the greatest educator, the everlasting source of guidance and knowledge for humanity. He taught the principles of morality and eternal values and enabled us to recognize our Creator. I have a deep sense of obligation to my parents, brothers, sisters, and son. Their unconditional love, care and confidence in my abilities helped me achieve this milestone in my life. For this and much more, I am forever in their debt. It is to them that I dedicate this book. In this arduous time, I also appreciate the patience and serenity of my wife, who brought joy to my life in so many different ways. It is indeed on account of her affections and prayers that I was able to achieve something in my life.

Shah Fahad

About the Editors

Dr. Shah Fahad is an assistant professor in the Department of Agronomy, University of Haripur, Khyber Pakhtunkhwa, Pakistan. He obtained his PhD in agronomy from Huazhong Agriculture University, China, in 2015. After doing his postdoctoral research in agronomy at the Huazhong Agriculture University (2015–2017), he accepted the position of assistant professor at the University of Haripur. He has published over 190 peer-reviewed papers with more than 160 research and 30 review articles on important aspects of climate change, plant physiology and breeding, plant nutrition, plant stress responses and tolerance mechanisms, and exogenous chemical priming-induced abiotic stress tolerance. He has also contributed 20 book chapters to various book editions published by Springer, Wiley-Blackwell and Elsevier. He has edited six book volumes, including this one, published by CRC Press, Springer and Intech Open. He has won the Young Rice International Scientist Award and the Distinguished Scholar Award in 2014 and 2015, respectively. He has worked on and is presently continuing to work on, a wide range of topics including climate change, greenhouse emission gasses, abiotic stresses tolerance, roles of phytohormones and their interactions in abiotic stress responses, heavy metals, and regulation of nutrient transport processes.

Dr. Osman Sönmez is a professor in the Department of Soil Science, Faculty of Agriculture, Erciyes University, Kayseri, Turkey. He obtained his MS and PhD in agronomy from Kansas State University, Manhattan, KS, USA from 1996–2004. In 2014, he accepted the position of associate professor at the University of Erciyes. Since 2014, he has worked in the Department of Soil Science, Faculty of Agriculture at Erciyes University. He has published over 90 peer-reviewed papers and research and review articles on soil pollution, plant physiology and plant nutrition.

Dr. Shah Saud received his PhD in turf grasses (horticulture) from Northeast Agricultural University, Harbin, China. He is currently working as a post-doctorate researcher in the Department of Horticulture, Northeast Agricultural University. Dr. Saud has published over 125 research publications in peer-reviewed journals. He has edited 3 books and written 25 book chapters on important aspects of plant physiology, plant stress responses and environmental problems in relation to agricultural plants. According to Scopus®, Saud's publications have received roughly 2,500 citations with an h-index of 24.

Dr. Depeng Wang completed his PhD in 2016 in the field of agronomy and crop physiology from Huazhong Agriculture University, Wuhan, China. Presently, he is a professor at the College of Life Science, Linyi University, Linyi, China. He is the principal investigator of Crop Genetic Improvement, Physiology & Ecology Center in Linyi University. His current research focuses on crop ecology, physiology and agronomy, with key characteristics associated with high yielding crop, the effect of temperature on crop grain yield and solar radiation utilization, morphological plasticity to agronomic manipulation in leaf dispersion and orientation, and optimal integrated crop management practices for maximizing crop grain yield. Dr. Wang has published over 36 papers in reputed journals.

Dr. Chao Wu engages in field crop cultivation and physiology, and plant phenomics. He completed his PhD in 2016 from Huazhong Agricultural University, Wuhan, China, and completed his post-PhD in 2019 from Nanjing Agricultural University, Nanjing, China. At present, he is an associate research fellow at the Guangxi Institute of Botany, Guangxi Zhuang Autonomous Region, and the Chinese Academy of Sciences, Guilin, China. He chairs the Natural Science Foundation of Jiangsu Province and the Postdoctoral Science Foundation research, and focuses mainly on physiological mechanisms of abiotic-stress tolerance (heat and drought) in crops and medicinal plants.

Dr. Muhammad Adnan is a lecturer in the Department of Agriculture at the University of Swabi (UOS), Pakistan. He completed his PhD (soil fertility and microbiology) from the Department of Soil and Environmental Sciences (SES), University of Agriculture, Peshawar, Pakistan and the Department of Plant, Soil and Microbial Sciences, Michigan State University, USA. He received his MSc and BSc (Hons.) in soil and environmental sciences from the Department of SES, University of Agriculture.

Dr. Veysel Turan is an assistant professor in the Department of Soil Science and Plant Nutrition, Bingöl University, Turkey. He obtained his PhD in soil science and plant nutrition from Atatürk University, Turkey in 2016. After doing his postdoctoral research in the Department of Microbiology, University of Innsbruck, Austria (2017–2018), he began working at Bingöl University. He has worked, and is presently working on, a wide range of topics including soil–plant interaction, heavy metal accumulation, bioremediation of soil by some plants, and soil amendment.

List of Contributors

Ishfaq Ahmad
Centre for Climate Research and Development (CCRD), COMSATS University
Islamabad, Pakistan

Syed Kamran Ahmad
Department of Entomology, School of Agriculture, ITM University Gwalior
M.P., India

Abid Ali
Department of Zoology, Abdul Wali Khan University
Mardan, Khyber Pakhtunkhwa, Pakistan

Mazhar Ali
Department of Environmental Sciences, COMSATS University Islamabad
Vehari Campus

Abdelrazzaq Al-Tawaha
Department of Crop Science, Faculty of Agriculture, Universiti Putra Malaysia,
Selangor, Malaysia

Abdel Rahman Al-Tawaha
Department of Biological Sciences, Al Hussein Bin Talal University
Ma'an Jordan

Amanullah
Department of Agronomy, The University of Agriculture
Peshawar, Pakistan

Muhammad Aqeel
Department of Botany, Govt. College Women University
Sialkot, Pakistan

Muhammad Arif
Department of Agronomy, The University of Agriculture
Peshawar, Pakistan

Dhriti Banerjee
Zoological Survey of India, Ministry of Environment, Forests and Climate Change (Govt. of India) 'M' Block
New Alipore, Kolkata

Usman Khalid Chaudhry
Department of Agricultural Genetic Engineering, Ayhan Şahenk Faculty of Agricultural Sciences and Technologies, Niğde Ömer Halisdemir University
Niğde, Turkey

Qurat ul Ain Farooq
College of Life Science and Bioengineering, Beijing University of Technology
Beijing, China

Olivet Delası Gleku
Department of Agricultural Genetic Engineering, Ayhan Şahenk Faculty of Agricultural Sciences and Technologies, Niğde Ömer Halisdemir University
Niğde, Turkey

Ali Fuat Gökçe
Department of Agricultural Genetic Engineering, Ayhan Şahenk Faculty of Agricultural Sciences and Technologies, Niğde Ömer Halisdemir University
Niğde, Turkey

Nayab Gul
Climate Change Centre, University of Agriculture
Peshawar

Zehranur Gülbahar
Department of Agricultural Genetic Engineering, Ayhan Şahenk Faculty of Agricultural Sciences and Technologies, Niğde Ömer Halisdemir University
Niğde, Turkey

Shanawar Hamid
Department of Agricultural Engineering, Faculty of Engineering, Khwaja Fareed University of Engineering and Information Technology
Rahim Yar Khan, Pakistan

Noor ul Haq
Department of Computer Science and BioInfor-

matics, Khushal Khan Khattak University, Karak
Khyber-Pakhtunkhwa, Pakistan

Imran
Department of Agronomy, The University of Agriculture
Peshawar, Pakistan

Nadia Iqbal
Department of Biochemistry and Biotechnology, The Women University
Multan, Pakistan

Muhammad Kashif Irshad
Institute of Pure and Applied Biology, Bahauddin Zakariya University
Multan, Pakistan

Waqar Islam
Institute of Geography, Fujian Normal University
Fuzhou, P.R. China

Hafiz Muhammad Rashad Javeed
Department of Environmental Sciences, COMSATS University Islamabad
Vehari Campus

Muhammad Daniyal Junaid
Department of Agricultural Genetic Engineering, Ayhan Şahenk Faculty of Agricultural Sciences and Technologies, Niğde Ömer Halisdemir University
Niğde, Turkey

Ebrar Karabulut
Department of Agricultural Genetic Engineering, Ayhan Şahenk Faculty of Agricultural Sciences and Technologies, Niğde Ömer Halisdemir University
Niğde, Turkey

Noreen Khalid
Department of Botany, Govt. College Women University
Sialkot, Pakistan

Shah Khalid
Department of Agronomy, The University of Agriculture
Peshawar, Pakistan

Tika Khan
Kunming Institute of Botany, Chinese Academy of Sciences
Kunming, China

Nasir Masood
Department of Environmental Sciences, COMSATS University Islamabad
Vehari Campus

Kailash Chandra Naga
ICAR-Central Potato Research Institute
Shimla, Himachal Pradesh, India

Huma Naz
Mohammad Ali Nazeer Fatima Degree College
Hardoi, U.P., India

Yasir Niaz
Department of Agricultural Engineering, Faculty of Engineering, Khwaja Fareed University of Engineering and Information Technology
Rahim Yar Khan, Pakistan

Ali Noman
Department of Environmental Science, Govt. College University Faislabad
Pakistan

Sibgha Noreen
School of Life Sciences, Lanzhou University, Lanzhou
Gansu Province, P.R. China

Muhammad Abbas Qazi
Kunming Institute of Botany, Chinese Academy of Sciences
Kunming, China

Adeela Sadaf
Department of Environmental Sciences, COMSATS University Islamabad
Vehari Campus

Muhammad Asad Salim
Kunming Institute of Botany, Chinese Academy of Sciences
Kunming, China

Tayeba Sanaullah
Institute of Pure and Applied Biology, Bahauddin Zakariya University
Multan, Pakistan

Jayita Sengupta
Zoological Survey of India, Ministry of Environment, Forests and Climate Change (Govt. of India) 'M' Block
New Alipore, Kolkata

Sedat Serçe
Department of Agricultural Genetic Engineering, Ayhan Şahenk Faculty of Agricultural Sciences and Technologies, Niğde Ömer Halisdemir University
Niğde, Turkey

Mohd Abas Shah
ICAR-Central Potato Research Station, Jalandhar
Punjab, India

Anamika Sharma
Western Triangle Agriculture Research Center, Montana State University Bozeman
USA

Sanjeev Sharma
ICAR-Central Potato Research Institute,
Shimla, Himachal Pradesh, India

Zeeshan Shaukat
Faculty of Information Technology, Beijing University of Technology
Beijing, China

Muhammad Mohsin Waqas
Center for Climate Change and Hydrological Modeling Studies, Water Management and Agricultural Mechanization Research Center, Department of Agricultural Engineering, Faculty of Engineering, Khwaja Fareed University of Engineering and Information Technology
Rahim Yar Khan, Pakistan

Hafiz Muhammad Wariss
Kunming Institute of Botany, Chinese Academy of Sciences
Kunming, China

Hafsa Zahid
Department of Zoology, Abdul Wali Khan University
Mardan, Khyber Pakhtunkhwa, Pakistan

Ismail Zeb
Department of Zoology, Abdul Wali Khan University
Mardan, Khyber Pakhtunkhwa, Pakistan

1

Agriculture Contribution toward Global Warming

Hafiz Muhammad Rashad Javeed[1], Nadia Iqbal[2], Mazhar Ali[1], and Nasir Masood[1]
[1]Department of Environmental Sciences, COMSATS University Islamabad, Vehari Campus
[2]Department of Biochemistry and Biotechnology, The Women University, Multan

1.1 Agriculture and Climate Change

Pakistan is known as an agricultural country as the agriculture sector contributes the main share of the economy of Pakistan. Agriculture continues to be the major provider of employment to the rural community and the basic needs of life for more than 70% of our population. Agriculture is supposed to be the backbone of our country's economy as it is the single largest sector having 21.8% share in Pakistan's economy. Agriculture contributes to the growth and development of our economy as it is the primary supplier of raw materials to industries and a large market as well for industrial products, that is, pesticides, fertilizers, tractors, and combines harvesters.

Substantially it contributes to foreign exchange. More than half of Pakistan's population belongs to rural areas, which are directly or indirectly linked to the agriculture industry for the basic necessities of life. Although all the sectors are susceptible to climatic changes, agriculture is the most. Within the past few decades, the agriculture sector is facing many difficulties, including climate change, water scarcity, and poor production of crop. While the impacts of climatic change are getting severe day by day, climate change and agriculture are linked as both take place on a global level. Climate change encounters agricultural productivities in several ways, that is, irregular patterns of rainfall, temperature fluctuations, heat waves, and changes in sea level (Adnan et al. 2018a,b; Ahmad et al. 2019; Akram et al. 2018a,b; Wang et al. 2018; Fahad and Bano 2012; Fahad et al. 2013; Farhat et al. 2020; Gul et al. 2020; Habib ur Rahman et al. 2017; Hammad et al. 2016, 2018, 2019, 2020a,b; Hussain et al. 2019, 2020; Ilyas et al. 2020; Jan et al. 2019; Kamarn et al. 2017; Khan et al. 2017a,b; Mubeen et al. 2020; Muhammad et al. 2019; Naseem et al. 2017; Rehman et al. 2020; Saleem et al. 2020a,b,c; Saud et al. 2013, 2014, 2016, 2017, 2020; Shafi et al. 2020; Shah et al. 2013; Subhan et al. 2020; Wahid et al. 2020; Wu et al. 2019, 2020; Yang et al. 2017; Zahida et al. 2017; Zafar-ul-Hye et al. 2020a,b; Zaman et al. 2017; Zamin et al. 2019). Further climate change is already negatively affecting crop production worldwide. Among the major determinants related to agricultural productivity, climate/weather could be the most significant factor on a global scale (Faostat 2010). Increasing average annual temperatures, seasonal variations, changing rainfall patterns, and floods and droughts are some of the main impacts of climate change.

Wheat, being the major staple of the South-Asian region, faces severe calamities caused by climate change. Due to seasonal variations, the growth period of wheat has been shortened, which ultimately have negative impacts on potential production and quality of wheat. Among these climatic variables that adversely affect wheat production temperature is an important variable to consider. Temperature affects the wheat crop throughout the season from sowing to harvesting. Rainfall also has a significant but positive role toward wheat productivity if it occurs in proper frequency and at critical growth stages of the wheat crop. Thus, the measure of the impacts of climate change on wheat productivity can provide us with some important perspectives, which may help us to combat the prevailing climate changes and a sufficient knowledge to adapt these climatic variations. By the end of the 21st century, there could be an increase in the average global temperature of 1.4°C to 5.8°C, resulting from increased concentrations of greenhouse gases (Houghton et al. 2001).

1.2 Intergovernmental Panel on Climate Change (IPCC)

According to a report on climate change by IPCC, the average air temperature would rise by 1.4°C to 5.8°C by the end of the 21st century. This rise in temperature would shorten the growing seasons of different crops due to changing growing patterns, for example, early flowering and fruit bearing, and as a result of enhanced respiration activities, there must be a decrease in the nutrient supply to the seed, which substantially lead the seeds to not be developed fully. Magrin et al. (2009) reported that the potential yields of wheat had been declining at increasing rates since the mid-20th century mainly due to minimum rises in temperature. Further temperature variations will lead to declined potential wheat yield for each 1°C rise in temperature (Fahad et al. 2016a,b,c,d). Zhu (2004) reported that the decreased yields of wheat crop could be the result of short growth periods. As increased temperatures accelerate the plant's metabolic activities so that they mature earlier, hence, they reduce the potential yields by reducing the time span of full development. Increased temperatures aggravate water stress. As water is an essential element for crop growth and potential yields, hot and humid weather conditions also contribute to the prevalence of insect pests, diseases, and weed infestations. As a result of global climate changes, by 2080, there would be a 14% decrease in winter wheat yields. You et al. (2005) reported that a 1% increase in temperature of wheat growing season reduced the potential yields by about 0.3%. Wajid et al. (2007) studied that drought has significantly reduced the potential wheat productions.

Globalization also has some adverse impacts on our environment. According to statistics reported by the World Bank, deforestation contributes up to 20% of the total global carbon emissions. The increasing global trades have also encouraged fishing, destruction of forestlands, and promotion of pollution. To cope with the adverse impacts of these environmental issues, there are serious financial and policy issues to be faced. As climate change is the global phenomenon, it affects all societies on our planet.

Global warming is the result of agricultural and industrial activities practised globally (Alharby and Fahad 2020; Fahad et al. 2014a,b, 2015a,b, 2016a,b,c,d, 2017, 2018, 2019a,b). So, the consequences are also global and all parts across the globe are predicted to experience changes resulting from global warming. One of the major attributes of these climatic variations is the adverse impact of climate change unevenly distributed worldwide. Many developing countries that don't contribute proportionally to the global greenhouse gas emissions such as Bangladesh will also sustain the critical issues resulting in floods from rising sea levels. Global warming can significantly cause the economic and social disruption worldwide. But the major issue is that these adverse effects of global warming will mainly affect the poor/underdeveloped countries, though the developed countries in temperate zones will experience less severe or moderate negative impacts of climate changes. Over the past decade, the Intergovernmental Panel on Climate Change (IPCC) has conducted several research, studies which showed that there has been an increase of about 0.6°C in average annual global temperatures since the industrial revolution took place. This elevated temperature is the result of enhanced concentrations of Greenhouse Gases (GHGs) in the atmosphere. These results of IPCC have been confirmed by Brohan et al. (2006), who reported the 20th century as the warmest century. During the past 11 years, 1990s has been declared as the warmest decade of millennium. In future, the global average temperature variations are expected to be increased from 1.4°C to 5.8°C (McCarthy et al. 2001). Considering the overall rise in average global temperatures, IPCC research predicts that large temperature and precipitation variations (either increase or decrease) will have significant direct and indirect impacts on global socio-economic and industrial sectors (water, agriculture, health, forestry, and biodiversity of flora and fauna), which could be positive or negative. However, the current global model simulations have shown that mostly the regional lands would be warmer than the expected global average annual temperatures (Giorgi and Bi 2005). If the issue of greenhouse gas emissions should not be under check, then these emissions could be grown substantially over the next century. Burning of fossil fuels and land use change contributes major share in greenhouse gas emissions (Stocker 2014). The ultimate results of this practice will be considerable average annual temperature rise and variations in rainfall patterns (Change et al. 2007). Although the global climate change puts significant impacts on various sectors, agriculture is the single largest sector facing adverse impacts of climate change (Cline 1996). Having a good understanding of these impacts provides considerable insights to formulate mitigation strategies. This understanding will also help in the adaptations of these climate changes. Various economic studies in the United States have quantified the climate change impacts on agriculture industry. Mathematical programming has been reported an effective approach to check crop switching in response to varying potential crop yields (Adams et al. 1990, 1999).

Global warming has become an emerging issue in the late 20th century as it has shown an increasing trend in the past few decades. As climate change affects various aspects of human life and different socio-economic sectors, it should be a major concern. The impacts of climate change and global warming on the agriculture sector will be more adverse and should be of more concern because agriculture directly affects food security and substantially affects global human populations. Due to the awareness of the impacts of climatic variations on ecosystems, biodiversity, economy, as well as global trade and humanity, make climate change and global warming attain the topmost priority at governmental, socio-economic, and community levels (Posas 2011). Most of the developing countries including Pakistan are economically based on the agriculture sector. Because the agriculture sector is directly exposed to natural phenomenon, so agriculture is the most adversely affected sector by changing climatic conditions compared to other sectors. Therefore, developing countries are more affected by climate changes as these are more vulnerable to climate variabilities than developed countries (Ali et al. 2019).

Pakistan's economy is highly dependent on the agriculture sector, which accounts for 25% share in total GDP of the country and provides employment to more than two-fifths of total rural labour force. Pakistan is located in a semiarid to arid region where rainfall is very low. In Pakistan, the major irrigation source for crop production is Indus River Irrigation System (IRS), which contributes about 90% of the irrigation requirements of the agriculture sector. According to IPCC fourth Assessment Report, by the end of this century, the global temperature would increase by 1°C to 6°C. This prediction of IPCC is especially relevant to Pakistan, as Indus River System (IRS) mainly relies on the Himalayan glaciers, which are supposed to melt in coming 50 years due to hot dry summers prior to monsoon season. And substantially in the near future, the Himalayan glaciers might disappear. Moreover, these climate changes also have negative impacts on precipitation patterns, leading to more intense and irregular patterns of rainfall resulting in heavy floods and drought periods in Pakistan. About 2.6 million acres of cultivated land have been lost in the recent floods and a decline of 6%–9% in the potential wheat yields has also been observed as well. The industrial sector of Pakistan heavily relies on the agricultural sector so the adverse climate change effects and impacts on agriculture directly affect the industrial sector and ultimately the economic sector suffers.

As the Gulf region is right next to Pakistan the climate changes that have adverse effects on Pakistan have direct impacts on Gulf countries as well. Recent research has concluded that in the coming 20–30 years, the atmospheric concentrations of greenhouse gases will increases to an extent, which would put considerable negative effects on global climate. Considering current increases, that is, 1.8°C and 4.0°C, by IPCC forecasts, by the end of this century, the average global surface temperatures will rise by 2.8°C. Natural environmental systems and natural climate would be changed by global warming, resulting in extreme weather conditions, ocean currents, rising sea levels, glaciers melting and changing rainfall patterns, which have direct and indirect ultimate impacts on the industrial and economic sectors. In spite of the advanced technologies and innovations including green revolution, weather and climate are the limiting variables for the potential crop yields worldwide. The forecasted fluctuations of temperature and rainfall patterns along with their direct or indirect impacts on natural resources, insect-pest, and disease infestations may affect the substantial crop yields. Studies on economic aspects of climate change have suggested that though global warming has increased crop yields by elevating temperatures to some extent, but in the long run, these slightly positive effects turn negative (Bruinsma 2003). Moreover, these climatic variations are unevenly distributed across the globe. In the lower latitude areas/regions, the agricultural sector suffers more from global warming (increasing temperature and irregular rainfall), due to their geographical location and limited resources to adapt these variations and also high dependency of their economy on the agricultural sector (Ullah 2017). Comparatively the high latitude regions could be less vulnerable to changing climate. If global warming continues at this extent and remains unchecked, during the last era of the 21st century, the global potential crop yields would decrease by 16%, while the developing countries would face an additional 19.7% reduction in agricultural production (Cline 2007). Agriculture shares a vital role in Southeast Asia and contributes more than 10% of the total GDP of the most regional economies. It also provides employment to over one-third of the labour force in the region. Similar to other developing countries of the world, almost a quarter of the poor in Southeast Asia belongs to rural community, majority of which is reliable to the agricultural sector for livelihood. Consequently, agricultural development has become the key factor to mitigate poverty in the region of Southeast Asia.

Furthermore, the increase in international trade from Southeast Asia's agricultural sector means that any climate change-related shocks to international markets for agricultural products will be easily transmitted to the region through trade channels. CGE model being economy-wide model best explains the interactions between industries, consumers, and governments in the global economies. This could be due to the fact that climate change had direct effects on the economy by affecting its agricultural production and indirect effects by altering agricultural productions of other countries.

1.3 Effect of Climate Change on Agricultural Productivity

Agricultural productivity is affected by climate change in different aspects, which originate from average annual temperature elevations, variable precipitation patterns, increase in the atmospheric CO_2 levels, and greenhouse gas emissions (Chijioke et al. 2011). Additionally, other constraints regarding temperature variations and irregular rainfall patterns have direct effects on the areas suitable for arid agriculture and crop water requirements, which could have effect on the agro-ecological potential of that area (Li et al. 2014). Rice cultivation practices contribute to the greater extents to the methane emissions, which is one of the major greenhouse gases (GHGs). Approximately 30% of methane and 11% of N_2O are emitted from the rice fields globally (Usepa 2006). Organic matter (OM) significantly affects soil ecosystems in various aspects of biological, chemical, and physical processes. The soil organic carbon (SOC) is one of the major indicators of soil quality and health. Soil organic carbon contributes to the large share of carbon in the global carbon cycle, and it represents the dynamic balance between deposition and loss of carbon input, respiration, erosion, and leaching through photosynthesis. The accumulation of organic carbon in the soil can convert atmospheric carbon dioxide into the soil and sequester atmospheric organic carbon pools, resulting in reduction in climate changes.

Therefore, the maintenance or increase in the soil organic carbon is a key factor to attain the desirable soil functions and for the regulation and sequestration of atmospheric carbon dioxide. In the agro-ecosystem, soil organic carbon balance is affected by cropping intensities, fertilization, irrigation, and crop rotations, in addition to organic matter. Among them, organic amendments such as green manuring, farmyard manure, and crop residues are widely accepted as the most viable soil management techniques for increasing soil organic carbon. To increase the soil organic carbon in temperate regions, the winter cover cropping is highly recommended cropping technique in mono-cultivation systems of rice. However, the greenhouse gas emissions might be increased by the application of cover crop residues during respiration and mineralization.

Methane, which has a global warming potential (GWP) of 25 times higher than carbon dioxide on the 100-year horizon, is produced mainly under very low conditions (less concentration) from organic matter decomposition. Cover crop biomass application significantly increases methane emissions during rice cultivation. Therefore, in order to provide an adequate management strategy for cover crops in rice paddy soils, the impact of cover cropping should be properly assessed by considering the risks caused by the C sequestrations and greenhouse gas emissions. Weather variations affect the agricultural sector

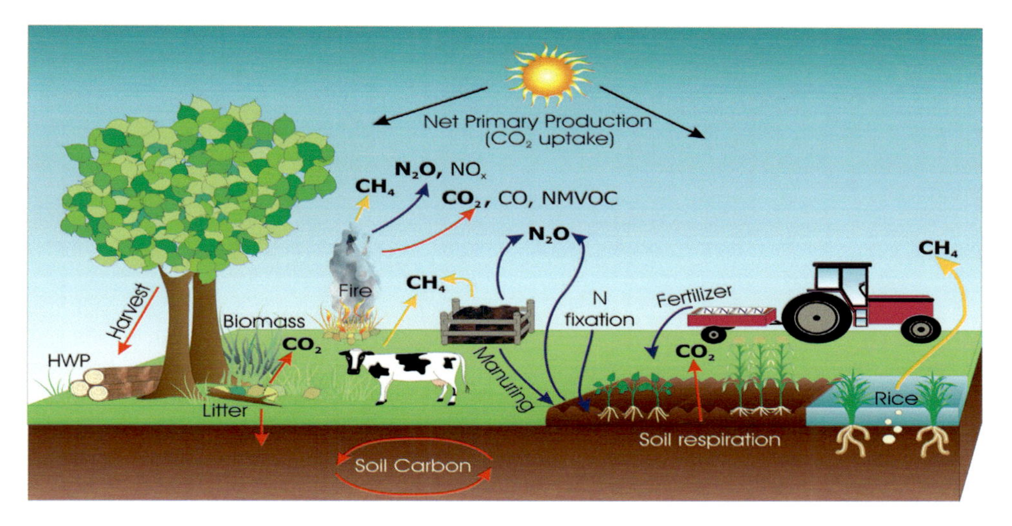

FIGURE 1.1 The main greenhouse gas emission sources, removals, and processes in managed ecosystems.

Source: https://climatepolicyinfohub.eu

in various ways. Temperature, radiation, rainfall, soil moisture, and carbon dioxide are important determinants of agricultural productivity and they have no linear relationships. And above the threshold levels of these climate variables, the crop yields are decreasing.

For example, the modelling studies discussed in recent IPCC reports suggest that crop temperatures in temperate regions, along with moderate increases in temperatures (1°C–3°C), as well as associated CO_2 increases and changes in precipitation, are expected to increase crop yields. Though, in areas with low latitudes, moderate temperatures (1°C–2°C) may have a negative yield effect on major grain crops. Warming above 3°C has a negative impact on all areas (Change ICJM et al. 2007). The impact of climate change on soil moisture is determined by changes in temperature and rainfall. With rising temperatures, both evaporation and precipitation are expected to increase. The net effect on water availability depends upon which energy is more effective. The IPCC report stated that by the middle of the 21st century, climate change in high latitudes and some humid tropics will increase water availability and also in some dry areas in the mid-latitudes and arid tropics (see Figure 1.1) (Change ICJM et al. 2007).

Some areas that are already drought prone will be exposed to more severe droughts. Increasing atmospheric CO_2 concentration has a positive effect on crop yields by promoting photosynthesis and reducing water loss through plant respiration. This carbon fertilization effect is strong in so-called C3 crops such as rice, wheat, soybeans, fine grains, legumes, and many trees with low rates of photosynthesis. For C4 crops such as corn, millet, sugarcane, and many grasses, these effects are minimal. Other factors, such as plant growth stages, or water and nitrogen use may also affect the effect of increased CO_2 on plant yields. Recent research, based on experiments conducted in the free air concentration enrichment method, compared to previous estimates from studies conducted under attached test conditions, found that C3 crops did not stimulate very low CO_2 and C4 crops on yield. Based on an analysis of recent data, the IPCC report states that C3 crops increase 10–25% and C4 crops increase 0–10% when CO_2 levels reach 550 ppm (Change ICJM et al. 2007).

Although many limiting factors were not included in the modelling and experiment analysis, considerable uncertainties still surrounded estimates of carbon pollution effects. Apart from temperature and carbon densities, some other environmental changes caused by global warming also affect agriculture. For example, patterns of pests and diseases may change with climate change, which may result in decreased agricultural production. In addition, agricultural productivity is affected by increased climate variability and the frequency of extreme events and extreme events such as droughts and floods. These are the difficulties in assessing the agricultural productivity impacts of climate change. Quantitative estimates of the agricultural impact of climate change are mainly based on three approaches: crop simulation models, agro-ecological zone (AEZ) models, and cross-sectional (Ricardian) models. Crop simulation models are based on controlled experiments in which crops are grown in agricultural or laboratory settings and simulate different environments and CO_2 levels to predict the yield responses of a particular crop variation to some environment and other variables of interest.

1.4 Impact of Climate Change on Insect Pests

Global warming increases pests populations, impacting the yield of major crops such as wheat, soybeans, and corn. Warmer temperatures for longer periods of time and faster growth rate for plants increase the metabolic rate and the number of reproductive cycles of the insect population. Insects that previously had only two reproductive cycles per year may receive an additional cycle due to population growth if warmer weather advances. Temperate locations and high latitudes are likely to cause unpredictable changes in insect populations.

Some experiments resulted that warm temperatures and increased CO_2 levels were repeated in one area of soybean, while the other remained in control. These studies found that soybeans with increased CO_2 levels increased faster and had higher yields, but Japanese beetles were attracted at significantly higher rate than the control area. Agricultural beetles with increased CO_2 also have more eggs on soybean plants and have a longer lifespan, indicating the potential for a rapidly spreading population. If this project continues, the area with the highest CO_2 levels will eventually yield lower than the control area. In areas with increased CO_2 levels, three genes in the soybean plant, which are responsible for chemical protection against pests, are reported as deactivated. One of these defences is the protein that inhibits the digestion of soy leaves in insects. Because this gene was inactive, beetles were able to digest much larger plant material than beetles in the control area. For this reason, long life expectancy and egg laying rate were observed in the experimental area.

There are some proposed solutions to the problem of pest population expansion. One proposed solution is to increase the number of pesticides used on future crops. It has the advantage of being relatively inexpensive and simple, but not ineffective. Many pesticides are immune to these pesticides. Another proposed solution is to use biological control agents. This involves placing rows of native vegetation between rows of crops. This solution is beneficial in its overall environmental impact. However, planting more native plants is not enough to prevent insect pests build immunity from pesticides. More room is needed to plant additional native plants, which will destroy additional acres of public land. The cost is also high with the use of pesticides.

1.5 Impact of Climate Change on Soil Erosion and Fertility

Warmer weather temperatures observed in previous decades are expected to lead to more powerful hydrological cycles, including more severe rainfall events. Erosion and soil erosion are likely to occur. Soil fertility is also affected by global warming. Increased erosion in the agricultural landscape can be caused by anthropogenic factors, with a loss of up to 22% of soil carbon in 50 years. However, the proportion of soil organic carbon is mediated by soil biology, which doubles as a narrow range. Soil organic carbon can be doubled as organic nitrogen in the soil, resulting in higher yields for plants, while providing higher yield potential. Imported fertilizer may reduce the nitrogen demand and provide an opportunity to replace costly fertilizer strategies.

Due to the extremes of climate that would result the increase in precipitations would probably result in greater risks of erosion, while at the same time providing soil with better hydration, according to the intensity of the rain. The possible evolution of the organic matter in the soil is a highly contested issue: while the increase in the temperature would induce a greater rate in the production of minerals, lessening the soil organic matter content, the atmospheric CO_2 concentration would tend to increase it.

1.6 Role of Agriculture Sector in Greenhouse Gas (GHG) Emissions

On one hand, the agricultural sector is highly vulnerable to climate change; on the other hand, it is an important source of greenhouse gas (GHG) emissions (Sun et al. 2019). The ability to convert atmospheric carbon dioxide into food through photosynthesis and partially compensate for the GHG generated from it is unique and different from elsewhere. CO_2 causes changes in temperature, precipitation, and changes in the radiation levels, which substantially affect the plants and animals and also change the demand and supply equation for water requirements (Jaiswal and Agrawal 2020). In addition, rising sea levels can disrupt the agricultural sector and increase the salinity of groundwater along with the coastlines. The effects of climate change on agriculture could be direct and indirect. All crops, animals, and regions are affected by climate change, but these effects may vary by species and region. It is fair to say that the global carbon cycle is widely used by agriculture. Bringing in extra land under agricultural practices can disrupt the carbon cycle by releasing more greenhouse gases than carbon sequestration through photosynthesis. Generally, it is accepted that agriculture contributes 14% of carbon dioxide equivalent per year to the global greenhouse gas emissions (Ghosh et al. 2019).

There are some reports that 14% increase in greenhouse gas emissions has been observed from crop and livestock production, which increased from 4.7 billion tonnes of CO_2 in 2001 to 5.3 billion tonnes of CO_2 in 2011 (Faostat and Production ACJR, Italy 2016). This development is mainly in developing countries due to the expansion of agricultural activities in these countries. Globally, agriculture has 54% of anthropogenic methane (CH_4) and 58% of nitrous oxide (N_2O) emissions. In 2014, South Asian countries together accounted for 8% of global agricultural greenhouse gases emissions (Faostat and Production ACJR, Italy 2016). The concept of the carbon footprint is derived from the environmental footprint given by Reese. The ecological footprint is a biologically productive land, and the sea area is essential for managing human populations expressed in terms of global hectares. The carbon footprint is part of the life cycle Assessment (LCA), which measures GHG. Nitrous oxide (N_2O) is one of the most potent greenhouse gases released mainly (almost 60% of global emissions) from the agricultural sector. Nitrogenous fertilizers have been identified as a potential source of N_2O, and out of every 100 kg of nitrogen fertilizer, 1.0 kg of nitrogen is released as nitrous oxide.

From this perspective, legumes with minimal nutrient requirements and some other cultural practices have little negative impacts on natural ecosystems. Improvements in soil organic carbon concentration and nitrogen and carbon cycle optimizations have the least adverse effects on the environment and agroecosystems. Periodic grazing and forage exclusion have been proved less effective compared to moderate grazing to reduce pure global warming potential (1.84 mg CO_2 ha^{-1} yr^{-1} on average), primarily as a result of reduced soil organic carbon accumulation. Integrated crop and livestock farming systems are increasingly being adopted mainly due to their ability to adjust forage intensity or forage allowance by storing rates (Carvalho et al. 2018). Overstocking adversely affects the lawns (shoot and root development) and consequently affects the soil organic matter and soil physical properties (Assmann et al. 2014). Significant allocation of carbon to the root systems of pasture species and increased carbon saturation deficit in the deep soil layers of tropical and subtropical soils may contribute to the impact of intercropped legumes on soil organic carbon stocks (Nath et al. 2018).

Organic fertilizers rich in nitrogen content and compost are also the ultimate sources of nitrous oxide emissions. Rice cultivation produces methane by anaerobic decomposition of organic matter (e.g. crop residues) in rice, which contributes 10% of the total agricultural greenhouse gas emissions. In addition to be a source of greenhouse gas emissions, farms also act as a potential ‘sink’ for carbon. In other words, to remove carbon from the atmosphere and store it in soil or timber. In this way, soil carbon sequestration and CH_4 and N_2O emissions in agricultural soils could be reduced. Biomass of agroforestry has carbon suppressing capacity. Mixing of timber leftovers with crops and livestock residues and incorporation of this mixture in the soil is also large and reversible process but might be subjected to uncertainties associated with carbon sequestration in the soils. The dataset used for this analysis was restricted to carbon stock changes in soil. To capture the full suppression capability of agroforestry interventions, the values of carbon sequestration in biomass were derived by Cardinael et al. (2018).

The challenges for the 21st century agriculture are different when considering the potential impact of climate change (Lozano et al. 2012). Agriculture, one of the largest source of emissions of greenhouse gases (GHGs), has been heavily affected by climate change. Increasing productions without crop expansion, reducing the environmental emissions such as land degradation and water depletion and addressing increased agricultural losses due to climate change are new challenges (Richards et al. 2019). India is the fourth-largest greenhouse gas emitter in the world, where agriculture accounts for 18% of the total national emissions. Anthropogenic greenhouse gas emissions, including the burning of rice residues, are a major contributor to global warming and climate change. The changing climate led the food security in South Asia complex, fragile and vulnerable, where the rice-wheat cropping system is one of the largest agricultural production systems. Intercultural operations like tillage systems with mulching/cover cropping have a significant impact on methane and nitrous oxide emissions.

Numerous field studies have shown that zero or no-tillage might cause methane emissions lower than the traditional tillage systems by maintaining methane oxidation efficiency. Numerous studies have shown that nitrous oxide gas emissions from no tillage practices might be lower or higher as compared to the traditional tillage systems (Choudhury et al. 2019). It is concluded that high nitrous oxide gas emissions under reduced tillage practices would be decreased over time. Global warming perverts the global carbon cycle, leading to the changes in ecosystem structures and functions. Although carbon dioxide (CO_2) is considered as the most important greenhouse gas, but the CO_2 fluxes are balanced by atmospheric fixation of carbon dioxide in plants as a net primary product. Thus, carbon dioxide contributes to less than 1% of the global warming potential of agriculture (Figure 1.2).

In the past half century, agroecosystems have had to increase their production to meet the challenges of food security. However, with some differences in various countries, the general trend in the

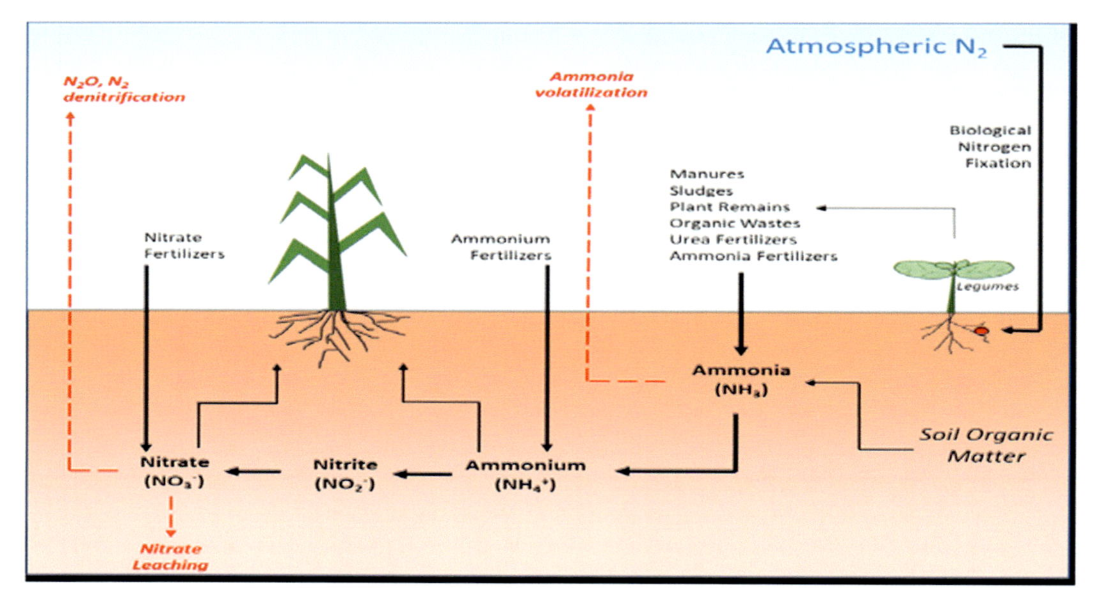

FIGURE 1.2 The effect of nitrogen fertilizer on water quality and the environment. It provides guidelines for managing soil fertility on farms to preserve water quality.

Source: https://content.ces.ncsu.edu

agriculture was to increase the per hectare potential yields and industrialization of the crop system (Agovino et al. 2019). Monoculture cropping system allows an easy spread of diseases and pests, hence reducing the resistance to climate change, which often puts additional stress on crop plants. Extensive mitigation techniques in agriculture have been proposed, which could be based on the evaluation of methods and technologies in various agricultural production systems in South Asia (Aryal et al. 2019). However, the performance of low-carbon agricultural production in China remains undeniable. Rice is being used by more than half of the world population as staple where China holds 30% share in global rice production (Cheng et al. 2020). Rice straw is the main by-product after harvesting of rice crop, and it contains larger amounts of organic matter and other soil nutrients, which are essential for plant growth. Mid-season drainage has proven to be an effective alternative to reducing CH_4 emissions from rice.

Therefore, the mitigation strategies of the negative impacts of the crop residues on different ecosystems are there. Various methods of crop residues management have been studied to minimize the adverse effects. Reduction in methane emissions and widespread greenhouse effect through the above measures and applications of the technology have not been suitable instead they increased the additional costs. Soil moisture content has also been found to be an important factor for methane and nitrous oxide gas emissions. Double-rice cropping system is a very common and important practice in China. Almost 17 million hectares of land is under this cropping system. The rice-wheat cropping systems contribute a major share to the production of greenhouse gases (Porter et al. 2017). Nitrogenous fertilizers are an important source of readily available nitrogen for the crop plants in exhaustive cropping systems. The application of appropriate nitrogen fertilizers increases the crop yield as it promotes radiation interception by increasing leaf area and photosynthetic rate of crop plants. Over the past half century, the nitrogen fertilizers might have increased per unit food production. Overuse and imbalanced use of nitrogen fertilizers for higher yields result in greater losses of nitrogen, causing various adverse effects and impacts on natural ecosystems and climate on local and global levels.

Leaching of nitrates negatively affects the quality of groundwater (Yilmaz 2019). The soils fertilized with nitrogen fertilizers are the main source of the emissions of nitrous oxide (N_2O) compared to upper regions, where paddy rice cultivation could be a major contributor to the global methane emissions on global scale. Substantially it contributes almost 11% to the total GHGs emissions (Mohammadi et al. 2016). Total non-CO_2 greenhouse gases emissions in the past decade were approximately 12% of the total global anthropogenic greenhouse gases emissions (Smith 2014). Application of inorganic fertilizers resulted in the release of nitrous oxide and methane with a rapid increase in their concentrations. Hence, inorganic/synthetic fertilizers could be ranked second among the major agricultural greenhouse gas emitters.

Nitrous oxide and methane emissions from the synthetic fertilizers have the largest absolute growth rate and are the second largest agricultural emissions after enteric fermentation (Kim et al. 2019). Among the various greenhouse gases emissions mitigation strategies, use of nitrogen inhibitors and slow-release nitrogen fertilizers or efficient nitrogen use fertilizers may reduce the methane. However, overuse or misuse of synthetic fertilizers in China is the major cause of increased environmental pollution (Stevens 2019). Recent studies have shown that soil methane emissions are reduced by the abundant use of biochar applied to paddy rice soils, which alters soil microbial ratio from methanogens to methanotrophs (Kim et al. 2017). The extensive use of synthetic fertilizers (Mostashari-Rad et al. 2019) and heavy agricultural machinery is also a major cause of rapid increase in greenhouse gas emissions. Burning of fossil fuels, farm machinery gas emissions, and chemical changes in the organic soil carbon in conventional cropping systems play an important role in serious environmental problems as these cause up to 50% carbon dioxide emissions (Liu et al. 2019).

In some cropping patterns, fertilizer compositions can also change N_2O emissions. For example, emissions can be two to four times higher than those after urea ammonium or urea transmitted. In the United States, the move towards using more urea in maize may help to reduce emissions of N_2O. Nitrification inhibitors such as nitropyrene, which delay soil ammonium's microbial transformations to nitrate, can delay the development of nitrate until plants utilize it more quickly. When plants need more fertilizers, using nitrogen fertilizer can also contribute to reducing N_2O emissions into the groundwater through leaching or into the atmosphere through volatilization. If the large quantity of fertilizer is used a few weeks after planting, rather than before planting, the chance increases that the N will end up in the crop instead of being lost to groundwater or the atmosphere. Adding N fertilizer to frozen fields in the fall or spreading manures often leads to a particularly high loss of nitrates and N_2Os. In those situations, the formulations for fertilizers are not aligned with crop demands.

Reduced or no tillage along with sustainable land use is an effective approach to reduce adverse effects of changing climate to some extent (Wang et al. 2019), because it reduces energy consumption and land degradation and thus is a cost-effective approach. In addition, tillage is an intensive energy consuming practice with a greater impact on the overall environmental system (Hellerstein and Vilorio 2019). Conventional tillage cultivation practices are not only energy demanding, but also cause changes in physical, chemical, and biological activities of soil, which could ultimately affect organic matter content and soil fertility status. Carbon dioxide is released from the soil mainly through plant and animal respiration processes. Various soil factors such as soil texture, temperature, soil humidity, soil pH, and available C:N ratio as well as rainfall affect the release of soil carbon dioxide.

Different researches concluded that use of biochar have also implemented negative, neutral, and positive effects on the soil emissions of carbon dioxide. Possibly the practices that mitigate CO_2 emissions could also help with adaptation to varying climatic conditions. Farmers might adapt to prevailing climate changes by switching to agro-forestry for carbon sequestration (Stokes and Howden 2010). To ensure sustainable agriculture, it is need of the hour to investigate the relationship between carbon emissions, economic development, and energy use in the agricultural sector. Biochar amendment in agricultural soils can reduce the greenhouse gas emissions due to reduction of nitrous oxide concentration (Cayuela et al. 2014). Therefore, there is a need to identify the possible mitigation approaches to mitigate farm soil CH_4 and N_2O and carbon dioxide gas emissions as well as maintain the potential grain yields of agricultural crops.

1.7 Role of Livestock Farming in Greenhouse Gas (GHG) Emissions

Livestock plays a vital role in the agricultural sector in developing nations; for example, it contributes 11% of the total GDP of Pakistan and 56% to the value addition in the agriculture sector. Global warming and climate change cause thermal stresses on animals and birds by increasing the environmental temperatures. Heat stress changes the physiological, biochemical, hormonal, and reproductive parameters of animal production (Schauberger et al. 2019). New strategies are therefore required to reduce thermal stress such as changes in housing and water management. In the past decades, animal husbandry has changed considerably. Small farmers with modern schemes have been replaced by restricted animal feeding operations with higher stock densities to meet the increasing demand for animal products in developed countries. Such complex schemes are economically successful but notable for their adverse environmental effects.

In contrast, the livestock sector has also been negatively affecting the global climate. The rearing of livestock/cattle requires considerable natural resources and plays a key role in global greenhouse gas emissions (Figure 1.3). Methane and nitrous oxide are the most significant greenhouse gases from animal farming. Methane, produced mainly by enteric fermentation, is a gas that is 28 times higher than carbon dioxide in the context of global warming. A molecule with a global warming effect of 265-fold greater than carbon dioxide is nitrous oxide (N_2O), produced from manure production and organic or chemical fertilizers. Carbon dioxide equivalent for the global warming potential is a standard unit. In addition to greenhouse gasses from enteric fermentation and the manure fermentation, feed production and the soil-based emissions of carbon dioxide and nitrous oxide (N_2O) are key hub for livestock industry.

Soil emissions are due to dynamics of soil carbon emissions, for example, decomposing waste plant, soil-organic matter mineralization, and soil usage shifts;, production of synthetic/inorganic fertilizers and pesticides; and the utilization of fossil fuel in agriculture on the farm (Widiawati 2019). Nearly 60% of the worldwide biomass harvested is used for feed or bedding in the animal farming systems. Greenhouse gas emissions from feed production account for 60–80% of egg, chicken, and pork, and also

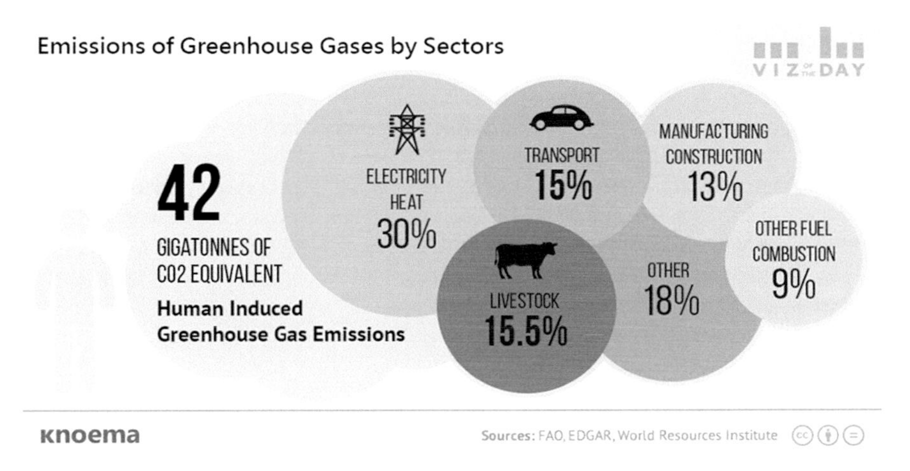

FIGURE 1.3 Global greenhouse gas emissions from livestock.

Source: FAO, EDGAR World Resources Institute

35–45% of milk and bovine production emission. Feed production emissions account for approximately 45% of the livestock sector. The use of manure as a fertilizer for food crops and the manure deposited on pastures cause a large amounts of nitrous oxide emissions, which represent approximately half of the greenhouse gas emissions (Bekuma et al. 2019). The more we turn a blind eye, the more we limit our ability to feed ourselves, protect waterways and habitats, and continue to make other use of our precious natural resources. Meat, eggs, and milk are the second-highest source of emissions and the highest of the transport of livestock, causing 14.5 per cent of the global emissions of greenhouse gases (GHGs). These are also a major cause of deforestation, loss of biodiversity, and wastewater as these use about 70% of agricultural land.

Moreover, air and water emissions can be attributed directly to the livestock industry, the major contributor to global water contamination. Additionally, the cattle industry is one of the major generators of deforestation and is synonymous with 75% of recent deforestation of the rainforest in the Brazilian Amazon. To date about a portion of loss of biodiversity has been linked with animal farming. In addition to the intensification of water and air contamination, globally produced agricultural species generate seven to nine times more waste as people. Pesticides and heavy metals are also released into water systems. The poultry farming has been identified as a large ammonia (NH_3) producer and, to a lesser degree, greenhouse gas production, mainly in national emission inventories. However, because the average emission factors for each type of animal are the basis of most national inventories, the factors influencing such emissions are not taken into consideration by breeding and manure management practices. The first step to enhance inventories and proposing mitigation options (e.g. best management practices and innovative systems) is to better understand the drivers of gaseous emissions and to identify the key factors in ammonia (NH_3) and greenhouse gas mitigations.

Additional public health risks are posed by concentrated animal farming activities to neighbouring communities, as viral conditions can extend from sick animals to humans and increased use of antibiotics encourages antibiotic resistance. Improper manure management from large volume plants threatens to spray faecal material that can reach nearby homes and cause breathing problems. Vegetable residues can also be transported through the soil to groundwater and contaminate rivers and streams by nitrates and pathogens. Every year animal-based food almost equivalent to one gigaton of carbon dioxide is wasted globally. If global meat and dairy demand continues to increase at its current speed, by the middle of the century, agriculture will consume about 70% of its permissible budget for all greenhouse gas emissions. Annual emissions from 49 gigatons of CO_2 today should be reduced by around 23 gigatons by 2050 in order to fulfil the global objective of limiting global warming at 2°C. Agriculture will invest 20 of them, leaving the world economy with just 3.

By adopting a plant-based diet, the greenhouse gas emissions intensity of the average western diet might be halved. Smarter livestock management, technology-enabled control of fertilizer applications, simplified improvements in field layout, and other more productive farming innovations can also reduce agricultural emissions. Proteins obtained from animal sources such as beef, chicken, pork, eggs, or milk/dairy typically contain adequate amounts of essential amino acids, which are also highly digestible for human development and growth. Nonetheless, in contrast, plant-based proteins can substitute one or several essential amino acids. The advancements in the processing and combination of plant protein sources with complementary protein sources can lead to the development of new sources of dietary protein with improved protein quality. The high-protein content of pulses (including beans, pea, soybean, and lentil) typically substitutes plants' protein sources that are readily processable and ultimately included in protein-enriched diets because of their high-protein content.

1.8 Conclusion

The effects of climate change are related to several processes dealing with many factors that have to be taken into consideration. Agriculture (including crop farming, husbandry, forest and fisheries) can be characterized as one system and environment as the other. This would contribute to a solution that is too fragmentary if the structures are treated separately. Human actions are now seen as likely to affect the atmosphere, one of the environmental elements. Weather, in turn, affects agriculture, the origins of all human and animal food. Environment affects agriculture. Nevertheless, the pattern and shortcomings of human societies and agriculture itself must be identified and taken into account in climate change impact studies. Global warming causes many human problems, but we human beings are causing global warming. But the global warming needs to reduce the gasoline, recycling and man, instead of increasing the earth's temperature, must contribute to reducing global warming. Their generation will start caring for the planet, because they will be impacted by global warming in the next decade. Global warming is therefore a major problem today.

REFERENCES

Adams RM, Rosenzweig C, Peart RM, Ritchie JT, McCarl BA, Glyer JD, Curry RB, Jones JW, Boote KJ, Allen Jr LH (1990) Global climate change and US agriculture. *Nature* 345:219.

Adams RM, McCarl BA, Segerson K, Rosenzweig C, Bryant KJ, Dixon BL, Conner R, Evenson RE, Ojima D (1999) *The economic effects of climate change on US agriculture*. Cambridge University Press, Cambridge.

Adnan M, Shah Z, Fahad S, Arif M, Alam M, Khan IA, Mian IA, Basir A, Ullah H, Arshad M, Rahman I-U, Saud S, Ishan MZ, Jamal Y, Amanullah, Hammad HM, Nasim W (2018) Phosphate-solubilizing bacteria nullify the antagonistic effect of soil calcification on bioavailability of phosphorus in alkaline soils. *Sci Rep* 8:4339. https://doi.org/10.1038/s41598-018-22653-7

Adnan M, Fahad S, Zamin M, Shah S, Mian IA, Danish S, Zafar-ul-Hye M, Battaglia ML, Naz RMM, Saeed B, Saud S, Ahmad I, Yue Z, Brtnicky M, Holatko J, Datta R (2020) Coupling phosphate-solubilizing bacteria with phosphorus supplements improve maize phosphorus acquisition and growth under lime induced salinity stress. *Plants* 9(900). doi: 10.3390/plants9070900

Agovino M, Casaccia M, Ciommi M, Ferrara M, Marchesano KJ (2019) Agriculture, climate change and sustainability: The case of EU-28. *Ecol Indic* 105:525–543.

Ahmad S, Kamran M, Ding R, Meng X, Wang H, Ahmad I, Fahad S, Han Q (2019) Exogenous melatonin confers drought stress by promoting plant growth, photosynthetic capacity and antioxidant defense system of maize seedlings. *PeerJ* 7:e7793. http://doi.org/10.7717/peerj.7793

Akram R, Turan V, Hammad HM, Ahmad S, Hussain S, Hasnain A, Maqbool MM, Rehmani MIA, Rasool A, Masood N, Mahmood F, Mubeen M, Sultana SR, Fahad S, Amanet K, Saleem M, Abbas Y, Akhtar HM, Waseem F, Murtaza R, Amin A, Zahoor SA, Uldin MS, Nasim W (2018a) Fate of organic and inorganic pollutants in paddy soils. In: Hashmi, MZ and Varma, A (eds.) *Environmental pollution of paddy soils*. Springer International Publishing, Cham, Switzerland, pp. 197–214.

Akram R, Turan V, Wahid A, Ijaz M, Shahid MA, Kaleem S, Hafeez A, Maqbool MM, Chaudhary HJ, Munis, MFH, Mubeen M, Sadiq N, Murtaza R, Kazmi DH, Ali S, Khan N, Sultana SR, Fahad S, Amin A, Nasim W (2018b) Paddy land pollutants and their role in climate change. In: Hashmi, MZ, Varma, A (eds.) *Environmental pollution of paddy soils*. Springer International Publishing, Cham, Switzerland, pp. 113–124.

Alharby AF, Fahad S (2020) Melatonin application enhances biochar efficiency for drought tolerance in maize varieties: Modifications in physio-biochemical machinery. *Agron J*: 1–22.

Ali S, Ying L, Nazir A, Ishaq M, Shah T, Ilyas A, Khan AA, Din IU, Tariq AJ (2019) The effect of climate change on economic growth: Evidence from Pakistan. *Pacific Int J* 2(3):21–30.

Aryal JP, Sapkota TB, Khurana R, Khatri-Chhetri AJ (2019) Climate change mitigation options among farmers in South Asia. *Environ Dev Sustain* 22:3267–3289.

Assmann JM, Anghinoni I, Martins AP, de Andrade SEVG, Cecagno D, Carlos FS, de Faccio Carvalho PC (2014) Soil carbon and nitrogen stocks and fractions in a long-term integrated crop–livestock system under no-tillage in southern Brazil. *Agric Ecosyst Environ* 190:52–59.

Bekuma A, Tadesse T, Galmessa U (2019) Review on negative impacts of livestock production on climate change and its mitigation strategies: A global issue. *World Sci News* 115:218–228.

Brohan P, Kennedy JJ, Harris I, Tett SF, Jones PD (2006) Uncertainty estimates in regional and global observed temperature changes: A new data set from 1850. *J Geophys Res-Atmos* 111(D12):12106.

Bruinsma JJF (2003) *World agriculture: Towards 2015/2030: An FAO perspective*. Earthscan, London.

Cardinael R, Umulisa V, Toudert A, Olivier A, Bockel L, Bernoux M (2018) Revisiting IPCC Tier 1 coefficients for soil organic and biomass carbon storage in agroforestry systems. *Environ Res Lett* 13(12):124020.

Carvalho PCdF, Peterson CA, Nunes PAdA, Martins AP, de Souza Filho W, Bertolazi VT, Kunrath TR, de Moraes A, Anghinoni I (2018) Animal production and soil characteristics from integrated crop-livestock systems: Toward sustainable intensification. *J Anim Sci* 96:3513–3525.

Cayuela M, Van Zwieten L, Singh B, Jeffery S, Roig A, Sánchez-Monedero M (2014) Biochar's role in mitigating soil nitrous oxide emissions: A review and meta-analysis. *Agric Ecosyst Environ* 191:5–16.

Change ICJM M, Chen Z, Marquis M, Averyt KB, Tignor M, Miller H (2007) *The physical science basis. Contribution of working group I to the fourth assessment report of the intergovernmental panel on climate change*. Cambridge University Press, Cambridge and New York, p. 996.

Cheng C, Yang X, Wang J, Luo K, Rasheed A, Zeng Y, Shang Q (2020) Mitigating net global warming potential and greenhouse gas intensity by intermittent irrigation under straw incorporation in Chinese double-rice cropping systems. *Paddy Water Environ* 18:99–109.

Chijioke OB, Haile M, Waschkeit C (2011) Implication of climate change on crop yield and food accessibility in Sub-Saharan Africa. *Centre for Development Research*. University of Bonn, Bonn.

Choudhury SR, Das A, Gupta S, Sharma R, Pathak S (2019) Agronomic management strategies for yield-scaled global warming potential under rice-wheat cropping system. *Current Journal of Applied Science and Technology*: 1–14. https://doi.org/10.9734/cjast/2019/v37i630310

Cline WR (1996) The impact of global warming of agriculture: Comment. *Am Econ Rev* 86:1309–1311.

Cline WR (2007) Global warming and agriculture: Impact estimates by country. *Peterson Institute*.

Fahad S, Bano A (2012) Effect of salicylic acid on physiological and biochemical characterization of maize grown in saline area. *Pak J Bot* 44:1433–1438.

Fahad S, Chen Y, Saud S,Wang K, Xiong D, Chen C,Wu C, Shah F, Nie L, Huang J (2013) Ultraviolet radiation effect on photosynthetic pigments, biochemical attributes, antioxidant enzyme activity and hormonal contents of wheat. *J Food, Agric Environ* 11(3&4):1635–1641.

Fahad S, Hussain S, Bano A, Saud S, Hassan S, Shan D, Khan FA, Khan F, Chen Y, Wu C, Tabassum MA, Chun MX, Afzal M, Jan A, Jan MT, Huang J (2014a) Potential role of phytohormones and plant growth-promoting rhizobacteria in abiotic stresses: Consequences for changing environment. *Environ Sci Pollut Res* 22(7):4907–4921. https://doi.org/10.1007/s11356-014-3754-2

Fahad S, Hussain S, Matloob A, Khan FA, Khaliq A, Saud S, Hassan S, Shan D, Khan F, Ullah N, Faiq M, Khan MR, Tareen AK, Khan A, Ullah A, Ullah N, Huang J (2014b) Phytohormones and plant responses to salinity stress: A review. *Plant Growth Regul* 75(2):391– 404. https://doi.org/10.1007/s10725-014-0013-y

Fahad S, Hussain S, Saud S, Tanveer M, Bajwa AA, Hassan S, Shah AN, Ullah A,Wu C, Khan FA, Shah F, Ullah S, Chen Y, Huang J (2015a) A biochar application protects rice pollen from high-temperature stress. *Plant Physiol Biochem* 96:281–287.

Fahad S, Nie L, Chen Y, Wu C, Xiong D, Saud S, Hongyan L, Cui K, Huang J (2015b) Crop plant hormones and environmental stress. *Sustain Agric Rev* 15:371–400.

Fahad S, Hussain S, Saud S, Hassan S, Chauhan BS, Khan F et al. (2016a) Responses of rapid viscoanalyzer profile and other rice grain qualities to exogenously applied plant growth regulators under high day and high night temperatures. *PLoS One* 11(7):e0159590. https://doi.org/10.1371/journal.pone.0159590

Fahad S, Hussain S, Saud S, Hassan S, Ihsan Z, Shah AN, Wu C, Yousaf M, Nasim W, Alharby H, Alghabari F, Huang J (2016b) Exogenously applied plant growth regulators enhance the morphophysiological growth and yield of rice under high temperature. *Front Plant Sci* 7:1250. https://doi.org/10.3389/fpls.2016.01250

Fahad S, Hussain S, Saud S, Hassan S, Tanveer M, Ihsan MZ, Shah AN, Ullah A, Nasrullah KF, Ullah S, Alharby NW, Wu C, Huang J (2016c) A combined application of biochar and phosphorus alleviates heat-induced adversities on physiological, agronomical and quality attributes of rice. *Plant Physiol Biochem* 103:191–198.

Fahad S, Hussain S, Saud S, Khan F, Hassan S, Jr A, Nasim W, Arif M, Wang F, Huang J (2016d) Exogenously applied plant growth regulators affect heat-stressed rice pollens. *J Agron Crop Sci* 202:139–150.

Fahad S, Bajwa AA, Nazir U, Anjum SA, Farooq A, Zohaib A, Sadia S, Nasim W, Adkins S, Saud S, Ihsan MZ, Alharby H,Wu C,Wang D, Huang J (2017) Crop production under drought and heat stress: Plant responses and Management Options. *Front Plant Sci* 8:1147. https://doi.org/10.3389/fpls.2017.01147

Fahad S, Ishan MZ, Khaliq AK, Daur I, Saud S, Alzamanan S, Nasim W, Abdullah MA, Khan IA, Wu C, Wang D, Huang J (2018). Consequences of high temperature under changing climate optima for rice pollen characteristics-concepts and perspectives. *Archives Agron Soil Sci.* doi:10.1080/03650340.2018.1443213

Fahad S, Adnan M, Hassan S, Saud S, Hussain S, Wu C, Wang, D, Hakeem KR, Alharby, HF, Turan V, Khan MA, Huang J. (2019a) Rice responses and tolerance to high temperature. In: Hasanuzzaman M, Fujita M, Nahar K, Biswas JK (eds.) *Advances in Rice Research for Abiotic Stress Tolerance*. Cambridge: Woodhead, pp. 201–224.

Fahad S, Rehman A, Shahzad B, Tanveer M, Saud S, Kamran M, Ihtisham M, Khan SU, Turan V, Rahman MHU (2019b) Rice responses and tolerance to metal/metalloid toxicity. In: Hasanuzzaman M, Fujita M, Nahar K, Biswas JK (eds.) *Advances in rice research for abiotic stress tolerance*. Cambridge, Woodhead, pp. 299–312.

Farhat A, Hafiz MH, Wajid I, Aitazaz AF, Hafiz FB, Zahida Z, Fahad S, Wajid F, Artemi C (2020) A review of soil carbon dynamics resulting from agricultural practices. *J Environ Manag* 268:110319.

Faostat F (2010) Disponível em. http://faostat.fao.org/site/567/default.aspx#ancor. Acessado em setembro.

Faostat F, Production ACJR, Italy (2016) *Food and agriculture organization of the United Nations*, 2010.

Gul F, Ahmed I, Ashfaq M, Jan D, Shah F, Li X, Wang D, Fahad M, Fayyaz M, Shah AS (2020) Use of crop growth model to simulate the impact of climate change on yield of various wheat cultivars under different agro-environmental conditions in Khyber Pakhtunkhwa, Pakistan. *Arabian J Geosci* 13:112. https://doi.org/10.1007/s12517-020-5118-1

Ghosh PK, Mahanta SK, Mandal D, Mandal B, Ramakrishnan S (2019) *Carbon management in tropical and sub-tropical terrestrial systems*. Springer, Cham, Switzerland.

Giorgi F, Bi X (2005) Updated regional precipitation and temperature changes for the 21st century from ensembles of recent AOGCM simulations. *Geophy Res Lett* 32. doi:10.1029/2005GL024288

Habib ur Rahman M, Ahmad A, Wajid A, Hussain M, Rasul F, Ishaque W, Islam MA, Shiela V, Awais M, Ullah A, Wahid A, Sultana SR, Saud S, Khan S, Shah F, Hussain M, Hussain S, Nasim W (2017) Application of CSM-CROPGRO-Cotton model for cultivars and optimum planting dates: Evaluation in changing semi-arid climate. *Field Crops Res.* http://dx.doi.org/10.1016/j.fcr.2017.07.007

Hammad HM, Farhad W, Abbas F, Fahad S, Saeed S, Nasim W, Bakhat HF (2016) Maize plant nitrogen uptake dynamics at limited irrigation water and nitrogen. *Environ Sci Pollut Res* 24(3):2549–2557. https://doi.org/10.1007/s11356-016-8031-0

Hammad HM, Abbas F, Saeed S, Shah F, Cerda A, Farhad W, Bernado CC, Wajid N, Mubeen M, Bakhat HF (2018) Offsetting land degradation through nitrogen and water management during maize cultivation under arid conditions. *Land Degrad Dev* 1–10. doi: 10.1002/ldr.2933

Hammad HM, Ashraf M, Abbas F, Bakhat HF, Qaisrani SA, Mubeen M, Shah F, Awais M (2019) Environmental factors affecting the frequency of road traffic accidents: a case study of sub-urban area of Pakistan. *Environ Sci Pollut Res.* https://doi.org/10.1007/s11356-019-04752-8

Hammad HM, Abbas F, Ahmad A, Bakhat HF, Farhad W, Wilkerson CJ W, Shah F, Hoogenboom G (2020a) Predicting kernel growth of maize under controlled water and nitrogen applications. *Int J Plant Prod.* https://doi.org/10.1007/s42106-020-00110-8

Hammad HM, Khaliq A, Abbas F, Farhad W, Shah F, Aslam M, Shah GM, Nasim W, Mubeen M, Bakhat HF (2020b) Comparative effects of organic and inorganic fertilizers on soil organic carbon and wheat productivity under arid region. *Communications in Soil Science and Plant Analysis.* doi: 10.1080/00103624.2020.1763385

Hellerstein D, Vilorio D (2019) *Agricultural resources and environmental indicators, 2019.* Economic Research Service, USDA, Washington, D.C.

Houghton JT, Ding Y, Griggs DJ, Noguer M, van der Linden PJ, Dai X, Maskell K, Johnson C (2001) *Climate change 2001: The scientific basis.* The Press Syndicate of the University of Cambridge, Cambridge.

Hussain S, Mubeen M, Ahmad A, Akram W, Hammad HM, Ali M, Masood N, Amin A, Farid HU, Sultana SR, Shah F, Wang D, Nasim W (2019) Using GIS tools to detect the land use/land cover changes during forty years in Lodhran district of Pakistan. *Environ Sci Pollut Res.* https://doi.org/10.1007/s11356-019-06072-3

Hussain MA, Fahad S, Sharif R, Jan MF, Mujtaba M, Ali Q, Ahmad A, Ahmad H, Amin N, Ajayo BS, Sun C, Gu L, Ahmad I, Jiang Z, Hou J (2020) Multifunctional role of brassinosteroid and its analogues in plants. Plant Growth Regul. https://doi.org/10.1007/s10725-020-00647-8

Ilyas M, Mohammad N, Nadeem K, Ali H, Aamir HK, Kashif H, Fahad S, Aziz K, Abid U, (2020) Drought tolerance strategies in plants: A mechanistic approach. *J Plant Growth Regulation.* https://doi.org/10.1007/s00344-020-10174-5

Jaiswal B, Agrawal M (2020) Carbon Footprints of Agriculture Sector. *Carbon Footprints.* Cham, Switzerland: Springer, pp. 81–99.

Jan M, Anwar-ul-Haq M, Shah AN, Yousaf M, Iqbal J, Li X, Wang D, Shah F (2019) Modulation in growth, gas exchange, and antioxidant activities of salt-stressed rice (Oryza sativa L.) genotypes by zinc fertilization. *Arabian J Geosci* 12:775. https://doi.org/10.1007/s12517-019-4939-2

Kamarn M, Wenwen C, Irshad A, Xiangping M, Xudong Z, Wennan S, Junzhi C, Shakeel A, Fahad S, Qingfang H, Tiening L (2017) Effect of paclobutrazol, a potential growth regulator on stalk mechanical strength, lignin accumulation and its relation with lodging resistance of maize. *Plant Growth Regul* 84:317–332. https://doi.org/10.1007/ s10725-017-0342-8

Khan A, Tan DKY, Munsif F, Afridi MZ, Shah F, Wei F, Fahad S, Zhou R (2017a) Nitrogen nutrition in cotton and control strategies for greenhouse gas emissions: A review. *Environ Sci Pollut Res* 24:23471–23487. https://doi.org/10.1007/s11356-017-0131-y

Khan A, Kean DKY, Afridi MZ, Luo H, Tung SA, Ajab M, Fahad S (2017b) Nitrogen fertility and abiotic stresses management in cotton crop: A review. *Environ Sci Pollut Res* 24:14551–14566. https://doi.org/10.1007/s11356-017-8920-x

Kim GW, Gutierrez-Suson J, Kim PJ (2019) Optimum N rate for grain yield coincides with minimum greenhouse gas intensity in flooded rice fields. *Field Crops Res* 237:23–31.

Kim J, Yoo G, Kim D, Ding W, Kang H (2017) Combined application of biochar and slow-release fertilizer reduces methane emission but enhances rice yield by different mechanisms. *Appl Soil Ecol* 117:57–62.

Li Y, Liao J, Guo H, Liu Z, Shen G (2014) Patterns and potential drivers of dramatic changes in Tibetan lakes, 1972–2010. *PloS One* 9:e111890.

Liu Y, Tang H, Muhammad A, Huang G (2019) Emission mechanism and reduction countermeasures of agricultural greenhouse gases – A review. *Greenh Gases: Sc Technol* 9:160–174.

Lozano R, Naghavi M, Foreman K, Lim S, Shibuya K, Aboyans V, Abraham J, Adair T, Aggarwal R, Ahn SY (2012) Global and regional mortality from 235 causes of death for 20 age groups in 1990 and 2010: A systematic analysis for the Global Burden of Disease Study 2010. *The Lancet* 380:2095–2128.

Magrin GO, Travasso MI, Rodríguez GR, Solman S, Núñez M (2009) Climate change and wheat production in Argentina. *Int J Glob Warm*, 1 (1):214–226.

McCarthy JJ, Canziani OF, Leary NA, Dokken DJ, White KS (2001) *Climate change 2001: impacts, adaptation, and vulnerability: contribution of working group ii to the third assessment report of the Intergovernmental Panel on Climate Change, 2*. Cambridge University Press, Cambridge.

Mohammadi A, Cowie A, Mai TLA, de la Rosa RA, Kristiansen P, Brandao M, Joseph S (2016) Biochar use for climate-change mitigation in rice cropping systems. *J Cleaner Prod* 116:61–70.

Mostashari-Rad F, Nabavi-Pelesaraei A, Soheilifard F, Hosseini-Fashami F, Chau K-w (2019) Energy optimization and greenhouse gas emissions mitigation for agricultural and horticultural systems in Northern Iran. *Energy* 186:115845.

Mubeen M, Ahmad A, Hammad HM, Awais M, Farid H, Saleem M, Sami ul Din M, Amin A, Ali A, Shah F, Nasim W (2020) Evaluating the climate change impact on water use efficiency of cotton-wheat in semi-arid conditions using DSSAT model. *J Water Climate Change*. doi:10.2166/wcc.2019.179/622035/jwc2019179.pdf

Nasim W, Ahmad A, Amin A, Tariq M, Awais M, Saqib M, Jabran K, Shah GM, Sultana SR, Hammad HM, Rehmani MIA, Hashmi MZ, Habib Ur Rahman M, Turan V, Fahad S, Suad S, Khan A, Ali S (2017) Radiation efficiency and nitrogen fertilizer impacts on sunflower crop in contrasting environments of Punjab. *Pakistan Environ Sci Pollut Res* 25:1822–1836. https://doi.org/10.1007/s11356-017-0592-z

Nath AJ, Brahma B, Sileshi GW, Das AK (2018) Impact of land use changes on the storage of soil organic carbon in active and recalcitrant pools in a humid tropical region of India. *Sci Total Environ* 624, 908–917.

Porter JR, Xie L, Challinor AJ, Cochrane K, Howden SM, Iqbal MM, Travasso M, Barros V, Field C, Dokken D (2017) Food security and food production systems. In: Field CB, Barros VR, Dokken DJ, Mach KJ, Mastrandrea MD, Bilir TE, Chatterjee M, Ebi KL, Estrada YO, Genova RC, Girma B, Kissel ES, Levy AN, MacCracken S, Mastrandrea PR, White LL (eds.) *Climate change 2014: impacts, adaptation, and vulnerability. Part A: Global and sectoral aspects. Contribution of working group II to the fifth assessment report of the Intergovernmental Panel on Climate Change*. Cambridge University Press, Cambridge, pp. 485–533.

Posas PJ (2011) Exploring climate change criteria for strategic environmental assessments. *Progress in Planning* 75:109–154.

Rehman M, Fahad S, Saleem MH, Hafeez M, Ur Rahman MH, Liu F, Deng G (2020) Red light optimized physiological traits and enhanced the growth of ramie (Boehmeria nivea L.). *Photosynthetica* 58 (4):922–931.

Richards M, Arsalan A, Cavatassi R, Rosenstock T (2019) IFAD research series 35 – Climate change mitigation potential of agricultural practices supported by IFAD investments: An ex ante analysis. https://www.ifad.org/documents/38714170/41066943/35_research.pdf.

Saleem MH, Fahad S, Adnan M, Mohsin A, Muhammad SR, Muhammad K, Qurban A, Inas AH, Parashuram B, Mubassir A, Reem MH (2020a) Foliar application of gibberellic acid endorsed phytoextraction of copper and alleviates oxidative stress in jute (Corchorus capsularis L.) plant grown in highly copper-contaminated soil of China. *Environ Sci Poll Res* https://doi.org/10.1007/s11356-020-09764-3

Saleem MH, Fahad S, Shahid UK, Mairaj D, Abid U, Ayman ELS, Akbar H, Analía L, Lijun L (2020b) Copper-induced oxidative stress, initiation of antioxidants and phytoremediation potential of flax (Linum usitatissimum L.) seedlings grown under the mixing of two different soils of China. *Environ Sci Poll Res* https://doi.org/10.1007/s11356-019-07264-7

Saleem MH, Rehman M, Fahad S, Tung SA, Iqbal N, Hassan A, Ayub A, Wahid MA, Shaukat S, Liu L, Deng G (2020c) Leaf gas exchange, oxidative stress, and physiological attributes of rapeseed (Brassica napus L.) grown under different light-emitting diodes. *Photosynthetica* 58 (3):836–845.

Saud S, Chen Y, Long B, Fahad S, Sadiq A (2013) The different impact on the growth of cool season turf grass under the various conditions on salinity and drought stress. *Int J Agric Sci Res* 3:77–84.

Saud S, Li X, Chen Y, Zhang L, Fahad S, Hussain S, Sadiq A, Chen Y (2014) Silicon application increases drought tolerance of Kentucky bluegrass by improving plant water relations and morph physiological functions. *Sci World J* 2014:1–10. https://doi.org/10.1155/2014/ 368694

Saud S, Chen Y, Fahad S, Hussain S, Na L, Xin L, Alhussien SA (2016) Silicate application increases the photosynthesis and its associated metabolic activities in Kentucky bluegrass under drought stress and post-drought recovery. *Environ Sci Pollut Res* 23(17):17647–17655. https://doi.org/10.1007/s11356-016-6957-x

Saud S, Fahad S, Yajun C, Ihsan MZ, Hammad HM, Nasim W, Amanullah Jr, Arif M, Alharby H (2017) Effects of nitrogen supply on water stress and recovery mechanisms in Kentucky bluegrass plants. *Front Plant Sci.* 8:983. doi:10.3389/fpls.2017.00983

Saud S, Fahad S, Cui G, Chen Y, Anwar S (2020) Determining nitrogen isotopes discrimination under drought stress on enzymatic activities, nitrogen isotope abundance and water contents of Kentucky bluegrass. *Sci Rep* 10:6415. https://doi.org/10.1038/s41598-020-63548-w

Schauberger G, Mikovits C, Zollitsch W, Hörtenhuber SJ, Baumgartner J, Niebuhr K, Piringer M, Knauder W, Anders I, Andre K (2019) Global warming impact on confined livestock in buildings: efficacy of adaptation measures to reduce heat stress for growing-fattening pigs. *Climatic Change* 156:567–587.

Shafi MI, Adnan M, Fahad S, Fazli W, Ahsan K, Zhen Y, Subhan D, Zafar-ul-Hye M, Brtnicky M, Datta R (2020) Application of single superphosphate with humic acid improves the growth, yield and phosphorus uptake of wheat (Triticum aestivum L.) in calcareous soil. *Agronomy* (10):1224. doi:10.3390/agronomy10091224

Shah F, Lixiao N, Kehui C, Tariq S, Wei W, Chang C, Liyang Z, Farhan A, Fahad S, Huang J (2013) Rice grain yield and component responses to near 2°C of warming. *Field Crop Res* 157:98–110.

Smith P (2014) Do grasslands act as a perpetual sink for carbon? *Global Change Biol* 20:2708–2711.

Stevens CJ (2019) Nitrogen in the environment. *Science* 363, 578–580.

Stocker T (2014) *Climate change 2013: the physical science basis: Working Group I contribution to the Fifth assessment report of the Intergovernmental Panel on Climate Change.* Cambridge University Press Cambridge.

Stokes C, Howden M (2010) *Adapting agriculture to climate change: preparing Australian agriculture, forestry and fisheries for the future.* CSIRO, Collingwood, Australia.

Subhan D, Zafar-ul-Hye M, Fahad S, Saud S, Brtnicky M, Hammerschmiedt T, Datta R (2020) Drought stress alleviation by ACC deaminase producing Achromobacter xylosoxidans and Enterobacter cloacae, with and without timber waste biochar in maize. *Sustainability* 12(6286). doi:10.3390/su12156286

Sun Q, Miao C, Hanel M, Borthwick AG, Duan Q, Ji D, Li H (2019) Global heat stress on health, wildfires, and agricultural crops under different levels of climate warming. *Environ Int* 128:125–136.

Ullah S (2017) Climate change impact on agriculture of Pakistan – A leading agent to food security. *Int J Environ Sci Nat Resour* 6:76–79.

Usepa E (2006) *Global Anthropogenic Non-CO_2 Greenhouse Gas Emissions: 1990–2020.* United States Environmental Protection Agency, Washington, DC.

Wahid F, Fahad S, Subhan D, Adnan M,, Zhen Y, Saud S, Manzer HS, Brtnicky M, Hammerschmiedt T, Datta R (2020) Sustainable management with mycorrhizae and phosphate solubilizing bacteria for enhanced phosphorus uptake in calcareous soils. *Agriculture* 10(334). doi:10.3390/agriculture10080334

Wajid A, Hussain K, Maqsood M, Ahmad A, Hussain A (2007) Influence of drought on water use efficiency in wheat in semi-arid regions of Punjab. *Soil and Environ* 26:64–68.

Wang D, Shah F, Shah S, Kamran M, Khan A, Khan MN, Hammad HM, Nasim W (2018) Morphological acclimation to agronomic manipulation in leaf dispersion and orientation to promote 'Ideotype' breeding: Evidence from 3D visual modeling of 'super' rice (Oryza sativa L.). *Plant Physiol Biochem.* https://doi.org/10.1016/j.plaphy.2018.11.010

Wang X, Qi J-Y, Zhang X-Z, Li S-S, Virk AL, Zhao X, Xiao X-P, Zhang H-L (2019) Effects of tillage and residue management on soil aggregates and associated carbon storage in a double paddy cropping system. *Soil and Tillage Res* 194:104339.

Widiawati Y (2019) Estimation of greenhouse gas (GHG) emission from livestock sector by using ALU tool: West Java case study. *IOP Conference Series: Earth and Environmental Science* 372:012044.

Wu C, Tang S, Li G, Wang S, Fahad S, Ding Y (2019) Roles of phytohormone changes in the grain yield of rice plants exposed to heat: A review. *PeerJ* 7:e7792. doi:10.7717/peerj.7792

Wu C, Kehui C, She T, Ganghua L, Shaohua W, Fahad S, Lixiao N, Jianliang H, Shaobing P, Yanfeng D (2020) Intensified pollination and fertilization ameliorate heat injury in rice (Oryza sativa L.) during the flowering stage. *Field Crops Res* 252:107795.

Yang Z, Zhang Z, Zhang T, Fahad S, Cui K, Nie L, Peng S, Huang J (2017) The effect of season-long temperature increases on rice cultivars grown in the central and southern regions of China. *Front Plant Sci* 8:1908. https://doi.org/10.3389/fpls.2017.01908

Yilmaz M (2019) The effects of different combinations of combined fertilizer doses on some turfgrass performances of turf mixture. *Pak J Bot* 51:1357–1364.

You L, Rosegrant MW, Fang C, Wood SJCD (2005) *Impact of global warming on Chinese wheat productivity*. Discussion Paper, International Food Policy Research Institute.

Zafar-ul-Hye M, Naeem M, Danish S, Fahad S, Datta R, Abbas M, Rahi AA, Brtnicky M, Holatko, Jiri, Tarar ZH, Nasir M (2020a) Alleviation of cadmium adverse effects by improving nutrients uptake in bitter gourd through cadmium tolerant rhizobacteria. *Environments* 7(54). doi:10.3390/environments7080054

Zafar-ul-Hye M, Tahzeeb-ul-Hassan M, Abid M, Fahad S, Brtnicky M, Dokulilova T, Datta RD, Danish S (2020b) Potential role of compost mixed biochar with rhizobacteria in mitigating lead toxicity in spinach. *Scientific Rep* 10:12159; https://doi.org/10.1038/s41598-020-69183-9.

Zahida Z, Hafiz FB, Zulfiqar AS, Ghulam MS, Fahad S, Muhammad RA, Hafiz MH, Wajid N, Muhammad S (2017) Effect of water management and silicon on germination, growth, phosphorus and arsenic uptake in rice. *Ecotoxicol Environ Saf* 144:11–18.

Zaman QU, Aslam Z, Yaseen M, Ishan MZ, Khan A, Shah F, Basir S, Ramzani PMA, Naeem M (2017) Zinc biofortification in rice: Leveraging agriculture to moderate hidden hunger in developing countries. *Arch Agron Soil Sci* 64:147–161. https://doi.org/10.1080/03650340.2017.1338343

Zamin M, Khattak AM, Salim AM, Marcum KB, Shakur M, Shah S, Jan I, Fahad S (2019) Performance of Aeluropus lagopoides (mangrove grass) ecotypes, a potential turfgrass, under high saline conditions. *Environ Sci Pollut Res* https://doi.org/10.1007/s11356-019-04838-3

Zhu B (2004) Climate change affecting crops. An article published in Chinese newspaper China daily, May 24, 2004, Beijing, China.

2

Climate Change and Climate Smart Plants Production Technology

Imran[1], Shah Fahad[2], Amanullah[1], Shah Khalid[1], Muhammad Arif[1], and Abdel Rahman Al-Tawaha[3]
[1]Department of Agronomy, The University of Agriculture, Peshawar, Pakistan
[2]Department of Agronomy, The University of Haripur, Pakistan
[3]Department of Biological Sciences, Al Hussein Bin Talal University, Ma'an, Jordan

2.1 Introduction

Food legumes are widely grown throughout the world, with a long cultivation history and in a diversity of cropping systems. Legumes are important food crops that can play a major role in achieving food security, nutrition, and human health, contributing to making agriculture more sustainable and helping mitigate and adapt to climate change (Imran et al., 2016). With the advantages of short growing periods and biological nitrogen fixation (BNF), food legumes play an important role in cropping systems (Baqa et al., 2015; Imran et al., 2016; Naveed et al., 2016; Samreen et al., 2016). Legumes can be integrated in rotation with cereals, tuber, and vegetable crops. Such systems provide a quality food harvest after nonleguminous crops and improve soil fertility on farms. According to FAO (1994), legumes are crop plants belonging to the *Leguminosae* family (commonly known as the pea family) that produce edible seeds and are used for human and animal consumption. However, those harvested for only dry grain are classified as legumes. Legume species used for oil extraction (e.g. soybean [*Glycine max* L.] and groundnut [*Arachis hypogaea* L.] and sowing purposes (e.g. clover [different species belonging to the genus *Trifolium* L.] and alfalfa [*Medicago sativa* L.], are not considered legumes. Likewise, legume species are not considered as legumes when they are used as vegetables (e.g. green peas [*Pisum sativum* L.] and green beans [*Phaseolus vulgaris* L.]. Consequently, when common bean (*Phaseolus vulgaris* L.) is harvested for dry grain, it is considered a legumes, but when the same species is harvested unripe (known as green beans), it is not treated as a legumes. Major constraints in legumes production are diseases and insect pest along with minimum attention of the farmers (Adnan et al., 2017; Baqa et al., 2015; Imran et al., 2016; Naveed et al., 2016; Samreen et al., 2016; Wani et al., 2012). Common insect and pest of mung bean are pods borer, whitefly, pods sucking bug, caterpillar, jassids, and aphids. Pigeon pea common diseases are rust leaf spot and Fuserium wilt, and occasionally sterility mosaic virus may also attack. This disease may spread and be transmitted by Eriophyid mite. The affected plant is totally or partially sterile and can not produce seed. Common insect pest of pigeon pea may be moth and pod fly. These diseases can reduce yield from 10% to 30%, depending upon the severity, pathogen, and weather conditions (Aise et al., 2011; Imran et al., 2016; Khan et al., 2019). Cow pea is also susceptible to various disease and insect pest. Common diseases of cow pea are aphids born by mosaic virus, fuserium wilt, leaf smut, and scab. General insect pest of cow might be beetles, aphids, pods borer, caterpillar, and white flies which seriously damage the crop. Certain insects can also provide access to disease organism and or transmit them directly to plants (Baqa et al., 2015; Imran et al., 2016;

Naveed et al., 2016; Samreen et al., 2016). Therefore, understanding the relationships between the insect and the crop will enable farmers to manage pests much better. Major insects that damage soybean crops are stem fly, white fly, green stink bug, cut worm, and larvae. Insects attack on all parts of the soybean plant and feed throughout the growing season. A new practice to control insect attack is based on knowledge of the economic injury levels of the consequential insects (Anchal et al., 1997; Akpalu et al., 2014). The economic injury level is the population of insects that is capable of producing an amount of economic damage which is at least equal to the cost of controlling the insects. Wise monitoring of major insects is required in order to effectively make decision relative to insecticide application (Ankomah et al., 1996; Imran et al., 2015, 2019; Imran, Amanullah, 2018).

In this context to obtain full farm productivity of legumes in unfavourable environments to increase food security and overcome hunger, intercropping or relay intercropping of food with other crops is another good choice. As ideal intercrops, food can increase crop yields and economic benefits of intercropping systems, and much more innovation will be needed in the future for adaptation to climate change and to meet the increasing demand for food for the overwhelming population.

2.2 Climate Change, Adaptability, and Ecology of Legumes Crop

Pulses, including two sorts, depend firstly on the growing period of winter season crops, such as chickpeas, lentils, lathyrus, broad bean and vetch, secondly, on summer season types, for example, soybeans and peas. They have many weals in our globe; they supply humans with food, a feed that also provides the soil with nitrogen fixation through bacterial nodes and increase the CO_2 concentration of the ambience (Baqa et al., 2015; Imran et al., 2016; Mousavi Derazmahalleh et al., 2019; Naveed et al., 2016; Samreen et al., 2016). Fabaceae, syn. Leguminosae, is the third largest family of angiosperms, including over 750 genera and 19,000 species ranging from small herbs to large trees. Studying the pulse species under stress conditions helps in choosing pasture breeding and soil preservation programs, mainly in arid and semi-arid regions. A study was undertaken to create biogeographical manners in the pulse flora of southern Africa so as to assist the selection of species with agricultural potency (Imran et al., 2015, 2019; Imran, Amanullah, 2018; Trytsman et al., 2016).

Acting as an important part in human nourishment, mammal feeding and soil fecundity improvement, pulse crops play a vast task of decreasing paucity and contributing to persistent economic progress in the globe. Embracing new varieties of pulses have latent endure under stress situations aid in expanding the planting zones which guide better the country's revenue. Leaves from pulse trees and shrubs can be a precious supplement for nourishment livestock, particularly through the desiccated time when other nurturing raw materials are insufficient or their alimentary value is low. The agronomic assessment at two agro-ecologically discrete regions in eastern Democratic Republic of the Congo (DRC) verified that diverse forage trees and shrub species tailored distinctly. The species with top forage superiority, *L. diversifolia* (especially accession ILRI 15551) and *L. leucocephala*, execute preferably with advanced soil productiveness (Imran et al., 2015, 2019; Imran, Amanullah, 2018; Katunga, 2014).

Legumes are generally developed rain fed and show distinct kinds of drought zones from terminal drought to sporadic drought (Purushothaman et al., 2013). Acclimation technologies are endeavours to reinstate the synchrony or to generate new management between ambiance and plant communities. Both chickpeas and pigeonpeas are important grain legumes developed for their protein-rich seeds used in human utilization, for their capacity to renovate and conserve soil fecundity by nitrogen fixation, and for their convenience to robust very well into different cropping manners (Upadhyaya et al., 2010). Heat stress through venereal progress can issue noteworthy yield loss. Chickpea (*Cicer arietinum* L.) is an important food legume and heat stress influence chickpea ontogeny over a range of milieu (Devasirvatham et al., 2012; Imran et al., 2015, 2019; Imran, Amanullah, 2018). Four accessions of *C. reticulatum* and one accession of *C. pinnatifidum* were found to be as impedance to water deficiency and heat stress (up to 41.8 °C) (Canci and Toker, 2009).

In chickpea, total root biomass in precocious development stages has been revealed to appreciably participate to seed product under incurable water deficit in middle and south India (Kashiwagi, 2015). Deeper root coordination and superior root desiccated mass could be allied with enhanced prevention to soil water restriction in drought tolerant legume than in drought-sensitive ones (Abdelhamid et al., 2010; Imran et al., 2014, 2015; Imran and Khan, 2015). Reproduction programs are ongoing to promote cultivars tailored to the spacious range of latitudes and with disease resistance (Siddique et al., 2012).

Conformation strategies that are debated include prior seeding of pulse crops, use of coldness pulses, crop succession within crop rotation, and modulations to the microclimate, such as forthright seeding into standing straw (Cutforth, 2007; Imran et al., 2014, 2015; Imran and Khan, 2015).

Prospect pulses genotypes will require to preserve or be smooth to be greater under the expected climate changes. The most significant way to enhance the granule yield of cool period food legumes per unit area under stress niches will be the enlargement of cultivars with conflict to numerous a biotic stresses in the circumstance of endurable incorporated pulses output, insect and pest executive, nutrients organization, and irrigation use if accessible (Toker and Yadav, 2010). Faba bean has a lot of proper imputes for sustainable recent cropping patterns but these outcomes propose that yield will be inadequate by anticipated climate change, presuppose the expansion of heat tolerant cultivars, or enhanced flexibility by other mechanisms such as earlier flowering periods (Bishop et al., 2016).

Faba bean yield will turn into ever more uneven with anticipated climate change and underline the requirement to recover heat stress resilience of faba bean floral advance and anthesis to endorse future (Bishop et al., 2016; Imran et al., 2014, 2015; Imran and Khan, 2015). In the study of Bishop et al. (2016), Faba bean yield was minimized by heat stress within the hotness range recognized to contrive yield decreasing in other output species, and as in other species. Replacement of imported soybean food by locally created pulses could have a convenient effect, mainly for pulses produced and used on-farm. Modest is identified about the natural adjustment of faba bean germplasm, or about the regions and impacts of genes that affect traits associated to drought adjustment, chiefly stomatal morphology and role as key manners for gas reciprocity between plant and ambiance (Khazaei, 2014). Katerji (2011) report that drought at flowering, early poding, and grain-filling phases decrease both grain and straw yields.

Promoting new varieties that can be tolerated for environment invasion mainly climate change will be sharing in raising the prime creation of pulses during minimize the stomata conductance, CO_2 improvement, universal warming tolerates. On the other, the nitrogen fixation outfitted by legume and C:N ratio can be enhanced.

In order to enhance crop execution, it is necessary to be aware of how crop species react and adjust to water deficit circumstances (Imran et al., 2014, 2015; Imran and Khan, 2015; Khazaei, 2014). Due to variability in quantity, interval, strength, and allocation of rainfall overall regions global, enhancement of new varieties is required to evade drought incident. With adjustment based on the genetic makeup and phenotype elasticity, species with high flexibility are more adapted to stress environment and can be survived for long period and generate high yield ingredients.

Drought is a very imperative ecological cause restraining the output of pulses crops as well other field crops globally. Climate change patterns estimate variableness in rainfall patterns, expand periods of winter and spring while drought at continents will effect areas of pulses planting. The planting required for varieties showing prior drought bearing and water use efficiency is fabricated in the enhancing stressed within pulses plant improving programs on selection of varieties preferable to tolerate expanded times of water deficit.

Modern molecular markers help in limiting quantitative trait loci that participated in selection of the favourable traits used in breeding programs. Amelioration that demanded for pulses depended on conductors of water deficit, for example, chlorophyll fluorescence, stomatal density, and root density. Genetic resource gathering and classification can have an important part in distinguishing pulses genotypes adapted to water-deficit-prone environments. In legumes, the influences of water deficit or other stresses on nitrogen fixation requires to be quantified alongside the influences on the plant itself (Abberton et al., 2008).

Direct gene transfer can be helpful in enhancing the yield and yield ingredients during genetic stability of legume crops under various ecology microclimates. Molecular biology has a wide range of capacity enforcements in pasture pulses breeding, including marker-assisted and genomics-assisted selection and the recognition of quantitative trait loci and elect genes for necessitate characters (Nichols et al., 2012).

On the other hand, classical breeding programs have an immense role in acclimatization of pulses based on introgression of genes. Nichols et al. (2012) pointed that new perennial pulses have recently been progressed to defeat the soil acidity and water logging output oblige of Lucerne and white clover and to minimize groundwater recharge and the dispersal of arid land salinity. Several studies have been undertaken on the adaptation. For any of these selected species to be prosperous on a marketable level, there are numerous other traits that need to be considered in their improvement. As a minimum, they would be demanded to have an extensive soil and climatic adaptation to help endorse a bulky seed market (Real et al., 2011). Real et al. (2011) and Imran (2018) stated that the three-year assessment at six regions in southern Australia recognized a set of 10 preference perennial forage legumes (*B. bituminosa* var. albomarginata; *C. australasicum*; *D. hirsutum*; *K. prostrata*; *L. bainesii*; *L. pedunculatus*; *L. cytisoides*; *M. sativa* subsp. sativa; *M. sativa* subsp. Caerulea; and M. sativa subsp. falcata) with appropriate adaptation, rendering, and insistence. Voisin et al. (2013) reported that even though the properties of pulses concerning the nitrogen cycle may raise the elasticity of output systems exposed with climate change, their involvement to extenuating climate change earns to be better determined at global levels. Ecology and adaption point of pulses depends on the emergent cropping systems and rotation to decrease the usage of pesticides and mineral and fertilizer and enhance the yield stability. Himanen (2016) concluded that intercropping can be availed as one adjustment strategy to reinforce the adaptive ability of Finnish farms. Promoting disease resistance, yield, and varieties with ameliorative time of flowering will share in financial system and beneficial business. Biodiversity farming and building of strong soils are assured, flourishing strategies to acclimatize to the augment in droughts presaged under climate change.

Abd El-Moneim and Cocks (1992) reported that the two selections, 347 and 311 of Lathyrus spp., generating from Syria and Turkey, respectively, joined both high pasture and seed yields with extensive adaptation and steadiness and could be deemed the most broadly adapted lines. Various entities have brightened to adapt to the latest technologies and techniques in provisions of both yield and firmness (Abd El-Moneim, 1988). Collaborative adaptation arrangement by farmers and researchers can construct approximately rationally and environmentally and economically resonance adaptation events as well as can motivate the precedence building of farm-adaptive ability (Himanen, 2016). Climate change adaptations exemplarily link to exploit on technological regenerations variegation strategies, changed organization, and the construction of public networks (Darnhofer, 2010). Participating farmers and pastoral stakeholders in limiting and boosting adaptation strategies for agriculture to expose climate change, with affirmation on the part of intercropping in structuring the adaptive strength of farms, are to have highly executions of climate prompt manners. Because climate change has a great part in food output for present and next generation that is predictable to turnoff more than three billion suffering from starvation, the next congress and strategic planning issued by governorates should take into consideration all factors that have direct influences on growth and allocation of the field crops focusing on the pulses crops, which contribute high percentage of food nutritive value.

2.3 Role of Legumes in Body Health Maintenance

Legumes (mung bean, mash bean, pigeon pea, cow pea, chickpea, field pea, lentil, and soybean) are hailed as the most protective bean (Aziz et al., 2016; Imran, Amanullah, 2018; Imran et al., 2015, 2016, 2017, 2018; Imran and Khan, 2015). It has the highest protein content amongst plant products. As animal protein contains all the essential amino acids, lacking in legumes protein, 'soy protein' refers to the protein found in soybeans. Soy is often used to replace the animal proteins in an individual's diet. Soybean belongs to family leguminacea but is mostly known as oilseed. Here it is important to understand soybean as a leguminous crop instead of oilseed (Devi et al., 2012). Soybean is the only vegetable food that contains all eight essential amino acids. Soybeans are processed to various soya products namely soy flour, soy milk, cottage cheese-like tofu, and fermented products like tempeh and miso.

2.4 Legumes in Diet Can Help in Cardiovascular Health Stimulation (Legumes and Soybean)

Soybean oil is a rich source of polyunsaturated fatty acids, including the two essential fatty acids called linoleic acid and linolenic acid. Substituting saturated fat sources with unsaturated fats may help lower your cholesterol levels (Devi et al., 2013). Elevated serum cholesterol levels lead to atherosclerosis and the increased risk for heart disease. These two essential fatty acids are precursors to hormones that regulate smooth muscle contractions and blood pressure. Because your heart is a muscle, there is significance in this health benefit. Soybeans are a source of soluble fibre, which can help lower cholesterol levels and potentially reduce the risk for heart disease (Gaur et al., 2012).

2.5 Legumes and Recovery of Digestive Health and Prevention of Colon Cancer

Mostly legumes and especially soybeans are a source of insoluble fibre, which adds bulk to stool and helps waste pass quickly through the digestive tract. Adding soybeans to your diet may prevent colon cancer, diverticular disease, constipation, and haemorrhoids. Soybeans have a high concentration of calcium, which is a mineral crucial to bone health (Imran, Amanullah, 2018; Imran et al., 2015, 2016, 2017, 2018; Imran and Khan, 2015; Siddique et al., 2012). Isoflavones.info says isoflavones can interfere as well as activate the activity of your own estrogen. Soybeans contain isoflavones, which are substances that have a chemical structure very similar to estrogen. During menopause, when estrogen levels drop, isoflavones bind to the estrogen receptors in cells and ease the symptoms of menopause, such as hot flashes. The isoflavones in soybeans increase bone density in women and offer protection against osteoporosis (Figure 2.1).

2.6 Climate Change and Legume Productivity and Profitability

The legumes production technology may be almost similar and undistinguishable throughout the world in this advance era of internet and makes the world global village but here are a few available technologies that can increase the productivity and profitability of legumes (Sinclair and Vadez, 2012).

2.6.1 Short-Duration, High-Yielding Varieties

Matching legumes crop maturity duration to available cropping window, including soil moisture availability, is a major strategy to avoid drought stress. Hence, emphasis in crop improvement programmes has been to develop high-yielding, short-duration cultivars which escape terminal drought (Imran et al., 2015). These short duration varieties provide opportunities for inclusion of a given crop or variety in the cropping systems with a narrow cropping window or new production niches.

The key factors for significant increase might be achieved with the following technique.

1. Introduction of high yielding, short duration, fusarium wilt resistant varieties
2. High adoption of improved cultivars and production-technologies-mechanized field operations and effective management of pod-borer
3. Availability of grain storage facilities to farmers at local level at affordable cost

2.6.2 Improved Varieties with Drought Tolerance

A wider dissemination of drought-tolerant material would provide sustenance to the livelihoods of farmers who are more vulnerable to shocks of crop failure. The drought-tolerant varieties can provide cost-effective long-term solutions against adverse effects of drought. Returns to investment in breeding for drought tolerance are likely to be higher compared to those in other drought management strategies

FIGURE 2.1 Tangible view of soybean experimental plots at ARI Agriculture Research Institute Mingora Swat with biofertilization of Trichoderma and PSB.

(Imran et al., 2015). On the other hand, even though the potential economic benefits of drought-tolerance breeding research are attractive, farmers may not benefit from it if appropriate institutional arrangements are not in place for multiplication and distribution of seeds of improved varieties. This is more so in the case of large seeded legumes whose seed requirement is very high (Varshney et al., 2010; Imran, Amanullah, 2018; Imran et al., 2015, 2016, 2017, 2018; Imran and Khan, 2015).

2.6.3 New Niches

Legume crops have great diversity of maturity durations that enable their cultivation in many niches and different production systems to increase production. A few examples are given below, but there are many more in other crops and niches that can be exploited successfully.

The main cropping systems involving major food could be the following:

Faba bean

a. Faba bean – rice – barley intercropping with green manure – cotton.
b. Faba bean – transplanting cotton – faba bean intercropping with green manure – maize intercropping with soybean, sweet potatoes, or rice.
c. Faba bean intercropping with green manure – cotton – faba bean or barley – transplanting cotton.
d. Faba bean intercropping with green manure – maize intercropping with cotton – Faba bean intercropping with green manure – maize intercropping soybean or sweet potatoes.
e. Faba bean intercropping with green manure – maize intercropping soybean – barley intercropping with green manure – cotton.
f. Faba bean intercropping with green manure – corn multiple cropping with rice – barley – cotton.

The main biennial cropping patterns, generally with two crops a year, are as follows

a. Faba bean – corn relay intercropping with sweet potatoes – wheat (canola) – peanut.
b. Wheat – cotton – faba bean – corn relay intercropping with sweet potatoes.
c. Wheat (barley) – corn relay intercropping with sweet potatoes – faba bean – peanut.
d. Faba bean – early rice – late rice – canola – semi-late rice.
e. Wheat – semi-late rice – faba bean – semi-late rice – barley.

Pea

a. The first year: peas – early rice – late rice.
b. The second year: barley (wheat) – early rice – late rice.
c. The third year: canola – early rice – late – rice.

Intercropping and relay intercropping

a. Pea – maize – maize.
b. Pea – canola – wheat.
c. Pea – wheat – potato.
d. Pea – barley – maize.

Mung bean

a. Wheat (canola) – mung bean – wheat – maize (millet).
b. Wheat – mung bean – wheat – mung bean (sweet potato).
c. Wheat – mung bean – cotton – wheat – mung bean (millet) (three crops in two years).
d. Wheat – mung bean – sweet potato – maize – wheat – mung bean (five crops in three years).

Adzuki bean

a. Adzuki bean – maize (sorghum) – millet.
b. Adzuki bean – maize – sorghum.
c. Adzuki bean – spring wheat – maize (flax) – maize.
d. Adzuki bean – wheat – wheat.
e. Adzuki bean – millet – maize – wheat.
f. Winter wheat (barley) – adzuki bean – winter wheat (barley) – adzuki.
g. Winter wheat (barley) – adzuki bean – cotton (maize, sorghum, or millet).
h. Winter wheat (barley) – summer maize intercropping with adzuki bean – spring maize intercropped with soybean.

i. Spring wheat – millet (broomcorn millet) – adzuki bean.
j. Winter wheat (barley) – maize intercropped with adzuki bean – winter wheat (barley) – maize intercropped with adzuki.
k. Winter wheat (barley) – sunflower intercropped with adzuki bean – winter wheat (barley) – sunflower intercropped with adzuki.

2.7 Advance Legumes Production Technology

2.7.1 Seed Inoculum

Mungbean (green gram) inoculation with various beneficial microbes (PSB, Trichoderma, *Rhizobium*, *Bradyrhizobium japonicum*, *Azospirillum*, *Azotobacter*, etc.) have been tested and its been noticed the highest green forage, dry matter, and crude protein (Figure 2.1). It has been reported that the highest number of plants m^{-2}, plant height (cm), number of leaves $plant^{-1}$, pods length (cm), numbers of seeds pod^{-1}, number of pods $plants^{-1}$, thousand seed weight, seed yield and harvest index were noted with the inoculation of bio fertilizer (PSB) (Sinclair and Vadez, 2012; Imran, Amanullah, 2018; Imran et al., 2015, 2016; Imran and Khan, 2015). The nitrogen-fixing bacteria (Rhizobia) that lives on legume roots in nodules are not native to most soils. The best way to introduce these bacteria is to inoculate the seed. Once introduced, the rhizobia population remains active in the soil for a long time. In the presence of the appropriate inoculant of Rhizobium japonicum, more nodules are formed on roots of legumes plant which can fix atmospheric nitrogen from the air that is almost as effective as nitrogen applied as fertilizer to promote growth and development of the plant (Imran, Amanullah, 2018; Imran and Khan, 2015, 2017).

2.7.2 Method of Inoculation

Inoculum is a black powder containing nitrogen fixing bacteria which are mixed with ground peat or some similar carrier and applied on seed just before planting time. Seeds are moistened with concentrated sugar solution; inoculant is applied @ 1250 gm per 100 kg seeds and then mixed thoroughly to have a uniform coating of inoculum on the seeds. This process should be done in shady place (Imran et al., 2016). The use of fungicide in case of seed treatment may interfere with inoculated seed and with symbioses of Rhizobium system. Thus, compatible fungicides (i.e. Benlate and Dithane Z-78 [Zineb]) with no toxicity to Rhizobia should be used. One must treat seed immediately before planting and use an inoculum dose a little higher than recommended.

2.7.3 Method of Sowing

The method of sowing should meet three objectives i) adequate and uniform depth of seed placement, ii) adequate seed numbers and uniform distribution, and iii) sufficient soil-seed contact for germination. Under existing conditions, legume seed must be planted with single row cotton drill or tractor drill because of their consistency in producing good stands (Imran et al., 2016).

2.7.4 Time of Sowing (*Kharif* Legumes/*Rabi* Legumes)

Variation in yield loss due to delay in the planting is the result of variation in weather and its influence on disease, plant height, flower abortion, amount of vegetative growth, and so forth (Imran et al., 2015, 2016). Therefore, the planting date is considered to be the one of the important factors for maximum crop yield. Late planting may result in significant decrease in seed yield (Figures 2.1–2.2).

2.7.5 Seed Rate

Optimum seed for *Kharif* or *Rabi* legumes at a depth of 3 to 5 cm with 30 to 45 cm row spacing or optimum space for each legume crop gives optimum population of 120,000 to 130,000 plants per acre.

FIGURE 2.2 Tangible view of mungbean (green gram) experimental plots at Agriculture Research Institute Mingora Swat with biofertilization of Trichoderma and PSB.

As much as 20 to 24 plants per metre of row are generally satisfactory. However, within a wide range of plant populations for a particular row spacing, legume yield do not vary significantly. However, low populations result in low poding height and excessive branches but good lodging resistance. Conversely, high plant populations result in increased lodging, high poding height and less branching. Generally, legumes in narrow rows show comparatively higher yielding because they capture more of the sun's energy which drives photosynthetic machinery of the plant. However, wide rows are used if the varieties are tall and bushy type.

2.7.6 Irrigation

Number of irrigations varies with climatic conditions, management practices and length of growing season. Moisture stress during flowering, pod filling, and seed development stages reduces yield. Usually four to six irrigations are required for *kharif* legumes and two to three irrigations for *Rabi* crop depending upon the rains.

Therefore, irrigation must be given at the following stages:

- Three weeks after germination
- Initiation of flowering

- Pod filling stage
- Seed development stage

2.7.7 Weed Management

Weeds complete with legume crops for nutrients, moisture, and light and thus reduce yield. The most effective measure for developing weed control in legumes vary, depending on types of weeds, degree of weed infestation, soil type, weather patterns, crop rotation, tillage methods, row spacing, and equipment available. A good weed control programme should include combination of preventive, cultural, mechanical, and chemical practices.

2.7.8 Preventive Measures

All these measures taken to prevent the introduction and spread of weeds include the use of weed free crop seed.

- Certified seed of legumes
- Clean planting/harvesting equipment
- Weed free soil, seed, and farm yard manure/organic residues

2.7.9 Cultural Control

2.7.9.1 Crop Rotation

Crop rotation reduces the weed populations of certain weeds common to a particular crop. It results in improved crop yield, quality, and improved soil conditions, and reduces the chance of plant disease and insect infestation.

2.7.9.2 Tillage Practices

Deep and dry plugging gives a substantial control of perennial weeds. By tillage method, weeds are buried with soil, thrown over and with disruption of the ultimate relationship between the weed, crop, and soil.

2.7.10 Mechanical (Physical) Control

Hand weeding is the most common practice and employed by the farmers. Two weeding are recommenced during the growing period for each leguminous crop. Mechanical weeding by rotary weeder controls the annual weeds effectively and economically.

2.7.11 Chemical Control

The experimental results have shown that proper use of pre-emergence herbicides, that is, pendimethalin 8 (Stomp), trifluralin (Treflan), and oxadiazon (Ronstar) can be applied after planting before the germination of legumes crop. These have been found very effective in controlling a wide range of grasses and broad leaf weeds.

2.7.12 Integrated Weed Control

This involves the use of two or more aforementioned weed control techniques. The combined use of more than one weed control technique is advantageous, because one technique rarely achieves complete eradication of weeds.

2.7.13 Diseases and Insect Pests of Legumes and Their Control

About 35 diseases of economic importance are known to affect leguminous crops throughout the world. Mostly on *kharif* legumes (mungbean), fungal disease of leaf spot, root rot and powdery mildews may attack in wet and warm weather while bacterial leaf blight and yellow mosaic virus are also serious diseases in Asian countries like Pakistan and India. Common insect and pest of mung bean are pods borer, whitefly, pods sucking bug, caterpillar, Jassids, and aphids. Pigeon pea common diseases are rust leaf spot, Fuserium wilt and occasionally sterility mosaic virus may also attack (Imran et al., 2016). This disease may spread and be transmitted by Eriophyid mite. The affected plant is totally or partially sterile and can not produce seed. Common insect pest of pigeon pea may be moth and pod fly. These diseases can reduce yield from 10 to 30%, depending upon the severity, pathogen, and weather conditions.

Cow pea is also susceptible to various disease and insect pest. Common diseases of cow pea are aphids-born mosaic virus, fuserium wilt, leaf smut, and scab. General insect pest of cow might be beetles, aphids, pods borer, caterpillar, and white flies, which damage the crop heavily. Certain insects can also provide access for disease organisms or transmit them directly to plants. Therefore, understanding the relationships between the insect and the crop will enable farmers to manage pests much better. Major insects that damage soybean crops are stem fly, white fly, green stink bug, cut worm, and larvae. Insects attack on all parts of the soybean plant and feed throughout the growing season. A new practice to control insect attack is based on knowledge of the economic injury levels of the consequential insects (Imran et al., 2016). The economic injury level is the population of insects that is capable of producing an amount of economic damage which is at least equal to the cost of controlling the insects. Wise monitoring of major insects is required in order to effectively make decisions relative to insecticide application.

In Pakistan, soybean major diseases have been observed and these are anthracnose, charcoal rot, purple seed stain, pod and stem blight, and bacterial blight/pustule. Careful diagnosis is very important for the disease control strategies, because more than one pathogen or a complex may be responsible for final loss in yield and seed quality. Thus, the collection of samples is essential throughout the growing season (Imran et al., 2016).

Major *Rabi* legumes like chick pea, lentil, and field pea might produce low yield and profitability due to various insect pest and disease attack. Common disease of chick pea are blight disease, the most dangerous enemy of chick, root rot, and wilting. Insects of chick pea are aphids, gram caterpillar, leaf folder, and pods borer which attack and reduce the yield from 30% to 70% relating to the severity. Reduction in yield could be overcome by planting disease and insect pest resistant cultivars, proper seed treatment, and crop rotation. Like chick pea, lentil crop is also susceptible to attack of aphids, gram caterpillar, leaf folder, pods borer, and weevils. This can be checked by crop rotation and appropriate pesticides (Imran et al., 2016). The most common disease of lentil are leaf rust, downy mildew, blight and fuserium wilt, whereas sometimes root rot or root wilting may occurred. Decrease in yield could be overcome by planting disease and insect pest resistant cultivars, proper seed treatment, and crop rotation. In another *Rabi* crop, field pea/pea, attack by fungus, such as powdery mildew, downy mildew, pods and leaf spots, grey mould, fuserium wilt, and root, rot attack may seriously damage the crop. The insects of pea include, pea moth, pea aphids, and weevil. The damage of the crop and yield reduction can be controlled by planting of disease-resistant cultivars, fungicide treatment, and crop rotation (Imran et al., 2016).

2.7.14 Disease Management Programme Should Include the Following Methods

2.7.14.1 Preventive Measures

- Avoid planting in wet and poorly drained soils to reduce chances of the development of soil borne diseases (root rot).
- Harvest seed soybeans as soon as they are mature.
- Keep the crop free from weeds because they may be the hosts to any diseases.
- Plant quality and healthy seed, free of mechanical damage.
- Store seeds at 8–10% moisture at 15 °C temperature to have more seed viability.

2.7.14.2 Cultural Control

- Follow crop rotation with non-leguminous crop.
- Plant early before the soil temperature rises.
- Plough down crop residues.
- Reduce plant population, increase row width and avoid high fertility to prevent a closed canopy, improve aeration, and increase drying in the canopy.

2.7.14.3 Seed Treatment

To protect from soil borne diseases, seed should be treated before planting with one of the available fungicides: Captan, Dithane M-45, Benlate, and Tecto @ 1.5–2 gm per 1 kg seed.

2.7.14.4 Foliar Spray

At the appearance of disease symptoms, after field survey, foliar spray with one of the systemic fungicides, Dithane M-45 @ 1 kg ha-1, Benlate and Tecto @ 120–150 gm ha-1 dissolved in 250 litter of water should be done after 10 to 15 days interval. This process should be repeated twice or thrice depending upon the severity of disease (Baqa et al., 2015; Imran et al., 2016; Naveed et al., 2016; Samreen et al., 2016).

2.7.14.5 Integrated Disease Control

- Use integrated control combining high tolerant cultivar, good drainage, complete tillage, seed treatment, and rotation with cereals.
- Use multirace resistance varieties or least susceptible cultivars and avoid tall, viney cultivars that may lodge.

2.7.14.6 Bio Control

Many insect predators, parasites, and pathogens occur in legume fields that help in keeping population of pest species below economic levels. However, experience has shown us that beneficial insects and pathogens do not do a complete job, and that chemical control becomes necessary.

2.7.14.7 Cultural Control

Early planted legumes tend to receive the majority of the overwhelming adult of flies and bugs, while late planted will alleviate some of the problems with these two insects and cutworm. Therefore, it is still more economical to plant early for high yields and control any potential insect problems with insecticides than it is to plant late for insect control (Imran et al., 2016).

2.7.14.8 Chemical Control

- For cutworm and termite, apply powder of BHC @ 7 kg per hectare or Dieldrin 20 EC @ 5–7 litres ha-1 mixed with irrigation water.
- For other insects especially flies, thrips and larvae spray, apply Dimecron 100% @ 600 ml ha-1 or Methyl-Parathion 50% @ 800 to 1,200 ml ha-1 Somicidin 20 EC @ 400 to 600 ml ha-1 dissolved in 250 litre of water. If attack is severe spray two times with an interval of 8–10 days.

2.7.15 Harvesting and Threshing

Various legume crops matures in 80 to 120 days depending upon growing season and the variety that was planted. Senescence is the decline in chemical activity associated with aging of plants and maturation is only loss of water from plants or seeds which are physiologically mature. Seeds are physiologically mature when they are no longer synthesizing food. Physiological loss of chlorophyll and acceleration of senescence is characteristics of dry, dehisent fruits (soybean pods). Ethylene and abscisic acid play an important role in abscission and dehiscence of pod and often capsule dry fruits (soybean). Oil and storage protein have reached their maximum dry weight. At this stage, seed moisture is 45–55%, pods and stems are yellow, and leaves are yellow or have dropped. Delay in harvesting not only reduces seed quality but it also reduces harvesting efficiency and increases shattering losses. As soon as the pods are dry enough to open easily, harvest it, thresh after drying within seven to ten days and threshed seed must be cleaned before storage or marketing.

2.7.16 Yield

Number of pods per plant is a function of spacing and intercepted light while leaf N is a principal factor in determining legumes seed yield. The average farmers yield ranges from 1,500 to 2,500 kg ha^{-1} of soybean, 1,200 to 1,800 of mungbean, and mashbean varies with seed size and crop variety. At Agricultural Research Institutes under high-level management practices, the yields range from 2,500 to 3,500 kg ha^{-1} in case of soybean crop.

2.7.17 Storage

Well dried seed should be stored at about 8–10% moisture content and 15 °C in tropical regions. To maintain dried seed at a low moisture level, two practices are feasible:

- To grow legumes for seed in an area where relative humidity is low
- To use moisture proof containers for seed storage

For long-term benefit and effectiveness, air conditioned storage in tropical and subtropical areas are more suitable. Adequate air-conditioned storage should be maintained at a temperature of 20 °C to 22 °C or less and a relative humidity of 60% or less when the storage period is of eight to nine months. Poor-quality legume seeds will deteriorate quickly in storage than high-quality seeds. In addition, the practice of 'carrying over' legume seeds should be discouraged because this crop does not store well and the quality of seed quickly diminishes during the second over wintering period.

2.7.18 Marketing

Beside government agencies, private agencies should be directed to procure the entire produce from the farmers with high cost and effective rate.

2.7.18.1 Advantages of Legume Planting

- It has a short-duration season and farmers could utilize rice, cotton, and rainfed fallow areas.
- It fits well in the existing cropping system without clashing with major crops.
- Economics often dictate crop sequence, but where choices are available, legumes should follow crops other than slegumes like cereals that make better use of the nitrogen provided by legume crops.
- Legume also provides good-quality protein which is high in one of the limiting amino acids lysine, and is useful as a supplement to other cereals.

2.8 Conclusion

For agronomic management of legumes under climate change stress conditions, a quick alternative need based on nanotechnology and biotechnology application is needed to develop new cultivars for the new environments. The successful and critical solution may be to develop resilient legume varieties, demonstrating great potential in difficult conditions or to combat with climate change to enhance food security. The efforts in this admiration is needed to breed lentils with vertical nodulation and nodule clusters associated with high nitrogen fixation. Legume varieties with mature leaf concentration and pea varieties with waxy leaf surfaces help to combat heat stress. The progress of additional efforts to screen germplasm for heat tolerance and disease-resistance in chickpeas, lentil, mung bean and soybean is witnessed. Almost all crops of human interest have advanced production technologies beyond their routine traditional production methods. Legume crops can contribute in alleviating hunger as well as malnutrition especially in the poorer countries. Fortunately, legume crops need no special attention unlike rice, wheat, potato, and most oil seeds. Also they can adapt to wide range of soil and climate and demand very little nutrient and water management. Legumes can also fix atmospheric nitrogen. Legumes production could be a great option economically, nutritionally, and ecologically. Some trends see the ability of fixing legumes to make good use of the elevated CO_2 to increase growth by utilising excess C for N fixation and nutrient uptake. This will require varieties and treatments that allow conversion of this growth to grain. The reduction of some moderate, but not severe, stresses (e.g. drought and early N, P, and K) under high CO_2 and a need to manage the increased demand put on these factors by the greater growth. There is potential need to rework the phenology of varieties for regions as temperature and water profiles change (Table 2.1).

TABLE 2.1

Several species of taxonomically similar legumes

Sub-Family	Tribe	Species
Papilionoideae	**Genisteae**	*Lupinus albus L.*
		Lupinus luteus L.
		Lupinus angustifolius L.
		Lupinus mutabilis Sweet
	Indigofereae	*Cyamopsis etragonoloba* (L.) Taub.
	Phaseoleae	*Canavalia ensiformis* (L.) DC.
		Mucuna pruriens (L.) DC.
		Cajanus cajan (L.) Huth
		Psophocarpus tetragonolobus (L.) DC.
		Sphenostylis stenocarpa (Hochst. ex A. Rich.) Harms
		Lablab purpureus (L.) Sweet
		Vigna angularis (Willd.) Ohwi and H. Ohashi
		Vigna radiata (L.) R. Wilczek
		Vigna mungo (L.) Hepper
		Vigna umbellata (Thunb.) Ohwi and H. Ohashi
		Vigna aconitifolia (Jacq.) Maréchal
		Vigna unguiculata (L.) Walp.
		Vigna subterranea (L.) Verdc.
		Phaseolus vulgaris L.
		Phaseolus lunatus L.
		Phaseolus coccineus L.
		Phaseolus acutifolius A. Gray
	Cicereae	*Cicer arietinum* L.
	Fabeae	*Vicia faba* L.
		Vicia sativa L.
		Lens culinaris Medik.
		Pisum sativum L.

REFERENCES

Abberton MT et al. (2008) *The genetic improvement of forage grasses and legumes to enhance adaptation of grasslands to climate change*. FAO, May 2008. http://www.fao.org/3/a-ai779e.pdf.

Abd El-Moneim AM (1988) Yield stability of selected forage vetches (Vicia spp.) under rainfed conditions in west Asia. *J Agric Sci* 111(2):295–301.

Abd El-Moneim AM, Cocks PS (1992) A adaptation and yield stability of selected lines of Lathyrus spp. Under rainfed conditions in West Asia. *Euphtyica* 66(1–2):69–89.

Abdelhamid M, Shokr M, Bekheta M (2010) Growth, root characteristics, and leaf nutrients accumulation of four faba bean (Vicia faba L.) cultivars differing in their broomrape tolerance and the soil properties in relation to salinity. *Commun Soil Sci Plant Anal* 41(22):2713–2728.

Adnan, M, Z Shah, S Fahad, M Arif, M Alam, IA Khan, IA Mian, A Basir, H Ullah, M Arshad, I-ur- Rehman, S Saud, MZ Ihsan, Y Jamal, Amanullah, H. M. Hammad & W Nasim (2017) Phosphate-solubilizing bacteria nullify the antagonistic effect of soil calcification on bioavailability of phosphorus in alkaline soils. *Sci Rep.* DOI:10.1038/s41598-017-16537-5.

Aise D, Erdal S, Hasanand A, Ahment M (2011) Effects of different water, phosphorus and magnesium doses on the quality and yield factors of soybean in Harran plain conditions. *Int J Phys Sci* 6(6):1484–1495.

Akpalu MM, Siewobr H, Oppong D, Akpalu SE (2014) Phosphorus application and rhizobia inoculation on growth and yield of soybean (Glycine max L. Merrill). *Am J Exp Agric* 4(6):674–685.

Ali I, Khan AA, Imran, Inamullah, Khan A, Asim M, Ali I, Zib B, Khan I, Rab A, Sadiq G, Ahmad N, Iqbal B (2019) Humic acids and Nitrogen levels optimizing productivity of green gram (Vigna radiate L.). *Rus Agric Sci* 45(1)43–47.

Anchal D, Kharwara PC, Rana SS, Dass A (1997) Response of gram varieties to sowing dates and phosphorus levels under on farm condition. *Himachal J Agric Res* 23:112–115.

Ankomah AB, Zapata F, Hardarson G (1996) Yield, nodulation and N2 fixation by cowpea cultivars at different phosphorus levels. *Biol Fert Soils* 22:10–15.

Anwar Shazma, Israeel, Iqbal B, Khan AA, Imran, Shah WA, Islam M, Khattak WA, Ikramullah, Akram W, Abbas W (2016) Nitrogen and phosphorus fertilization of improved varieties for enhancing phenological traits of wheat. *Pure and Appl Biol* 5(3):511–519. http://dx.doi.org/10.19045/bspab.2016.50065.

Aziz ALA, Ahiabor BDK, Opoku A, Abaidoo RC (2016) Contributions of rhizobium inoculants and phosphorus fertilizer to biological nitrogen fixation, growth and grain yield of three soybean varieties on a fluvic luvisol. *Am J Exp Agric* 10(2):1–11.

Baqa S, Khan AZ, Inamullah, Imran, Khan AA, Anwar S, Iqbal B, Khan S, Usman A (2015) Influence of farm yard manure and phosphorus application on yield and yield components of wheat. *J. Pure Appl Biol* 4(4):458–464.

Bishop KA, Leakey ADB, Ainsworth EA (2016) How seasonal temperature or water inputs affect the relative response of C3 crops to elevated [CO_2]: A global analysis of open top chamber and free air CO_2 enrichment studies. *Food Energy Secur* 3(1):33–45. doi: 10.1002/fes3.44.

Canci H, Toker C (2009) Evaluation of annual wild Cicer species for drought and heat resistance under field conditions. *Genet Resour Crop Evol.* 56:1. https://doi.org/10.1007/s10722-008-9335-9.

Cutforth HW (2007) Adaptation of pulse crops to the changing climate of the northern great plains. *Agron J* 99:1684–1699.

Darnhofer L (2010) Adaptiveness to enhance the sustainability of farming systems: *Rev Agron Sustain Dev* 30:545–555.

Devasirvatham V et al. (2012) High temperature tolerance in chickpea and its implications for plant improvement. *Crop Pasture Sci* 63(5):419–428.

FAO (1994) Definition and classification commodities. 4. *Pulses and derived products*. http://www.fao.org/es/faodef/fdef04e.htm.

Devi KN, Singh LNK, Devi TS, Devi HN, Singh TB, Singh KK, Sing WM (2012) Response of soybean to sources and levels of phosphorus. *J Agric Sci* 4(6):44–53.

Devi KN, Singh TB, Athokpam HS, Singh NB, Shamurailatpam D (2013) Influence of inorganic, biological and organic manures on nodulation and yield of soybean and soil properties. *Aust J Sci* 7(9):1407–1415.

Gaur PM, Jukanti AK, Srinivasan S, Gowda CLL (2012) Chickpea (Cicer arietinum L.). In: Bharadwaj DN (ed.) *Breeding of field crops*. Agrobios, Jodhpur, India. pp. 165–194.

Himanen SJ (2016) Engaging farmers in climate change adaptation planning: Assessing intercropping as a means to support farm adaptive capacity. *Agriculture* 6(34). doi:10.3390/agriculture6030034.

Imran (2015a) Effect of germination on proximate composition of two maize cultivars. *Biol Agric Healthc.* https://www.iiste.org/Journals/index.php/JBAH/article/view/20194/20669.

Imran (2015b) Influence of nitrogen levels and decapitation stress on biological potential of rapeseed (Brassica Napus L) under water difficit condition of Swat-Pakistan. *J Nat Sci Res.* https://iiste.org/Journals/index.php/JNSR/article/view/20084/20339.

Imran (2017) Climate change is a real fact confronting to agricultural productivity. *Int J Environ Sci Nat Res* 3(3):555613. DOI: 10.19080/IJESNR.2017.03.555613.

Imran (2018a) Climate change is threat toward agronomy (base of food, fiber system), and food security. *Food Nutr J: FDNJ-160.* DOI: 10.29011/2575-7091.100060.

Imran (2018b) Ecological environmental variability influence growth and yield potential of rice under northern climatic scenario. *Rus Agric Sci* 44(1):18–24.

Imran (2018c) Organic matter amendments improve soil health, productivity and profitability of maize and soybean. *Annu Rev Res* 1(3):555564.

Imran (2018d) Physiological and morphological traits of agronomic crops influenced by climate change. *Mod Concept Dev Agron* 1(4). MCDA.000524.

Imran (2018e) Phosphorous fertilization influenced weeds attributes and phenological characteristics of mungbean cultivars (Vigna Radiata L.). *Rus Agric Sci* 44(3):229–234.

Imran, Amanullah (2018) Review. Global impact of climate change on water, soil resources and threat towards food security: Evidence from Pakistan. *Adv Plants Agric Res* 8(5):350–355. DOI: 10.15406/apar.2018.08.00349.

Imran, Amanullah, Bari A, Ali R (2018) Peach sources, phosphorous and beneficial microbes enhance productivity of soybean. *Soy Res* 16(2):39–48.

Imran, Amanullah, Khan AA, Bari A, Ud Din R, Ali R, Ahmed N, Ali A, Maula F, Ahmad F, Naveed S, Ullah I, Zada H, Khan GR (2019) Production statistics and modern technology of maize cultivation in Khyber Pakhtunkhwa Pakistan. *Plant Sci Arch* 2 (2):1–12.

Imran, Bari A, Ali R, Ahmad N, Ahmad Z, Khattak MI, Ali A, Ahmad F, Khan I, Naveed S. (2017) Traditional rice farming accelerate CH4 & N2O emissions functioning as a stronger contributors of climate change. *Int J Environ Sci Nat Res.* DOI: 10.19080/ARTOAJ.2017.09.555765.

Imran, Hussain I, Khattak I, Rehman A, Ahmad F, Shah ST (2015) Roots nodulation, yield and yield contributing parameters of mungbean cultivars as influenced by different phosphorous level in swat-Pakistan. *J Pure Appl Biol* 4(4):557–567.

Imran, Khan AA (2015a) Biochar application and shoot cutting duration (days) influenced growth, yield and yield contributing parameters of brassica Napus L. *J Biol Agric Healthc.* ISSN 2224-3208 (Paper) ISSN 2225-093X (Online).

Imran, Khan AA (2015b) Effect of transplanting dates on yield and yield components of various rice genotypes in hilly area cold climatic region of khyber pakhtunlhwa-Pakistan. *J Biol Agric Healthc.* ISSN 2224-3208 (paper) issn 2225-093x (Online).

Imran, Khan AA (2015c) Grain yield and phenology of maize cultivars influenced by various phosphorus sources. Food Sci and Quality Management. ISSN 2224-6088 (Paper) ISSN 2225-0557.

Imran, Khan AA (2015d) Influence of compost application and seed rates on production potential of late sown maize on high elevation in Swat -Pakistan. *J Environ Eearth Sci* ISSN 2224-3216 (Paper) ISSN 2225-0948.

Imran, Khan AA (2015e) Phenological Charateristics of Brassica Napus L. as influenced by biochar application and shoot cutting duration (days). *Civ Environ Res.* ISSN 2224-5790 (Paper) ISSN 2225-0514 (Online).

Imran, Khan AA (2017) Canola yield and quality enhanced with sulphur fertilization. *Rus Agric Sci* 43:113. doi: 10.3103/S1068367417020100.

Imran, Khan AA, Ahmad F, Irfanullah (2015) Nitrogen levels, tillage practices and irrigation timing influenced yeild, yeild components and oil contents of canola. *Civ Environ Res.* ISSN 2224-5790 (Paper) ISSN 2225-0514 (Online).

Imran, Khan AA, Akhtar K, Zaheer S, Faisal S, Ali S (2015) Rice seedling characteristics of various genotypes influenced by different sowing dates in Swat-Pakistan. *J Environ Earth Sci.* ISSN 2224-3216 (Paper) ISSN 2225-0948.

Imran, Khan AA, Ahmad F (2015) Phenology of various rice genotypes as affected by different transplanting dates under cold climatic region of Khyber Pakhtunkhwa-Pakistan. *J Environ Earth Sci.* ISSN 2224-3216 (Paper) ISSN 2225-0948.

Imran, Khan AA, Ahmad, F, Irfanullah (2015) Influence of hydrated calcium sulphate ($CaSO_4.2H_2O$) and nitrogen levels on water infiltration rate and maize varieties productivity in rainfed area of Swat, Pakistan. *Chem Environ Res*. ISSN 2224-3224 (Print) ISSN 2225-0956 (Online).

Imran, Khan AA, Inamullah I, Ahmad F. (2016) Yield and yield attributes of Mungbean (Vigna radiata L.) cultivars as affected by phosphorous levels under different tillage systems. *Cogent Food Agric* 2:1151982.

Imran, Khan AA, Inamullah, Luqman (2015) Weeding stages and their effect on yield and yield components of rice in upper Swat, Pakistan. *Pak J Weed Sci Res* 21(4):555–563.

Imran, Khan AA, Irfanullah, Ahmad F (2014) Production potential of rapeseed (brassica napus l.) As influenced by different nitrogen levels and decapitation stress under the rainfed agro-climatic condition of swat Pakistan, *J Glob Innov Agric Soc Sci*. ISSN (Online): 2311-3839; ISSN (Print): 2312-5225.

Imran, Khan AA, Khan IU, Naveed S (2016) Weeds density and late sown maize productivity influenced by compost application and seed rates under temperate environment. *Pak J Weed Sci Res* 22(1):169–181.

Imran, Khan AA, Ullah I, Zada H, Ahmad F, Shah ST, Usman A, Ullah I (2015) Yield and yield attributes of rapeseed cultivars as influence by sulfur level under swat valley conditions. *J Pure Appl Biol* 4(3):296–301.

Imran, Khan AA, Zada H, Irfanullah, Ahmad F (2015) Graine yield and yield components of wheat cultivar 'Siran 2010' as affected by phosphorous levels under rain fed condition. *J Nat Sci Res*. ISSN 2224-3186 (Paper) ISSN 2225-0921 (Online).

Imran, Khattak I, Hussain I, Rehman A, Anwar S, Ahmad F, Khan AA, Zada H (2015) Growth and yield of maize hybrids as effected by different sowing dates in Swat Pakistan. *J Pure Appl Biol* 5(1):114–120.

Imran, Maula F, Uzair M, Zada H (2015) Farmers income enhancement through off-season vegetables production under natural environment in Swat-Pakistan *J Environ Earth Sci*. ISSN 2224-3216 (Paper) ISSN 2225-0948.

Imran, Jamal N, Alam A, Khan AA (2017) Grain yield, yield attributes of wheat and soil physio-chemical charicteristics influenced by biochar, compost and inorganic fertilizer application. *Agric ResTechol*. DOI: 10.19080/ARTOAJ.2017.10.555795.

Imran, Naveed S, Khan AA, Khattak I (2015) Impact of phosphorus levels and seed rates on growth and yield of late sown maize on high elevation in Swat, Pakistan. *J Agric Res* 28 4:406–413.

Imran, Uzair M, Maula F, Vacirca M, Farfaglia S, Khan MN (2015) Introduction and promotion of off-season vegetables production under natural environment in hilly area of upper swat-Pakistan. *J Biol Agric Healthc* 5(11). ISSN 2224-3208.

Imran, Zada H, Naveed S, Khattak I, Ahmad S (2016) Variable rates of phosphorous application influenced phenological traits of green gram (Vignaradiata L.). *J Agric Res* 1(3):1–5.

Iqbal A, Amanullah, Ali A, Iqbal M, Ikramullah, Imran (2017) Integrated use of phosphorus and organic matter improve fodder yield of moth bean (Vigna aconitifolia (Jacq.) under irrigated and dryland conditions of Pakistan. *J AgriSearch* 4(1):10–15.

Jukanti AK, Gaur PM, Gowda CLL, Chibbar RN (2012) Nutritional quality and health benefits of chickpea (Cicer arietinum L.): A review. *Br J Nutr* 108:S11–S26.

Kashiwagi J (2015) Scope for improvement of yield under drought through the root traits in chickpea (Cicer arietinum L.). *Field Crops Res* 175:54–74.

Katerji N (2011) Faba bean productivity in saline–drought conditions. *Eur J Agron* 35(1):2–12.

Katunga MMD (2014) Agro-ecological adaptation and participatory evaluation of multipurpose tree and shrub legumes in mid altitudes of Sud-Kivu, D. R. Congo. *Am J Plant Sci* 5:2031–2039.

Khan A, Fahad S, Khan A, Saud S, Adnan M, Wahid F, Noor M, Nasim W, Hammad HM, Bakhat HF, Ahmad S, Rehman MH, Wang D, Sonmez O (2019) Managing tillage operation and manure to restore soil carbon stocks in wheat–maize cropping system. *Agron J* 111:1–10.

Khazaei H (2014) Leaf traits associated with drought adaptation in faba bean (Vicia faba L.). Doctoral thesis. Helsinki, p. 59.

Mousavi Derazmahalleh M et al. (2019) Adapting legume crops to climate change using genomic approaches. *Plant Cell Environ* 42:6–19.

Muhammad, B, Adnan M, Munsif F, Fahad S, Saeed M, Wahid F, Arif M, Amanullah, Wang D, Saud S, Noor M, Zamin M, Subhan F, Saeed B, Raza MA, Mian IA (2019) Substituting urea by organic wastes for improving maize yield in alkaline soil. *J Plant Nutr*. https://doi.org/10.1080/01904167.2019.1659344.

Naveed S, Ibrar M, Khattak I, Khan I, Imran, Khan MI, Khan H 2016 Anthelmintic, antilice, insecticidal, cytotoxic and phytotoxic potential of ethanolic extracts of two wild medicinal plants Iphiona Grantioides and Plucheaarguta. *J Woulfenia* 23(11):13–25.

Nichols PGH et al. (2012) Temperate pasture legumes in Australia - their history, current use and future prospects. Crop & Pasture *Science* 63:691–725.

Purushothaman R et al. (2013) Root anatomical traits and their possible contribution to drought tolerance in grain legumes. *Plant Prod Sci* 16 (1):1–8.

Real D et al. (2011) Evaluation of perennial forage legumes and herbs in six Mediterranean environments. *Chil J Agric Res* 71(3):357–369.

Samreen U, Ibrar M, Lalbadshah, Naveed N, Imran, Khatak I (2016) Ethnobotanical study of subtropical hills of Darazinda, Takht-e-Suleman range F.R.D.I. Khan, Pakistan. *Pure Appl Biol* 5(1):149–164. http://dx.doi.org/10.19045/bspab.2016.50020.

Siddique KHM, Johansen C, Turner NC, Jeuffroy MH, Hashem A, Sakar D, Gan Y, Alghamdi SS (2012) Innovations in agronomy for food legumes. A review. *Agron Sustain Dev* 32:45–64.

Sinclair TR, Vadez V (2012) The future of grain legumes in cropping systems. *Crop & Pasture Science.* http://dx.doi.org/10.1071/CP12128.

Toker C, Yadav SS (2010) Legumes cultivars for stress environments, in: Yadav S, Redden R (Eds.) *Climate Change and Management of Cool Season Grain Legume Crops*. Dordrecht:Springer, pp. 351–376.

Trytsman M et al. (2016) Diversity and biogeographical patterns of legumes (Leguminosae) indigenous to southern Africa. *PhytoKeys* 70:53–96.

Upadhyaya HD et al. (2010) Phenotyping chickpeas and pigeon peas for adaptation to drought. In: Drought phenotyping in crops: From theory to practice. *Generation Challenge Programme*, pp. 347–355.

Varshney RK, Thudi M, May GD, Jackson SA (2010) Legume genomics and breeding. *Plant Breed Rev* 33:257–304.

Voisin A-S et al. (2013) Legumes for feed, food, biomaterials and bioenergy in Europe: A review. *Agronomy for Sustainable Development.* https://hal.archives-ouvertes.fr/hal-00956058/document.

Wani SP, Sarvesh KV, Krishnappa K, Dharmarajan BK, Deepaja SM (2012) Bhoochetana: Mission to boost productivity of rainfed agriculture through science-led interventions in Karnataka. International Crops Research Institute for the Semi-Arid Tropics, Andhra Pradesh, India.

3

Climate Change and Plant Growth – South Asian Perspective

Muhammad Daniyal Junaid, Usman Khalid Chaudhry, and Ali Fuat Gökçe
Department of Agricultural Genetic Engineering, Ayhan Şahenk Faculty of Agricultural Sciences and Technologies, Niğde Ömer Halisdemir University, Niğde, Turkey

3.1 Introduction

Food security and global agriculture are severely threatened by climate change; if it will not be given proper attention, sustainable agricultural development will be at stake (IPCC, 2007). In the present scenario, developing nations have already been suffering from the drastic effects of climate change. There are significant changes in climate over the past 50 years, which are the result of human activities, that is, emission of greenhouse gases and intense industrialization (IPCC, 2001). South Asia covers diverse climatic zones and landscapes from the Karakorum range of Pakistan to the islands of Sri Lanka and the Maldives. Presently, the region is experiencing a series of climate change knocks, including soil erosion, glacial melting, rising sea level, forest fires, and contamination of ground water. Monsoon patterns have been disturbed and they have aggravated related hazards more frequently in recent years. Since the geography of this region and population patterns are severely random, South Asia is extremely vulnerable to several climate changes.

Climate change influences a big human population across South Asia in various ways. Due to large number of population and lower human development in South Asia, this region is vulnerable to environmental changes (Sterrett, 2011). Almost 70% population of this region lives in rural areas and depends on agriculture as prime income source; climate change has endangered this rural population (Oxfam, 2009). Moreover, more than half of the world's total population among which high number of people are poor live in this region who depend upon climatic conditions for their living (agriculture, forestry, and fishing). In South Asian economy, the agriculture sector has a share of 25% (Fischer et al., 2005). There is 0.7°C global increase in temperature over the past 30 years (Stern, 2006), and global sea level has also raised by 1.88 cm (IPCC, 2007). This situation will adversely affect South Asia due to the geography, less resources, and high agricultural dependency (Stern, 2006).

A slight change in climate can alter the natural habitat of a plant and affect their composition and functioning even in fertile areas (Adnan et al., 2018, 2020; Ahmad et al., 2019; Akram et al., 2018a,b; Wang et al., 2018; Gul et al., 2020; Farhat et al., 2020; Habib ur Rahman et al., 2017; Hammad et al., 2016, 2018, 2019 2020a,b; Hussain et al., 2019, 2020; Ilyas et al., 2020; Jan et al., 2019; Kamarn et al., 2017; Khan et al., 2017a,b; Mubeen et al., 2020; Muhammad et al., 2019; Naseem et al., 2017; Rehman et al., 2020; Saleem et al., 2020a,b,c; Saud et al., 2013; Saud et al., 2014, 2016, 2017, 2020; Shafi et al., 2020; Shah et al., 2013; Subhan et al., 2020; Wahid et al., 2020; Wu et al., 2019; Wu et al., 2020; Yang et al., 2017; Zafar-ul-Hye et al., 2020a,b; Zahida et al., 2017; Zaman et al., 2017; Zamin et al., 2019). It is important to understand the basis of climate changes, and its relationship with the plant growth and functioning, to sustain biodiversity and to develop in a sustainable way. Especially, in South Asia, it is the most populous and water scarce region in the world. This chapter focuses on the exploitation of knowledge related to the impacts of climate change on South Asian agriculture and the plants responses these changes.

3.2 Climate Change in South Asia

Meteorological projections show that South Asia is extremely vulnerable to drought, precipitation fluctuations, and similar drastic events; these factors are graphically presented in Figure 3.1. The extent of these factors vary with the different locations and populations as these climatic variations are heterogenous. In tropical parts of South Asia, rice and wheat growth will be negatively influenced by rising temperatures similarly, variations in soil moistures occur due to rise in temperatures and pest disease occurrence may increase (Sivakumar and Stefanski, 2010). At present, South Asia is facing severe climate changes, some of the effects of climate changes in the region on plants and their growth have been discussed next.

3.2.1 Enhanced Temperatures

Increased temperatures increase water demand in plants and alter their photosynthetic pathways and growth stages (Alharby and Fahad, 2020; Fahad and Bano, 2012; Fahad et al., 2013, 2014a,b, 2015a,b, 2016a,b,c,d, 2017, 2018, 2019a,b). CO_2 effects are also affected by beyond a certain range of high temperature, that is, in wheat increase in productivity with the increase of 0.8°C and decline with the increase of temperature beyond 1.5°C have been observed (Xiao et al. 2005). In India only, the increase in temperature by 2°C and precipitation by 7% will cause the damage of 8.4% to net revenues (Kumar and Parikh, 2001). In South Asia, it is predicted that yields will reduce by 6–9% with 1°C elevation in average temperature in arid, semiarid, and sub-humid areas (Sultana and Ali, 2006).

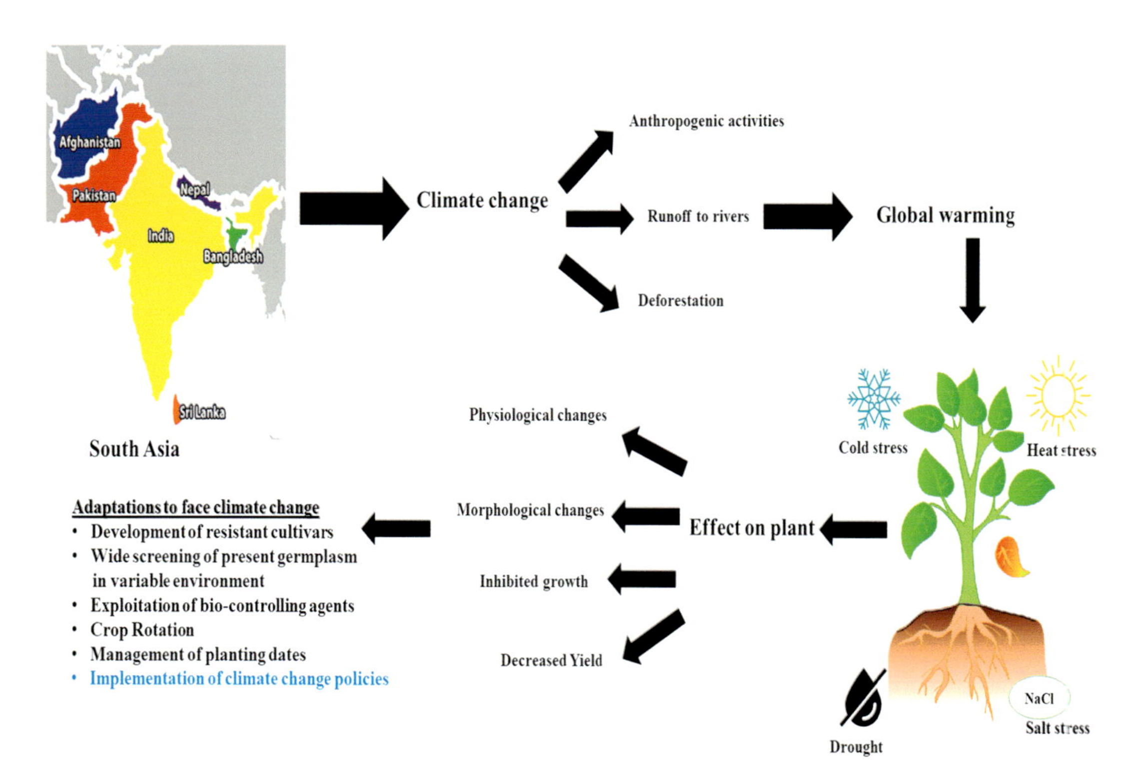

FIGURE 3.1 Various climatic factors affecting plant growth in South Asia.

Especially, important cash crops like mango, cotton, and sugarcane will decline gradually with the increase of 0.3°C (MoE Ministry of Environment, 2003). It is predicted that 0.5°C increase will reduce 6% of total rice yield in Sri Lanka. Ultimately, dried up land affect the production of tea, rubber, coconut, and cotton, which are Sri Lanka's key products (MENR, Ministry of Environment and Natural Resources, 2000). In Pakistan, climate change has already shifted the sowing and harvesting times and disturbed the growth stages of major cereal crops. Drylands and mountain areas are more prone to climate change adversaries than other areas and ecosystem of these areas have disturbed at large scale (Gitay et al., 2001; Hassan et al., 2005). In Sri Lanka, it is estimated that 7–8% tropical rain forests will turn into dry forests due to increased temperature and low rainfall (Somaratne and Dhanapala, 1996); hence, there are more chances of wild fires in combination with droughts (Laurance and Williamson, 2001).

3.2.2 Precipitation and Water

By 2050, there will be 2.5 billion people of South Asia who will be affected by droughts and water scarcity (HDR, 2006). Farming community in drylands is already facing problems like pests, diseases, and soil depletion, and these factors when couple with drought may cause drastic effect on agriculture (Stige et al., 2006).

South Asian economy's lifelines are the big rivers covering a vast area throughout Indian-subcontinent. Himalayan-Hindu Kush mountains are the source of water for World's half of the population as it constitutes largest nine rivers of South Asia, including Indus, Ganges, and Brahmaputra. These rivers are also major source for the agriculture in South Asia. Glaciers are melting at alarming rate, filling the glacial lakes of Nepal and Bhutan (UNEP United Nations Environment Programme, 2007). Groundwater and rainwater are the main source of freshwater for the Maldives, and they are vulnerable to climate change as sea levels are increasing and islands of Maldives are lying low, causing sea water entering into fresh water (Ministry of Environment and Construction, 2005). Meteorological projections depict that until the end of the 21st century, there will be decreased precipitations in northern and southern sub-tropics (IPCC, 2007), and this may severely affect the agricultural production of South Asia (Christensen et al., 2007). In Nepal, 80% of overall country's water is used for only irrigation, and studies showed that the reduction in winter snowfalls will decrease the water runoff. Similar scenario has already happened in Pakistan where areas rely on mountain sources for irrigation (Subbiah, 2001).

3.2.3 Anthropogenic Climate Change

Human activities, deforestation, cultivation, use of excessive pesticides and fertilizers, extraction of underground minerals, irrigation, and drainage have caused a lot of drastic changes on earth. These changes have changed the soil's chemical, hydrological, and thermal properties. Augmentation of irrigation practices in those areas that were dry earlier have changed land cover and pattern of solar radiation absorption, this increased the energy segregating into vapours (Adegoke et al., 2003). Studies have shown that the reduction in forests and growing pastures on those areas have changed the amount of radiant energy which cause air heating and reduced evaporation (Von Randow et al., 2004). Industrial, domestic, and mobility activities coupling with agricultural activities have resulted into CO_2, CH_4, and N_2O gases which result into extreme radiations in the environment (IPCC, 2001). Since industrialization era, combustion of fossil fuels and land use has significantly contributed in the increase (31%) of CO_2 concentration (IPCC, 2001). Similarly, the amount of methane in the environment is increased up to 151% roughly, within past 250 years and nitrous oxide has increased up to 17% (IPCC, 2001). Furthermore, fossils fuel combustion has also elevated sulphate emissions. CO_2 is also an important abiotic factor required for plant growth, any change in its availability may also affect the plant biology and living systems. Urbanization has also influenced plant growth by changing the atmospheric CO_2 (Ziska et al., 2003). There is censorious requirement to understand the fundamentals of the plant growth mechanisms, in order to see how plants react to these changes.

3.2.4 Agriculture Affected by Climate Change in South Asia

It will not be wrong to say that GDP and employment of South Asian countries primarily depend upon agriculture. Agriculture there is mainly dependent upon precipitations, that is, success of monsoons because three-fifths of cropped areas are arid (Kelkar and Bhadwal, 2007). Yield of various crops has been reduced due to elevated temperatures and fluctuating weather conditions adding up with melting glaciers, scarcity of resources and other factors (Cruz et al., 2007). Recent studies suggested that reduction in the productivity of cereal crops is likely to happen in upcoming years as a result of climate change; however, different regions will show different responses to climate change. Crop yield projections stressed that crop yields will increase in East and South East Asia up to 20% and they are likely to decrease up to 30% in Central and South Asia. By 2050, there will be 8% yield decrease in rice and 32% in wheat, studies also revealed that increase in 0.5°C winter temperature could cause damage to the end yield by 0.45 tons/ha. Losses to major South Asian countries due to climate changes are presented in Table 3.1. Indian crops will see decrease in yields up to 2–5% if temperature rises from 0.5°C to 1.5°C. Along with the productivity, climate change also reduces the area of production. A total of 28 mha area in South Asia is dependent upon precipitation and it will require 10% more moisture with increase of 1°C temperature. In dry and grasslands of South Asia, there are numerous short lived, fast growing species which will be drastically affected by climate change (Mitchell and Csillag, 2001). In South Asia, natural grasslands are fragile and can decline with increased evaporations along-with elevated temperatures (Lu and Lu, 2003). Not only is the agriculture, livestock also threatened by thermal stress (Easterling et al., 2007). Mangrove forests in Pakistan along Indus are important for shrimp production, wood, and fodder. Climate change will force people to move to urban areas for employment (MoE Ministry of Environment, 2003). Its impacts on major South Asian countries are shown in Table 3.2.

Hotter climates will increase the probability of extreme events or pests, where crops and livestock would be damaged by diseases and pests. Insect populations will increase when there will be higher temperatures and extended seasons of growth; warmer winters cannot effectively kill insect populations thus they will increase. This will make our agriculture more vulnerable to the outburst of diseases and insect-pests (Cruz et al., 2007). Increased temperatures can also hinder the ability of farmers to work, so rural populations which depend upon traditional agriculture become vulnerable and the livelihoods would be at risk.

The agricultural sector is the major employment source for South Asian countries, reduction in cultivable land and reduced plant production will directly affect the people. Under moderate climate change situation there will be 0.2MT crop loss only due to salinity (Habibullah et al., 1999). If temperature increases severely (4°C), there will be potential reduction in rice, wheat, and potato production (Karim et al. 1996). These problems couple with other sectors will ultimately cause adverse effects on economy and threaten food security.

3.3 Plant Responses to Climate Change

A slight change in climate can alter the natural habitat of a plant and affects its composition and functioning even in fertile areas. It is important to understand the basis of plant responses, growth and

TABLE 3.1

Climate change impacts on agriculture in South Asia

Country	Impacts
Pakistan	Losses of 6–9% in wheat are expected with 1°C increase in temperature
India	Adverse losses 32% & 42% in rice and wheat crops for 2.5°C and 41% & 52% for 4.9°C
Bangladesh	Wheat and rice can face 30 and 50% yield losses with 4°C increase in temperature
Nepal	Mostly people gather foods by migrating as Nepal do not produces sufficient food so climate change can hinder the distribution of food

Source: ISET Institute for Social and Environmental Transition (2008).

TABLE 3.2

Climate change impacts in South Asia

Countries	Temperature	Precipitation	Rivers
Pakistan	It is expected that in next 100 years, there will be an increase of 3–5°C. Temperature is rising in northern and southern Pakistan, but this increase is higher in winter than summer.	Fluctuations in monsoons. There is increased (20%) rain fall in Southern Pakistan and less (5%) in northern region of Pakistan.	Glaciers play key role in Indus flow. Indus basin is expected to suffer from reduced water flow, up to 15%.
India	For past 100 years, it has been observed that there is constant higher temperature. There is an increase of 0.48°C within past century. According to the projections, there will be 2.5–4.9°C increase in temperature until the end of this century.	Increased annual rainfall and increased droughts are expected. Meteorological data predict that rainy days will decline, whereas there are expectations of increased precipitations in central India (10 to 30%).	Ganga basin is expected to face high flows in monsoon. Snow melting will increase the water flow in ganga (85%). It is expected that in next 50 years, this flow will decrease by 30%.
Bangladesh	Bangladesh has similar warming rate as compared to India.	Hydrological cycle will get disturbed by improper rainfalls. It causes droughts and area will be rainfall deficient.	Higher flow of the rivers which comes into country from outside can affect significantly on the coastal region. Storms and cyclones are expected to become intense.
Nepal	There is 0.06°C increase per year in average temperature. In northern region it is even higher.	Alterations in precipitations have been observed. Rainfalls will become intense in near future. On western mid-hills, lower rainfalls are expected. Mountainous region's rainfall can't be generalized as it is under effect of microclimate.	Melting of glaciers is expected which can cause over flow in the rivers. It can cause flooding of the nearby regions and fluctuate water availability.

Source: ISET Institute for Social and Environmental Transition (2008).

functioning, to climate change in order to sustain biodiversity and to develop in a sustainable way. Mechanisms in plants to face environmental changes have always been evolved with time. Understanding of these mechanisms can help to improve crops and affectively face climate change. Plants have been categorized on the basis of photosynthetic metabolic pathways as C_3, C_4, and CAM plants. This classification is done generally on the basis of the enzymes that involve in photosynthetic fixation of CO_2, Rubisco, PEP carboxylase. More than 90% world plant biomass consists of C_3 plants (i.e. wheat, rice, fruits, and vegetables), C_4 (i.e. maize, sugarcane, and sorghum) and Cam (i.e. pineapple). In C_3 plants, the photorespiration rates are higher while in C_4 plants photorespiration is negligible.

3.3.1 Responses of Plants to CO_2

Different plant species show different responses to environmental factors depending upon their nature. It has been previously observed that C_3 and C_4 plants show better productivity in biomass under increased CO_2 level. In some C_3 species, NPP (net primary production) response was observed negative under dry and ambient CO_2 conditions to increased temperatures and vice versa. Contrarily, in C_4 plants, NPP response found positive especially in high precipitation levels (Chen et al., 1996). Drought, heat, and frost have lower effects on plant growth under high atmospheric CO_2 conditions, as plant accumulates high concentrations of membrane stability sugars and secondary carbon rich compounds, that is, tannins (Niinemets, 2010).

Various C_3 plants have showed positive productivity influenced by higher CO_2 concentration, that is, high grain yield in rice was observed, as it started its flowering phase earlier than usual under elevated CO_2 (Uprety et al., 2000). Contrarily, reduction in maize and wheat productivity has been seen under high CO_2 concentration because of decreased length of growth and vernalization period in later (Alexandrov and Hoogenboom, 2000). Cotton has showed significant increase (48%) in yield under free-air CO_2 enrichment (FACE) (Easterling and Apps, 2005). Non-uniform plant responses to CO_2 can be due to the availability of other factors like nutrients, water, and temperature (Chen et al., 1995).

3.3.2 Effects of Climate Change on Respiration and Photosynthesis

Response of photosynthesis and respiration to temperature vary among the plants. Respiration is a 'process of oxidation of food and breakdown of sugars and production of energy for plant growth, reproduction and maintenance'; its rate is highly influenced by carbohydrates and adenylate and has greatly vulnerable to the fluctuating temperatures. Mitochondrial respiration has key role in plant growth and survival (Atkin et al., 2005), it is controlled by the tissue sucrose content (Farrar and Williams, 1991). Elevated CO_2 have positive effect in the growth and assimilation of plants, but long-term exposure to elevated CO_2 reduces the rate of plant-respiration (Farrar and Williams, 1991). In cold climatic regions, plants show that transpiration has improved with short-term increase in temperature, whereas plants growing in tropics don't show this increase in transpiration rate (Atkin and Tjoelker, 2003). This can be due to the acclimatization of the tropical plants to higher temperatures.

Optimum temperature for photosynthesis differs in different species and their growth conditions. Plants that are well adapted to high temperatures have high optimum temperatures for photosynthesis (Battaglia et al., 1996). Photosynthesis – 'process in green plants for formation of sugar molecules by CO_2 uptake from atmosphere, sunlight and water' – in C_3 plants is highest at temperature range between 20°C and 32°C, whereas respiration rate increases within the range of 15°C–40°C. C_3 plants are more distributed among cooler regions than C_4 plants and this difference is not reflected in photosynthetic response. Temperature can strongly affect photosynthesis (Berry and Raison, 1982; Medlyn et al., 2002). Effect of temperature on photosynthesis as it is a biochemical process depends upon its various processes and their overall mechanism (Farquhar and von Caemmerer, 1982). Requirement of temperature by cellular processes increases as their activity improves in moderate temperatures (Medlyn et al., 2002). As many key enzymes in plants are sensitive to high temperatures, the photosynthesis activity reduces at increased temperatures and may become severely minimized when exposed to very high temperature (Berry and Björkman, 1980).

Studies depicted that there is interaction of various climatic factors on photosynthesis, such as temperature and CO_2 interaction. The response of plant species to temperature changes with the increase in CO_2 concentrations. Under low CO_2 concentration in cell, rubisco activity reduces the photosynthesis activity (Kirschbaum and Farquhar, 1984). Assimilation rate and the regeneration of RuBP are also highly influenced at higher CO_2 concentration and temperature (Medlyn et al., 2002).

3.3.3 Plant Responses to Dry Conditions

Plant growth related to yield is sensitive to low water potential depending upon the plant parts (Westgate and Boyer, 1985). Growth of plant parts which contribute to yield is directly related to solute accumulation in expanding parts (Sharp and Davies, 1979). When soil water level reduces, uptake of water by plant roots also decreases and it causes lower water potential in plant cells, which limits the growth of shoots and roots (Sharp et al., 2004).

Studies have shown that under soil dryness, plants can remunerate through osmosis and maintain their growth (Greacen and Oh, 1972; Meyer and Boyer, 1972). However, ability of osmotic adjustments varies between different cultivars (Morgan, 1977).

Under dry conditions, plants take soil water variably, and osmotic adjustment is difficult in such genotypes who has poor potential of water extraction from soil. Moreover, physiological and biochemical phenomenon are also affected by dry conditions of soil.

Studies exhibited that there is disastrous effect on photosynthetic activity as chloroplasts get damaged by dry conditions and hot weather (Pearcy et al., 1977). In extreme environments, plant production is highly associated with photosynthetic output. Water deficits are also influenced by the changes in chloroplast activities, that is, photophosphorylation and photosystem II (Boyer, 1976). Lower water potentials cause depressed photosynthetic activity which ultimately closes the stomata and alters activity of chloroplast. Reduction in photosynthate cause inhibition of yield (McPherson and Boyer, 1977). Some studies also depicted that if plants exposed to low leaf water potential, then their photosynthetic behaviour can be amended (Ackerson and Hebert, 1981).

3.3.4 Plant Responses to Altered Precipitations

Recent years have showed that raining patterns throughout the world has changed due to climate change. Ultimately, they are causing irregularity in the availability of soil moisture. Plant productivity and our ecosystem are greatly influenced by the availability of moisture in the soil (Davies and Gowing, 1999). Effects on plant productivity can be due to collective factors. If there is no adequate water supply to the plant, it cannot uptake necessary moisture and nutrients from soil (Stevens et al., 2004). Changes in moisture either in plants or in soil significantly affect plant processes (Brzostek et al., 2012). Oscillations in soil water content and fluctuating precipitations affect physiology and functioning of plants (Volder et al., 2013). Plant water relations and hydraulic architecture along with above ground primary productivity are heavily influenced by increased precipitations (Reyer et al., 2012). Under high precipitation, plant also becomes unable to take up enough water because of the partial pressure of oxygen on roots (Jackson and Weinand, 1995). Excessive plant water condition causes fluctuations in hormonal production, accumulation of toxic metabolites, and chemical changes in plants (Else and Jackson, 1998). With alterations in precipitations each season, phenology, leaf development, wood development, tiller, and grass development of plants may get influenced (Volder et al., 2010). Plant show variations in responses depending upon the intensity of precipitations (Gerten et al., 2008). In near future, it is expected that variations in soil water content will increase rapidly (Smith, 2011). Intense precipitations can lead to deep drainage of water, where shallow soils drain more quickly and, in turn, plant growth will be reduced due to water logging (Knapp et al., 2008).

3.3.5 Plant Responses to Light

Plant growth is primarily affected by light, its quality, intensity, and duration. Rate of photosynthesis varies with the changes in wavelengths of light. Light intensity varies with the changes in time and season as distance of light in atmosphere and angle of incidence change; moreover, it also changes with the altering humidity and clouds (Manske, 2011). In temperate zones, plants highly depend upon the duration of sunlight for their normal growth activities and phenological development. Changes in photoperiod lengths affect physiological processes as they affect flowering and cause resistance to hardening in cold season. Photoperiod triggers vegetative growth and reproductive initiation but can be amended by precipitation and temperature (Dahl, 1995). Short-day plants and long-day plants response to photoperiod differently, for example, short-day plants respond to increase length of night period and vice versa (Leopold, 1975).

3.3.6 Plant Growth Responses to Temperature

Plant growth is highly influenced by the temperature, that is, low temperatures cause lower plant growth in response to production of assimilates (Korner, 2006). Crop productivity potential has positive influence under a specific range of increase in local temperature (1–3°C). However, productivity is negatively affected beyond this range (IPCC, 2007). This may be caused due to changes in vernalization, phenological phases, and photosynthetic and transpiration rates (Qaderi and Reid, 2009).

Studies show that increase in soil temperature influences nutrient availability, leaching of minerals, and decomposition of organic matter. Where temperature is limiting factor, net primary production increases with the favourable moisture and substrate availability. Slight increase in temperature raises the availability of nutrients in soil and also affects weathering.

3.4 Potential Positive Effects of Climate Change

Along with the decreased agricultural productivity, climate changes have potential positive effects on plant growth. Atmospheric CO_2 levels have potential positive effects on some plants; for example, as they reduce transpiration rates and increase growth rate, CO_2 fertilization can increase agricultural productivity. Especially, water use efficiency of crop plants increases significantly under higher CO_2 levels. Plants such as potato, rice, wheat, and soybean (C_3 plants) get benefit from extra CO_2. Tropical originated plants (C_4 plants) also get benefit from CO_2 increase, but less as compared to C_3 plants. Different geographic regions of the world could get benefit from increased temperature, as in those areas where agriculture is restricted or hindered due to low temperatures, the land mass for agricultural production would potentially increase and the maturation time required for crop production would also decrease. Higher rate of rainfalls in some areas of world may also provide more water for irrigation, enabling higher yields. Studies have revealed that ability of plants to accumulate solutes (which results in improved yields) increases in water scarce conditions (Morgan, 2000). In wheat, it has been observed that solute regulation could be responsible for proper plant functioning, that is, in reproductive plant parts, delayed accumulation of harmful concentrations of ABA (abscisic acid). Under water scarce conditions, plant yields can be improved by the manoeuvring the processes which can alter plant water relations (Richards, 2006). Soil drying can help in improved yields as it assists the movement of solutes between organs. It has been depicted in previous studies that high nitrogen in the soil showed lower yield as compared to the crop grown under low soil nitrogen, it is because N causes delay in senescence and non-senescence plant stem traps a large amount of carbohydrates. Less soil water enables these carbohydrates to travel in different parts of plant, resulting into higher yield (Yang et al., 2000).

Hydraulic relations of the plant directly are influenced by the environmental variations, where plant growth gets affected as cell turgor is a cause for growth and cell walls stretching. Changes in moisture content in cell indirectly effects metabolism and hinders growth (Kaiser, 1987). Cell wall properties can affect plant growth when they get disturbed by any kind of stress, these can alter solute concentrations in cell wall. It has been observed that there is up regulation of certain putative enzymes (expansins and XET) through root tip under drought conditions (Wu et al., 2001).

3.5 Adaptations in South Asia to Face Climate Change

Diseases and insect-pest problems arising due to climate change should be addressed adequately. Mitigation is not possible by taking one step, it needs a series of changes and improvements. Development of such cultivars which can cope with insect-pests and also produce economic yield is need of time. It is the most environment friendly mitigation tactic against climate change. For this purpose, wide screening of germplasm should be done to search out durable resistance in the breeding lines/accessions against diseases and insect-pests. Additionally, the changes in pests should be monitored, either there are any changes happening in pathogens or not. Development of resistant cultivars which can replace susceptible cultivars would be appropriate technique to fight diseases and insect-pests. Secondly, insect pests can be somewhat controlled using Bio-Control Agents.

Adaptation is an important strategy against changing climatic conditions (Pielke, 2007). It is change in species in response to any change and to fit in environment. It can be at any level, from a single individual to international level. Selection of management practices in South Asia plays key role in incidence of insect pest incidence. So, adoption of suitable management practices can help to control insect pests. IPM (integrated pest management) is also an environment friendly and economic practice to use. Modern techniques, that is, drip irrigation system and sprinkler, for efficient water management should be introduced from local farm levels. Crop rotation is also necessary for the soil management and controlling the population of various pathogens. Similarly, use of mulches and integrated nutrient management (INM) helps in sustainable development and conservation of soil. Agricultural production should be focused in reducing losses and conserving priceless natural resources. In South Asia, in response to climate change, plantation dates can be shifted and with adequate crop rotation the destructive effects of climate change can be minimized, as we know South Asian is a developing region which is food insecure region with lack of proper infrastructure to produce new cultivars and techniques against climate change (Lobell et al., 2008).

Countries in South Asia have developed National Action Plans for adaptations to cope with climate change. National Adaptation Programs of Action (NAPA) was formed in Bangladesh and Nepal in 2005 and 2010, respectively. Some countries like Pakistan and India also have constituted national climate change strategies. Similarly, Sri Lanka developed climate change policy for 2011–2015. The main target of these policies was the welfare of society and their protection from the hazards of climate change. Pakistan is involved in UNFCCC processes since 1992. In 2011, the Pakistani government approved first draft of national climate change policy. On the other hand, the Indian government released their first action plan against climate change in 2008 (GoI Government of India, 2006). In 2010, Nepal finalized country's strategy to respond potential challenges by climate change (NPC, 2010). Most South Asian countries are extremely vulnerable to climate change because of high rates of population growth, unavailability of proper facilities, and lowest human development. There are still ambiguities in the strategies to face climate change due to restricted economic facilities and technology in South Asia.

3.6 Conclusion

South Asia is the most populous region in the world, facing food insecurity and lack of resources management. With land degradation this makes this region more vulnerable to climate change. About 70% population of major south Asian countries directly or indirectly depend upon agriculture, and despite of this, there is lack of proper facilities, planning, and sustainable development in the region. There is crucial need to comprehend the effects of changing climate, its potential hazards, and possible positive effects along with the responses of plants.

Climate change can manipulate our natural ecosystem, and we are already observing its effects on our environment. Scientists and the meteorological data conclude that world's agriculture will be severely affected by extreme meteorological events such as heat waves, altered precipitations, floods, droughts, and increased temperatures. Human contribution in climate change is enormous since heavy industrialization and urbanization has triggered global warming through gaseous emissions. South Asia has a diverse ecosystem and a lot of species are endangered. There is 28 mha area which is dependent upon the rainfall for irrigation. Due to climate change, these species are under severe danger; moreover, heat is also a major threat for plant growth. Climate change will force people to move from rural areas to urban areas which it will cause more pressure on cities. Diseases and insect pests are getting more adapted to the higher temperatures so agriculture is more threatened by biotic stresses also. Major cereal crops like rice and wheat will face serious reduction in yields with increase in temperature.

Along with the negative effects of climate change, there are also positive effects of climate change on agriculture. Some plant species have shown better growth under increased CO_2 and temperature. Landmass for agriculture production is also increasing with the increase in temperature. In South Asia, countries are making climate change policies and developing a proper mechanism to face climate change effectively, but still a lot is needed to be done.

REFERENCES

Ackerson RC, Hebert RR (1981) Osmoregulation in cotton in response to water stress: I. Alterations in photosynthesis, leaf conductance, translocation, and ultrastructure. *Plant Physiol* 67:484–488.

Adegoke JO, Pielke Sr RA, Eastman J, Mahmood R, Hubbard KG (2003) Impact of irrigation on midsummer surface fluxes and temperature under dry synoptic conditions: A regional atmospheric model study of the US High Plains. *Mon Weather Rev* 131:556–564.

Adnan M, Zahir S, Fahad S, Arif M, Mukhtar A, Imtiaz AK, Ishaq AM, Abdul B, Hidayat U, Muhammad A, Inayat-Ur R, Saud S, Muhammad ZI, Yousaf J, Amanullah, Hafiz MH, Wajid N (2018) Phosphate-solubilizing bacteria nullify the antagonistic effect of soil calcification on bioavailability of phosphorus in alkaline soils. *Sci Rep* 8:4339. https://doi.org/10.1038/s41598-018-22653-7.

Adnan M, Fahad S, Muhammad Z, Shahen S, Ishaq AM, Subhan D, Zafar-ul-Hye M, Martin LB, Raja MMN, Beena S, Saud S, Imran A, Zhen Y, Martin B, Jiri H, Rahul D (2020) Coupling Phosphate-Solubilizing Bacteria with Phosphorus Supplements Improve Maize PhosphorusAcquisition and Growth under Lime Induced Salinity Stress. *Plants* 9(900). doi: 10.3390/plants9070900.

Ahmad S, Kamran M, Ding R, Meng X, Wang H, Ahmad I, Fahad S, Han Q (2019) Exogenous melatonin confers drought stress by promoting plant growth, photosynthetic capacity and antioxidant defense system of maize seedlings. *PeerJ* 7:e7793. http://doi.org/10.7717/peerj.7793.

Akram R, Turan V, Hammad HM, Ahmad S, Hussain S, Hasnain A, Maqbool MM, Rehmani MIA, Rasool A, Masood N, Mahmood F, Mubeen M, Sultana SR, Fahad S, Amanet K, Saleem M, Abbas Y, Akhtar HM, Waseem F, Murtaza R, Amin A, Zahoor SA, Ul Din MS, Nasim W (2018a) Fate of organic and inorganic pollutants in paddy soils. In: Hashmi MZ, Varma A (eds.) *Environmental pollution of paddy soils*. Springer International Publishing, Cham, Switzerland, pp. 197–214.

Akram R, Turan V, Wahid A, Ijaz M, Shahid MA, Kaleem S, Hafeez A, Maqbool MM, Chaudhary HJ, Munis, MFH, Mubeen M, Sadiq N, Murtaza R, Kazmi DH, Ali S, Khan N, Sultana SR, Fahad S, Amin A, Nasim W (2018b) Paddy land pollutants and their role in climate change. In: Hashmi MZ, Varma A (eds.) *Environmental pollution of paddy soils*. Springer International Publishing, Cham, Switzerland, pp. 113–124.

Alexandrov VA, Hoogenboom G (2000) Vulnerability and adaptation assessments of agriculturalcrops under climate change in the Southeastern USA. *Theor Appl Climatol* 67:45–63.

Alharby AF, Fahad S (2020) Melatonin application enhances biochar efficiency for drought tolerance in maize varieties: Modifications in physio-biochemical machinery. *Agron J:* 1–22.

Atkin OK, Tjoelker MG (2003) Thermal acclimation and the dynamic response of plant respiration to temperature. *Trends Plant Sci* 8:343–351.

Atkin OK, Bruhn D, Hurry VM, Tjoelker MG (2005) Evans Review No. 2: The hot and the cold: unravelling the variable response of plant respiration to temperature. *Funct Plant Biol* 32:87–105.

Battaglia M, Beadle C, Loughhead S (1996) Photosynthetic temperature responses of Eucalyptus globulus and Eucalyptus nitens. *Tree Physiol* 16:81–89.

Berry J, Björkman O (1980) Photosynthetic response and adaptation to temperature in higher plants. *Annu Rev Plant Physiol* 31:491–543.

Berry JA, Raison JK (1982) Responses of macrophytes to temperature. In: Lange, OL, Nobel, PS, Osmond, CB, Ziegler, H (eds.) *Physiological plant ecology I. Responses to the physical environment, encyclopedia of plant physiology, new series*, Vol. 12 A. Springer-Verlag, Berlin, Heidelberg, and New York, pp. 277–338.

Boyer JS (1976) Water deficits and photosynthesis. *Water Deficits and Plant Growth* 4:153–190.

Brzostek ER, Blair JM, Dukes JS, Frey SD, Hobbie SE, Melillo JM, Mitchell RJ, Pendall E, Reich PB, Shaver GR, Stefanski A (2012) The effect of experimental warming and precipitation change on proteolytic enzyme activity: positive feedbacks to nitrogen availability are not universal. *Glob Change Biol* 18:2617–2625.

Chen M, Dickinson RE, Zeng X, Hahmann AN (1996) Comparison of precipitation observed over the continental United States to that simulated by a climate model. *J Clim* 9:2233–2249.

Chen XM, Alm DM, Hesketh JD (1995) Effects of atmospheric CO_2 concentration on photosynthetic performance of C3 and C4 plants. *Biotronics* 24:65–72.

Chikaraishi Y, Naraoka H, Poulson SR (2004) Hydrogen and carbon isotopic fractionations of lipid biosynthesis among terrestrial (C3, C4 and CAM) and aquatic plants. *Phytochemistry* 65:1369–1381.

Christensen JH, Hewitson B, Busuioc A, Chen A, Gao X, Held R, Jones R, Kolli RK, Kwon WK, Laprise R, Magaña Rueda V (2007) Regional climate projections. In: *Climate change, 2007: The physical science basis. Contribution of working group I to the fourth assessment report of the Intergovernmental Panel on Climate Change*. Cambridge University Press, Cambridge, pp. 847–940.

Cruz RV, Harasawa H, Lal M, Wu S, Anokhin Y, Punsalmaa B, Honda Y, Jafari M, Li C, Ninh NH (2007) Climate change 2007: Impacts, adaptation and vulnerability. Contribution of working group II to the fourth assessment report of the Intergovernmental Panel on Climate Change.

Dahl BE (1995) Developmental Morphology of Plants. In: Bedunah DJ, Sosebee RE (eds.) *Wildland plants: Physiological ecology and developmental morphology*. Society for Range Management, Denver, pp. 22–58.

Davies WJ, Gowing DJG (1999) Plant responses to small perturbations. In: Physiological plant ecology: 39th symposium of the British Ecological Society 39: 67. Cambridge University Press, Cambridge.

Easterling W, Apps M (2005) Assessing the consequences of climate change for food and forest resources: A view from the IPCC. In: Salinger J, Sivakumar MVK, Motha RP (eds.) *Increasing climate variability and change: Reducing the vulnerability of agriculture and forestry*. Springer, Dordrecht, pp. 165–189.

Easterling WE, Aggarwal PK, Batima P, Brander KM, Erda L, Howden SM, Kirilenko A, Morton J, Soussana JF, Schmidhuber J, Tubiello FN (2007) Food, fibre and forest products. In: Parry ML, Canziani OF, Palutikof JP, van der Linden PJ, Hanson CE (eds.) *Climate change 2007: Impacts, adaptation and vulnerability. Contribution of working group II to the fourth assessment report of the Intergovernmental Panel on Climate Change*. Cambridge University Press, Cambridge, pp. 273–313.

Else MA, Jackson MB (1998) Transport of 1-aminocyclopropane-1-carboxylic acid (ACC) in the transpiration stream of tomato (Lycopersicon esculentum) in relation to foliar ethylene production and petiole epinasty. *Funct Plant Biol* 25:453–458.

Fahad S, Bano A (2012) Effect of salicylic acid on physiological and biochemical characterization of maize grown in saline area. *Pak J Bot* 44:1433–1438.

Fahad S, Chen Y, Saud S, Wang K, Xiong D, Chen C,Wu C, Shah F, Nie L, Huang J (2013) Ultraviolet radiation effect on photosynthetic pigments, biochemical attributes, antioxidant enzyme activity and hormonal contents of wheat. *J Food, Agric Environ* 11(3&4):1635–1641.

Fahad S, Hussain S, Bano A, Saud S, Hassan S, Shan D, Khan FA, Khan F, Chen Y, Wu C, Tabassum MA, Chun MX, Afzal M, Jan A, Jan MT, Huang J (2014a) Potential role of phytohormones and plant growth-promoting rhizobacteria in abiotic stresses: Consequences for changing environment. *Environ Sci Pollut Res* 22(7):4907–4921. https://doi.org/10.1007/s11356-014-3754-2.

Fahad S, Hussain S, Matloob A, Khan FA, Khaliq A, Saud S, Hassan S, Shan D, Khan F, Ullah N, Faiq M, Khan MR, Tareen AK, Khan A, Ullah A, Ullah N, Huang J (2014b) Phytohormones and plant responses to salinity stress: A review. *Plant Growth Regul* 75(2):391–404. https://doi.org/10.1007/s10725-014-0013-y.

Fahad S, Hussain S, Saud S, Tanveer M, Bajwa AA, Hassan S, Shah AN, Ullah A,Wu C, Khan FA, Shah F, Ullah S, Chen Y, Huang J (2015a) A biochar application protects rice pollen from high-temperature stress. *Plant Physiol Biochem* 96:281–287.

Fahad S, Nie L, Chen Y, Wu C, Xiong D, Saud S, Hongyan L, Cui K, Huang J (2015b) Crop plant hormones and environmental stress. *Sustain Agric Rev* 15:371–400.

Fahad S, Hussain S, Saud S, Hassan S, Chauhan BS, Khan F et al. (2016a) Responses of rapid viscoanalyzer profile and other rice grain qualities to exogenously applied plant growth regulators under high day and high night temperatures. *PLoS One* 11(7): e0159590. https://doi.org/10.1371/journal.pone.0159590.

Fahad S, Hussain S, Saud S, Hassan S, Ihsan Z, Shah AN,Wu C, Yousaf M, Nasim W, Alharby H, Alghabari F, Huang J (2016b) Exogenously applied plant growth regulators enhance the morphophysiological growth and yield of rice under high temperature. *Front Plant Sci* 7:1250. https://doi.org/10.3389/fpls.2016.01250.

Fahad S, Hussain S, Saud S, Hassan S, Tanveer M, Ihsan MZ, Shah AN, Ullah A, Nasrullah KF, Ullah S, AlharbyH NW, Wu C, Huang J (2016c) A combined application of biochar and phosphorus alleviates heat-induced adversities on physiological, agronomical and quality attributes of rice. *Plant Physiol Biochem* 103:191–198.

Fahad S, Hussain S, Saud S, Khan F, Hassan S, Jr A, Nasim W, Arif M, Wang F, Huang J (2016d) Exogenously applied plant growth regulators affect heat-stressed rice pollens. *J Agron Crop Sci* 202:139–150.

Fahad S, Bajwa AA, Nazir U, Anjum SA, Farooq A, Zohaib A, Sadia S, NasimW, Adkins S, Saud S, Ihsan MZ, Alharby H, Wu C,Wang D, Huang J (2017) Crop production under drought and heat stress: Plant responses and management options. *Front Plant Sci* 8:1147. https://doi.org/10.3389/fpls.2017.01147.

Fahad S, Ishan MZ, Khaliq AK, Daur I, Saud S, Alzamanan S, Nasim W, Abdullah MA, Khan IA, Wu C, Wang D, Huang J (2018) Consequences of high temperature under changing climate optima for rice pollen characteristics-concepts and perspectives. *Archives Agron Soil Sci.* doi: 10.1080/03650340.2018.1443213.

Fahad S, Adnan M, Hassan S, Saud S, Hussain S, Wu C, Wang D, Hakeem KR, Alharby, HF, Turan V, Khan MA, Huang J. (2019a) Rice responses and tolerance to high temperature. In: Hasanuzzaman M, Fujita M, Nahar K, Biswas, JK (eds.) *Advances in rice research for abiotic stress tolerance.* Woodhead, London, pp. 201–224.

Fahad S, Rehman A, Shahzad B, Tanveer M, Saud S, Kamran M, Ihtisham M, Khan SU, Turan V, Rahman MHU (2019b) Rice responses and tolerance to metal/metalloid toxicity. In: Hasanuzzaman M, Fujita M, Nahar K, Biswas JK (eds.) *Advances in rice research for abiotic stress tolerance.* Woodhead, London, pp. 299–312.

Farhat A, Hafiz MH, Wajid I, Aitazaz AF, Hafiz FB, Zahida Z, Fahad S, Wajid F, Artemi C (2020) A review of soil carbon dynamics resulting from agricultural practices. *J Environ Manag* 268 (2020):110319.

Farquhar GD, von Caemmerer S (1982) Modelling of photosynthetic response to environmental conditions. In: Lange OL, Nobe PS, Osmond CB, Ziegler H. (eds.) Physiological plant ecology II. *Water relations and carbon assimilation, Encyclopedia of plant physiology, Vol.12B.* Springer-Verlag, Berlin, Heidelberg, and New York, pp. 549–588.

Farrar JF, Williams ML (1991) The effects of increased atmospheric carbon dioxide and temperature on carbon partitioning, source-sink relations and respiration. *Plant Cell Environ* 14:819–830.

Fischer G, Shah M, Tubiello F, Van Velhuizen H (2005) Socio-economic and climate change impacts on agriculture: An integrated assessment, 1990–2080. *Philos T R Soc B* 360:2067–2083.

Gerten D, Luo Y, Le Maire G, Parton WJ, Keough C, Weng E, Beier C, Ciais P, Cramer W, Dukes JS, Hanson PJ (2008) Modelled effects of precipitation on ecosystem carbon and water dynamics in different climatic zones. *Glob Change Biol* 14:2365–2379.

Gitay H, Brown S, Easterling W, Jallow B, Antle J, Apps MJ, Beamish R, Chapin T, Cramer W, Frangi J, Laine J (2001) Ecosystems and their goods and services. In: McCarthy JJ, Canziani OF, Leary NA, Dokken DJ, White KS (eds.) Climate change 2001: Impacts, adaptation, and vulnerability. *Contribution of working group II to the third assessment report of the intergovernmental panel on climate change.* Cambridge University Press, Cambridge, pp. 237–342.

GoI (Government of India) (2006) *Energy for the future: Making development sustainable.* Ministry of Environment and Forests, GoI and TERI Press, New Delhi.

Greacen EL, Oh JS (1972) Physics of root growth. *Nat New Biol* 235:24.

Gul F, Ahmed I, Ashfaq M, Jan D, Shah F, Li X, Wang D, Fahad M, Fayyaz M, Shah AS (2020) Use of crop growth model to simulate the impact of climate change on yield of various wheat cultivars under different agro-environmental conditions in Khyber Pakhtunkhwa, Pakistan. *Arabian J Geosci* 13:112. https://doi.org/10.1007/s12517-020-5118-1.

Habib ur Rahman M, Ahmad A, Wajid A, Hussain M, Rasul F, Ishaque W, Islam MA, Shiela V, Awais M, Ullah A, Wahid A, Sultana SR, Saud S, Khan S, Shah F, Hussain M, Hussain S, Nasim W (2017) Application of CSM-CROPGRO-Cotton model for cultivars and optimum planting dates: Evaluation in changing semi-arid climate. *Field Crops Res* http://dx.doi.org/10.1016/j.fcr.2017.07.007.

Habibullah M, Ahmed AU, Karim Z (1999) Assessment of foodgrain production loss due to climate induced enhanced soil salinity. *In: Vulnerability and adaptation to climate change for Bangladesh.* Springer, Dordrecht, 55–70.

Hammad HM, Farhad W, Abbas F, Fahad S, Saeed S, Nasim W, Bakhat HF (2016) Maize plant nitrogen uptake dynamics at limited irrigation water and nitrogen. *Environ Sci Pollut Res* 24(3):2549–2557. https://doi.org/10.1007/s11356-016-8031-0.

Hammad HM, Abbas F, Saeed S, Shah F, Cerda A, Farhad W, Bernado CC, Wajid N, Mubeen M, Bakhat HF (2018) Offsetting land degradation through nitrogen and water management during maize cultivation under arid conditions. *Land Degrad Dev* 1–10. doi: 10.1002/ldr.2933

Hammad HM, Ashraf M, Abbas F, Bakhat HF, Qaisrani SA, Mubeen M, Shah F, Awais M (2019) Environmental factors affecting the frequency of road traffic accidents: a case study of sub-urban area of Pakistan. *Environ Sci Pollut Res.* https://doi.org/10.1007/s11356-019-04752-8.

Hammad HM, Abbas F, Ahmad A, Bakhat HF, Farhad W, Wilkerson CJ W, Shah F, Hoogenboom G (2020a) Predicting kernel growth of maize under controlled water and nitrogen applications. *Int J Plant Prod.* https://doi.org/10.1007/s42106-020-00110-8.

Hammad HM, Khaliq A, Abbas F, Farhad W, Shah F, Aslam M, Shah GM, Nasim W, Mubeen M, Bakhat HF (2020b) Comparative effects of organic and inorganic fertilizers on soil organic carbon and wheat productivity under arid region. *Communications in Soil Science and Plant Analysis.* doi: 10.1080/00103624.2020.1763385.

Hassan R, Scholes R, Ash N (eds.) (2005) *Ecosystems and human wellbeing: Volume 1: Current state and trends.* Island Press, Washington, D.C., p. 917.

HDR (2006) Human development report 2006. *Beyond scarcity: Power, poverty and the global water crisis.* United Nations Development Programme, New York.

Hussain S, Mubeen M, Ahmad A, Akram W, Hammad HM, Ali M, Masood N, Amin A, Farid HU, Sultana SR, Shah F, Wang D, Nasim W (2019) Using GIS tools to detect the land use/land cover changes during forty years in Lodhran district of Pakistan. *Environ Sci Pollut Res.* https://doi.org/10.1007/s11356-019-06072-3.

Hussain MA, Fahad S, Sharif R, Jan MF, Mujtaba M, Ali Q, Ahmad A, Ahmad H, Amin N, Ajayo BS, Sun C, Gu L, Ahmad I, Jiang Z, Hou J (2020) Multifunctional role of brassinosteroid and its analogues in plants. *Plant Growth Regul.* https://doi.org/10.1007/s10725-020-00647-8.

IPCC (2001) *Climate change 2001: Impacts, adaptation and vulnerability.* Cambridge University Press, Cambridge.

IPCC (2007) Climate change 2007 – Impacts, adaptation and vulnerability. In: Parry ML, Canziani OF, Palutikof JP, van der Linden PJ, Hanson CE (eds.) *Contribution of working group II to the fourth assessment report of the Intergovernmental Panel on Climate Change.* Cambridge University Press, Cambridge, United Kingdom, and New York.

Ilyas M, Mohammad N, Nadeem K, Ali H, Aamir HK, Kashif H, Fahad S, Aziz K, Abid U, (2020) Drought tolerance strategies in plants: A mechanistic approach. *J Plant Growth Regulation.* https://doi.org/10.1007/s00344-020-10174-5.

ISET (Institute for Social and Environmental Transition) (2008) *Climate adaptation in Asia: Knowledge gaps and research issues in South Asia, Full report of the South Asia Team.* ISET-International and ISET-Nepal, Kathmandu, Nepal.

Jackson IJ, Weinand H (1995) Classification of tropical rainfall stations: a comparison of clustering techniques. *Int J Climatol* 15:985–994.

Jan M, Anwar-ul-Haq M, Shah AN, Yousaf M, Iqbal J, Li X, Wang D, Shah F (2019) Modulation in growth, gas exchange, and antioxidant activities of salt-stressed rice (Oryza sativa L.) genotypes by zinc fertilization. *Arabian J Geosci* 12:775. https://doi.org/10.1007/s12517-019-4939-2.

Kaiser WM (1987) Effects of water deficit on photosynthetic capacity. *Physiol Plant* 71:142–149.

Kamarn M, Wenwen C, Irshad A, Xiangping M, Xudong Z, Wennan S, Junzhi C, Shakeel A, Fahad S, Qingfang H, Tiening L (2017) Effect of paclobutrazol, a potential growth regulator on stalk mechanical strength, lignin accumulation and its relation with lodging resistance of maize. *Plant Growth Regul* 84:317–332. https://doi.org/10.1007/ s10725-017-0342-8.

Karim Z, Hussain SG, Ahmed M (1996) Assessing impacts of climatic variations on foodgrain production in Bangladesh. *In climate change vulnerability and adaptation in Asia and the Pacific.* Springer, Dordrecht, 53–62.

Kelkar U, Bhadwal S (2007) South Asian regional study on climate change impacts and adaptation: Implications for human development. http://hdr.undp.org/sites/default/files/kelkar_ulka_and_bhadwal_suruchi.pdf.

Khan A, Tan DKY, Munsif F, Afridi MZ, Shah F, Wei F, Fahad S, Zhou R (2017a) Nitrogen nutrition in cotton and control strategies for greenhouse gas emissions: A review. *Environ Sci Pollut Res* 24:23471–23487. https://doi.org/10.1007/s11356-017-0131-y.

Khan A, Kean DKY, Afridi MZ, Luo H, Tung SA, Ajab M, Fahad S (2017b) Nitrogen fertility and abiotic stresses management in cotton crop: A review. *Environ Sci Pollut Res* 24:14551–14566. https://doi.org/10.1007/s11356-017-8920-x.

Kirschbaum MUF, Farquhar GD (1984) Temperature dependence of whole-leaf photosynthesis in Eucalyptus pauciflora Sieb. Ex Spreng *Aust J Plant Physiol* 11:519–538.

Knapp AK, Beier C, Briske DD, Classen AT, Luo Y, Reichstein M, Smith MD, Smith SD, Bell JE, Fay PA, Heisler JL (2008) Consequences of more extreme precipitation regimes for terrestrial ecosystems. *Bioscience* 58:811–821.

Korner C (2006) Significance of temperature in plant life. In: Morison JIL, Morecroft MD (eds.) *Plant growth and climate change*. Blackwell Publishing, New York, pp. 48–69.

Kumar KK, Parikh J (2001) Indian agriculture and climate sensitivity. *Glob Environ Chang* 11:147–154.

Laurance WF, Williamson GB (2001) Positive feedbacks among forest fragmentation, drought, and climate change in the Amazon. *Conserv Biol* 15:1529–1535.

Leopold AC (1975) Aging, senescence, and turnover in plants. *BioScience* 25:659–662.

Lobell DB, Burke MB, Tebaldi C, Mastrandrea MD, Falcon WP, Naylor RL (2008) Prioritizing climate change adaptation needs for food security in 2030. *Science* 319:607–610.

Lu XR, Lu XY (2003) Climate tendency analysis of warming and drying in grassland of Northeast Qingzang Plateau of China. *Grassl Sci* 24:8–13.

Manske LL (2011) Range plant growth and development are affected by climatic factors. https://www.ag.ndsu.edu/DickinsonREC/grazing-handbook-files/topic1-volume1-ch2-report1.pdf.

McPherson HG, Boyer JS (1977) Regulation of grain yield by photosynthesis in maize subjected to a water deficiency 1. *Agron J* 69:714–718.

Medlyn BE, Dreyer E, Ellsworth DE, Forstreuter M, Harley PC, Kirschbaum MUF, LeRoux, X, Loustau D, Montpied P, Strassemeyer J, Walcroft A, Wang KY (2002) Temperature response of parameters of a biochemically-based model of photosynthesis. II. A review of experimental data. *Plant Cell Environ* 25:1167–1179.

MENR (Ministry of Environment and Natural Resources) (2000) *Initial national communication under the United Nations Framework Convention on Climate Change*. MENR, Government of Sri Lanka, Colombo, Sri Lanka.

Meyer RF, Boyer JS (1972) Sensitivity of cell division and cell elongation to low water potentials in soybean hypocotyls. *Planta* 108:77–87.

Ministry of Environment and Construction (2005) State of the environment – Maldives 2005. https://www.environment.gov.mv/v2/wp-content/files/publications/20170202-pub-soe-2016.pdf.

Mitchell SW, Csillag F (2001) Assessing the stability and uncertainty of predicted vegetation growth under climatic variability: northern mixed grass prairie. *Ecol Model* 139:101–121.

MoE (Ministry of Environment) (2003) *Pakistan's initial national communication on climate change*. MoE, Government of Islamic Republic of Pakistan, Islamabad.

Morgan JM (1977) Differences in osmoregulation between wheat genotypes. *Nature* 270:234.

Morgan JM (2000) Increases in grain yield of wheat by breeding for an osmoregulation gene: relationship to water supply and evaporative demand. *Aust J Agric Res* 51:971–978.

Mubeen M, Ahmad A, Hammad HM, Awais M, Farid H, Saleem M, Sami ul Din M, Amin A, Ali A, Shah F, Nasim W (2020) Evaluating the climate change impact on water use efficiency of cotton-wheat in semi-arid conditions using DSSAT model. *J Water Climate Change*. doi: 10.2166/wcc.2019.179/622035/jwc2019179.pdf.

Muhammad B, Adnan M, Munsif F, Fahad S, Saeed M, Wahid F, Arif M, Jr. Amanullah, Wang D, Saud S, Noor M, Zamin M, Subhan F, Saeed B, Raza MA, Mian IA (2019) Substituting urea by organic wastes for improving maize yield in alkaline soil. *J Plant Nutrition*. doi: 10.1080/01904167.2019.1659344.

Nasim W, Ahmad A, Amin A, Tariq M, Awais M, Saqib M, Jabran K, Shah GM, Sultana SR, Hammad HM, Rehmani MIA, Hashmi MZ, Habib Ur Rahman M, Turan V, Fahad S, Suad S, Khan A, Ali S (2017) Radiation efficiency and nitrogen fertilizer impacts on sunflower crop in contrasting environments of Punjab. *Pak Environ Sci Pollut Res* 25:1822–1836. https://doi.org/10.1007/s11356-017-0592-z.

Niinemets Ü (2010) Responses of forest trees to single and multiple environmental stresses from seedlings to mature plants: Past stress history, stress interactions, tolerance and acclimation. *Forest Ecol Manag* 260:1623–1639.

NPC (2010) Food Security Monitoring Task Force 2010, The Food Security Atlas of Nepal, National Planning Commission, Government of Nepal.

Oxfam GB (2009) South Asia regional vision and change strategy external context analysis: Poverty and inequality in South Asia.

Pearcy RW, Berry JA, Fork DC (1977) Effects of growth temperature on the thermal stability of the photosynthetic apparatus of Atriplex lentiformis (Torr.) Wats. *Plant Physiol* 59:873–878.

Pielke Jr, RA (2007) Future economic damage from tropical cyclones: Sensitivities to societal and climate changes. *Philos T R Soc A* 365:2717–2729.

Qaderi MM, Reid DM (2009) Crop responses to elevated carbon dioxide and temperature. In *Climate Change and Crops*. Springer, Berlin and Heidelberg, pp. 1–18.

Rehman M, Fahad S, Saleem MH, Hafeez M, Muhammad Habib ur Rahman, Liu F, Deng G (2020) Red light optimized physiological traits and enhanced the growth of ramie (Boehmeria nivea L.). *Photosynthetica* 58 (4):922–931.

Reyer CP, Leuzinger S, Rammig A, Wolf A, Bartholomeus RP, Bonfante A, De Lorenzi, F, Dury M, Gloning P, Abou Jaoudé R, Klein T (2013) A plant's perspective of extremes: terrestrial plant responses to changing climatic variability. *Glob Change Biol* 19:75–89.

Richards RA (2006) Physiological traits used in the breeding of new cultivars for water-scarce environments. *Agric Water Manag* 80:197–211.

Saleem MH, Fahad S, Adnan M, Mohsin A, Muhammad SR, Muhammad K, Qurban A, Inas AH, Parashuram B, Mubassir A, Reem MH (2020a) Foliar application of gibberellic acid endorsed phytoextraction of copper and alleviates oxidative stress in jute (Corchorus capsularis L.) plant grown in highly copper-contaminated soil of China. *Environ Sci Poll Res*. https://doi.org/10.1007/s11356-020-09764-3.

Saleem MH, Fahad S, Shahid UK, Mairaj D, Abid U, Ayman ELS, Akbar H, Analía L, Lijun L (2020b) Copper-induced oxidative stress, initiation of antioxidants and phytoremediation potential of flax (Linum usitatissimum L.) seedlings grown under the mixing of two different soils of China. *Environ Sci Poll Res*. https://doi.org/10.1007/s11356-019-07264-7.

Saleem MH, Rehman M, Fahad S, Tung SA, Iqbal N, Hassan A, Ayub A, Wahid MA, Shaukat S, Liu L, Deng G (2020b) Leaf gas exchange, oxidative stress, and physiological attributes of rapeseed (Brassica napus L.) grown under different light-emitting diodes. *Photosynthetica* 58 (3):836–845.

Saud S, Chen Y, Long B, Fahad S, Sadiq A (2013) The different impact on the growth of cool season turf grass under the various conditions on salinity and drought stress. *Int J Agric Sci Res* 3:77–84.

Saud S, Li X, Chen Y, Zhang L, Fahad S, Hussain S, Sadiq A, Chen Y (2014) Silicon application increases drought tolerance of Kentucky bluegrass by improving plant water relations and morph physiological functions. *SciWorld J* 2014:1–10. https://doi.org/10.1155/2014/368694.

Saud S, Chen Y, Fahad S, Hussain S, Na L, Xin L, Alhussien SA (2016) Silicate application increases the photosynthesis and its associated metabolic activities in Kentucky bluegrass under drought stress and post-drought recovery. *Environ Sci Pollut Res* 23(17):17647–17655. https://doi.org/10.1007/s11356-016-6957-0x.

Saud S, Fahad S, Yajun C, Ihsan MZ, Hammad HM, Nasim W, Amanullah Jr, Arif M and Alharby H (2017) Effects of nitrogen supply on water stress and recovery mechanisms in Kentucky bluegrass plants. *Front Plant Sci* 8:983. doi: 10.3389/fpls.2017.00983.

Saud S, Fahad S, Cui G, Chen Y, Anwar S (2020) Determining nitrogen isotopes discrimination under drought stress on enzymatic activities, nitrogen isotope abundance and water contents of Kentucky bluegrass. *Sci Rep* 10:6415. https://doi.org/10.1038/s41598-020-63548-w.

Shafi MI, Adnan M, Fahad S, Fazli W, Ahsan K, Zhen Y, Subhan D, Zafar-ul-Hye M, Brtnicky M, Datta R (2020) Application of single superphosphate with humic acid improves the growth, yield and phosphorus uptake of wheat (Triticum aestivum L.) in calcareous soil. *Agronomy* 10:1224. doi:10.3390/agronomy10091224.

Shah F, Lixiao N, Kehui C, Tariq S, Wei W, Chang C, Liyang Z, Farhan A, Fahad S, Huang J (2013) Rice grain yield and component responses to near 2°C of warming. *Field Crop Res* 157:98–110.

Sharp RE, Davies WJ (1979) Solute regulation and growth by roots and shoots of water-stressed maize plants. *Planta* 147:43–49.

Sharp RE, Poroyko V, Hejlek LG, Spollen WG, Springer GK, Bohnert HJ, Nguyen HT (2004) Root growth maintenance during water deficits: physiology to functional genomics. *J Exp Bot* 55:2343–2351.

Sivakumar MV, Stefanski R (2010) Climate Change in South *Asia. In: Faiz MA, Rahman AHMM, Islam KI, Siva Kumar MVK, Lal R (eds.) Climate change and food security in South Asia*. Springer, Dordrecht, pp. 13–30.

Smith MD (2011) An ecological perspective on extreme climatic events: a synthetic definition and framework to guide future research. *J Ecol* 99:656–663.

Somaratne S, Dhanapala AH (1996) Potential impact of global climate change on forest distribution in Sri Lanka. *Water Air Soil Poll* 92:129–135.

Stern N (2006) *Stern's review on economics of climate change.* Cambridge University Press, Cambridge.

Sterrett C (2011) Review of climate change adaptation practices in South Asia. Oxfam Research Report. https://www.oxfam.org/en/research/review-climate-change-adaptation-practices-south-asia.

Stevens CJ, Dise NB, Mountford JO, Gowing DJ (2004) Impact of nitrogen deposition on the species richness of grasslands. *Science* 303:1876–1879.

Stige LC, Stave J, Chan KS, Ciannelli L, Pettorelli N, Glantz M, Herren HR, Stenseth NC, (2006) The effect of climate variation on agro-pastoral production in Africa. *Proc Natl A Sci* 103:3049–3053.

Subbiah (2001) *Climate variability and drought in Southwest Asia.* Asian Disaster Preparedness Center, Bangkok.

Subhan D, Zafar-ul-Hye M, Fahad S, Saud S, Brtnicky M, Hammerschmiedt T, Datta R (2020) Drought stress alleviation by ACC deaminase producing Achromobacter xylosoxidans and Enterobacter cloacae, with and without timber waste biochar in maize. *Sustainability* 12(6286). doi:10.3390/su12156286.

Sultana H, Ali N (2006) Vulnerability of wheat production in different climatic zones of Pakistan under-climate change scenarios using CSM-CERES-Wheat Model, Paper presented in the Second International Young Scientists' Global Change Conference, Beijing, 7–9 November 2006, organized by START (the global change System for Analysis, Research and Training) and the China Meteorological Administration.

UNEP (United Nations Environment Programme) (2007) GEO yearbook 2007: An overview of our changing environment. *UNEP, Nairobi*, p. 86.

Uprety DC, Kumari S, Dwivedi N, Mohan R (2000) Effect of elevated CO_2 on the growth and yield of rice. *Indian J Plant Physiol* 5:105–107.

Volder A, Tjoelker MG, Briske DD (2010) Contrasting physiological responsiveness of establishing trees and a C4 grass to rainfall events, intensified summer drought, and warming in oak savanna. *Glob Change Biol* 16:3349–3362.

Volder A, Briske DD, Tjoelker MG (2013) Climate warming and precipitation redistribution modify tree–grass interactions and tree species establishment in a warm-temperate savanna. *Glob Change Biol* 19:843–857.

Von Randow C, Manzi AO, Kruijt B, De Oliveira PJ, Zanchi FB, Silva RL, Hodnett MG, Gash JHC, Elbers JA, Waterloo MJ, Cardoso FL (2004) Comparative measurements and seasonal variations in energy and carbon exchange over forest and pasture in South West Amazonia. *Theor App Climatol* 78:5–26.

Wahid F, Fahad S, Subhan D, Adnan M, Zhen Y, Saud S, Manzer HS, Brtnicky M, Hammerschmiedt T, Datta R (2020) Sustainable management with mycorrhizae and phosphate solubilizing bacteria for enhanced phosphorus uptake in calcareous soils. *Agriculture* 10 (334). doi:10.3390/agriculture10080334.

Wang D, Shah F, Shah S, Kamran M, Khan A, Khan MN, Hammad HM, Nasim W (2018) Morphological acclimation to agronomic manipulation in leaf dispersion and orientation to promote 'Ideotype' breeding: Evidence from 3D visual modeling of 'super' rice (Oryza sativa L.). *Plant Physiol Biochem.* https://doi.org/10.1016/j.plaphy.2018.11.010.

Westgate ME, Boyer JS (1985) Osmotic adjustment and the inhibition of leaf, root, stem and silk growth at low water potentials in maize. *Planta* 164:540–549.

Wu Y, Thorne ET, Sharp RE, Cosgrove DJ (2001) Modification of expansion transcript levels in the maize primary root at low water potentials. *Plant Physiol* 126:1471–1479.

Wu C, Tang S, Li G, Wang S, Fahad S, Ding Y (2019) Roles of phytohormone changes in the grain yield of rice plants exposed to heat: A review. *PeerJ* 7:e7792. doi:10.7717/peerj.7792.

Wu C, Kehui C, She T, Ganghua L, Shaohua W, Fahad S, Lixiao N,Jianliang H, Shaobing P, Yanfeng D (2020) Intensified pollination and fertilization ameliorate heat injury in rice (Oryza Sativa L.) during the flowering stage. *Field Crops Res* 252:107795.

Xiao X, Zhang Q, Hollinger D, Aber J, Moore III B (2005) Modeling gross primary production of an evergreen needleleaf forest using MODIS and climate data. *Ecol Appl* 15:954–969.

Yang J, Zhang J, Huang Z, Zhu Q, Wang L (2000) Remobilization of carbon reserves is improved by controlled soil-drying during grain filling of wheat. *Crop Sci* 40:1645–1655.

Yang Z, Zhang Z, Zhang T, Fahad S, Cui K, Nie L, Peng S, Huang J (2017) The effect of season-long temperature increases on rice cultivars grown in the central and southern regions of China. *Front Plant Sci* 8:1908. https://doi.org/10.3389/fpls.2017.01908.

Zafar-ul-Hye M, Naeem M, Danish S, Fahad S, Datta R, Abbas M, Rahi AA, Brtnicky M, Holatko, Jiri, Tarar ZH, Nasir M (2020a) Alleviation of cadmium adverse effects by improving nutrients uptake in bitter gourd through cadmium tolerant rhizobacteria. *Environments* 7(54). doi:10.3390/environments7080054.

Zafar-ul-Hye M, Tahzeeb-ul-Hassan M, Abid M, Fahad S, Brtnicky M, Dokulilova T, Datta R D, Danish S (2020b) Potential role of compost mixed biochar with rhizobacteria in mitigating lead toxicity in spinach. *Sci Rep* 10:12159. https://doi.org/10.1038/s41598-020-69183-9.

Zahida Z, Hafiz FB, Zulfiqar AS, Ghulam MS, Fahad S, Muhammad RA, Hafiz MH, Wajid N, Muhammad S (2017) Effect of water management and silicon on germination, growth, phosphorus and arsenic uptake in rice. *Ecotoxicol Environ Saf* 144:11–18.

Zaman QU, Aslam Z, Yaseen M, Ishan MZ, Khan A, Shah F, Basir S, Ramzani PMA, Naeem M (2017) Zinc biofortification in rice: leveraging agriculture to moderate hidden hunger in developing countries. *Arch Agron Soil Sci* 64:147–161. https://doi.org/10.1080/03650340.2017.1338343.

Zamin M, Khattak AM, Salim AM, Marcum KB, Shakur M, Shah S, Jan I, Fahad S (2019) Performance of Aeluropus lagopoides (mangrove grass) ecotypes, a potential turfgrass, under high saline conditions. *Environ Sci Pollut Res* https://doi.org/10.1007/s11356-019-04838-3.

Ziska LH, Gebhard DE, Frenz DA, Faulkner S, Singer BD, Straka JG (2003) Cities as harbingers of climate change: common ragweed, urbanization, and public health. *J Allergy Clin Immunol* 111:290–295.

4

Climate Change and Indoor Agriculture – The Environment for Plants[*]

Olivet Delası Gleku, Zehranur Gülbahar, Ebrar Karabulut, and Sedat Serçe
Department of Agricultural Genetic Engineering, Ayhan Şahenk Faculty of Agricultural Sciences and Technologies, Niğde Ömer Halisdemir University, Niğde, Turkey

4.1 Introduction

4.1.1 General Climate Change Issues

The rapid change in climatic issues over the next 50 years will play a major role in agriculture. For every 1°C increase in atmospheric temperature, it is valued that 10% of the agricultural land area will be lost to other land uses as population rises (Despommier, 2011; Fahad et al., 2013, 2014a,b, 2015a,b, 2016a,b,c,d, 2017, 2018, 2019a,b; Hesham and Fahad, 2020). Climate change is a long-term alteration in the earth's climate, especially a change due to a rise in the mean ambient temperature and it is as a result of global warming (Fahad and Bano, 2012; Fahad et al., 2013, 2014a,b, 2015a,b, 2016a,b,c,d, 2017, 2018, 2019a,b; Hesham and Fahad, 2020).

Global warming is the gradual rise in the mean temperature of the earth's atmosphere due to the rising amount of the sun's heat being released to the earth which is trapped in the atmosphere and not emitted back into space (Michael Shafer, 2019).

When fewer amounts of greenhouse gases are caught up in the atmosphere, more temperature escapes into space; however, when more greenhouse gases are caught up in the atmosphere, less temperature escapes into space, which is the situation of our planet resulting in warmer atmospheric temperature. Trapped greenhouse gases keep going up after the 18th century. In the 1950s, CO_2 level was at 300 ppm but since then, CO_2 level keeps increasing and currently higher than 400 ppm. Evidence provided by NASA indicates that since the Industrial Revolution, the CO_2 present in the earth's atmosphere has increased (Michael Shafer, 2019).

Greenhouse gases exist in nature and are necessary for human survival and that of other earth organisms. They absorb some of the sun's warmth preventing it from being reflected back into space and creating habitable temperatures for the earth. After more than a century and a half of industrialization, deforestation, and large-scale agriculture, amounts of greenhouse gases in the atmosphere have increased to extreme levels not recorded in three million years. As populations, economies, and standards of living grow, so does the successive additions of greenhouse gas (GHGs) emissions.

Here are some basic well-established methods and principles of science:

- The amount of GHGs in the earth's atmosphere is directly connected to the mean global temperature on earth.

---

[*] **Contributions**: All authors made equal contributions to this chapter.

- The amount has been increasing gradually, and mean worldwide temperatures along with it, since the era of the Industrial Revolution.
- The most abundant GHG, rated for about two-thirds of GHGs, is carbon dioxide (CO_2), which is greatly produced by burning fossil fuels.

Climate change is a major issue for discussion in our time and we are at a moment of decision making. From unpredictable weather patterns that threatened food production to rising sea levels that increase the risk of catastrophic flooding, the footprints of climate change span globally and are unparalleled in scale. Without taking far-reaching actions today, remodelling these foot prints in the future will be more demanding and expensive.

Temperature is increasing resulting in droughts, glacier melt and sea level rises and natural disasters such as floods occur. Many major coastal cities and countries are threatened by floods. In Turkey, several deltas are estimated to be flooded (Karaca et al., 2008).

Some scientists consider that there are not adequate proofs of climate change while many have no doubt that global warming is naturally happening. However, it is a fact that our activities support this situation (Table 4.1). Researchers concluded that there is more than 95% likelihood that human activities over the past 50 years have added to the warming up of our planet (Karaca et al., 2008). While electricity and heat generation represent 25% (e.g. fossil fuel), wrong agricultural practices take 24% as the second part (misuse of pesticides and unconscious fertilization, etc.) before the industry.

The Intergovernmental Panel on Climate Change (IPCC) is an international climate change regulatory authority instituted in 1988 by the World Meteorological Organization (WMO) and the United Nations Environment Programme (UNEP) to evaluate climate change based on the most recent science. The IPCC in 2007 promulgated that 2°C is the acceptable global carbon emission limit which is the cause of global warming, but in 2018, they released a new tolerance limit of 1.5°C. The IPCC furnished the world with a special report on the influence of global warming of 1.5°C in October 2018. It found out that restricting global warming to 1.5°C would require swift, far-reaching and unparalleled changes in all facets of society such as transformation in land, energy, industry, buildings, transport, and cities. There needs to be a drop in the emissions of human-caused worldwide average CO_2 by about 45% from the 2010 levels by 2030, reaching 'net zero' around 2050. This implies that any persisting emissions would need balancing by removing CO_2 from the air. With direct benefits to humans and the natural ecosystems, it was found that restricting global warming to 1.5°C compared to 2°C would go hand in hand with ensuring a well-grounded and equitable society. Several influences of climate change would be abated by bringing the global warming cut off point to 1.5°C in contrast to having 2°C, or more. For instance, by 2100, global sea-level rise would reduce by 10 cm with global warming of 1.5°C as compared with 2°C. The possibility of an Arctic Ocean free of sea ice in summer would occur once per century with global warming of 1.5°C, compared with at least once per decade at 2°C. Coral reefs would decrease by 70–90% with global warming of 1.5°C, whereas virtually all (>99%) would be lost with 2°C.

TABLE 4.1

The role of human activities contributing to global warming (Modified from IPCC, 2014)

Electricity and heat production	25%
Agriculture, forestry, and other land use	24%
Industry	21%
Transportation	14%
Other energy sources	10%
Buildings	6%

4.2 Climate Change and Agriculture

A global issue that is becoming apparent is the long-term decrease in agricultural land per capita. Statistics on the world's population growth from the United Nations Food and Agriculture Organization (FAO) opened up the projected decrease of fertile and productive land per person by 2050 to one-third of the accessible amount at 1970 (FAO, 2016). This decline is prognosticated to persist due to the upshot of climate change, the increasing amount of geographic dry lands, the depletion in freshwater supply, and growth of population (Fedoroff, 2015). While utilizable agricultural lands are reducing, the world's population is roughly calculated to reach almost 10 billion by 2050, and according to FAO, there is a need for producing 70% more food than is currently produced to be able to adequately feed the world.

4.3 Effects of Climate Change on Agriculture

Today, the consequences of increasing climate change, temperatures, changes in the order of precipitation, and increasing meteorological disorders are interfering with all vital human activities. While climate change affects the health, transportation, trade sectors, the agricultural sector is one sector that experiences far-reaching impacts. In as much as it is being affected by climate change, agriculture is also a sector that greatly contributes to climate change because it adds to the 'greenhouse gases' global flow through activities such as the destruction of forest lands in order to convert them into agricultural areas which leads to the release of greenhouse gases, causing a rise in the overall greenhouse gas emissions (Adnan et al., 2018, 2020; Ahmad et al., 2019; Akram et al., 2018a,b; Wang et al., 2018; Gul et al., 2020; Farhat et al., 2020; Habib ur Rahman et al., 2017; Hammad et al., 2016, 2018, 2019, 2020a,b; Hussain et al., 2019, 2020; Ilyas et al., 2020; Jan et al., 2019; Kamarn et al., 2017; Khan et al., 2017a,b; Mubeen et al., 2020; Muhammad et al., 2019; Naseem et al., 2017; Rehman, 2020; Saleem et al., 2020a,b,c; Saud et al., 2013; Saud et al., 2014, 2016, 2017, 2020; Shafi et al., 2020; Shah et al., 2013; Subhan et al., 2020; Wahid et al., 2020; Wu et al., 2019; Wu et al., 2020; Yang et al., 2017; Zafar-ul-Hye et al., 2020a,b; Zahida et al., 2017; Zaman et al., 2017; Zamin et al., 2019).

According to Demir (2013), the presumed rise in temperature over the next 30 years will cause notable problems especially in the summer months (0.6–0.8°C). While climate change negatively affects agricultural production in Africa, South America, South Asia, the Middle East, and Australia, in North America, Europe, and North Asia, agricultural production is affected positively by the same. In a research conducted by Demir (2009), it was observed that the distribution of species would be affected to a great extent in countries such as France, Algeria, and Spain by 2050 and that 80% of the diversity of these areas would be lost. While the temperature rises and CO_2 intensity increases as a resultant effect of climate change, it appears to have a positive impact on the level of agricultural production in some regions in the short term. But these constituents may cause a decrease in the quality of product and quantity of production in the long term (Akalın, 2014; Gürkan et al., 2017). Annual precipitation and temperature distribution in agriculture are of great importance as to which crop to grow. With global warming, both annual precipitation and temperature distribution can change throughout the world. Therefore, which crop will be grown will change accordingly. Decreased production amounts due to climate change results in economic losses. According to Dellal (2012), it is estimated that the amount of crop production will decrease by 8.18% in wheat, 2.24% in barley, 9.11% in corn, 4.53% in cotton, and 12.89% in sunflower due to the decrease in yield as a resultant of climate change in 2050 (Gürkan et al., 2017).

Changes in the rainfall patterns due to climate change will have an impact on agriculture. Due to the inconsistencies in precipitation regimes, the southern latitudes of developing countries are anticipated to be in a more unfavourable situation than the northern latitudes. It is, however, hoped that the accumulated mass of carbon dioxide in the atmosphere will positively add to the growth of certain agricultural products. Plants like rice and wheat (with higher CO_2 concentration and lower temperature, lower ability to use light intensity, temperate zone plants) will be influenced positively by the increased amount of CO_2, whereas plants such as corn and sugar cane, which are largely grown in African and Latin American countries (require lower carbon dioxide concentration, higher temperature and lower water, and also require seasonal drought-resistant, initially binding organic molecules containing four carbon atoms) will be adversely influenced by the increasing amount of carbon dioxide (Akalın, 2014; Doğan and Tüzer, 2011). There is an expected decline in the global food supply in the future due to pressure on agricultural production from the influence of climate change. Global predictions are that grain production will decline from 20% to 30% in the future, without the CO_2 fertilization effect. Supposing there is no CO_2 fertilizer impact in developing countries, there is a decline in agricultural products compared to 2000 in 2050. In these countries, the highest crop losses are expected to be in rice and wheat grown by irrigation systems.

It is seen that developed countries are less affected by these impacts of climate change on average compared to developing countries, and even climate change has a positive impact on the quantity of some agricultural products in developed countries (IFPRI, 2009). With CO_2 fertilization, without effectual modification and genetic advancement, each degree centigrade rise in the mean global temperature would reduce the mean global yield of wheat by 6.0%, rice by 3.2%, corn by 7.4% and soybean by 3.1% (Zhao et al., 2017).

4.4 Report of the Influence of Climate Change on Some Agricultural Crops

4.4.1 Sunflower

Sunflower is an important oil plant and holds the largest share in Turkey in the production of vegetable oil. Sunflower, which can be grown in a wide geography and adapt to different climatic conditions, is affected by changing climatic conditions just like all other plants. The Turkey Statistical Institute (TSI) considered the period from 2006 to 2016, according to this source only in 2007 sunflower production decreased by 19.7% yielding less amount of sunflower oil in the country compared to 2006. For example, the sunflower vegetation period was taken into consideration for Tekirdağ, Turkey, a nine-month drought monitoring system (WMD) analysis was conducted from December to August. According to the WMD results in 2007, there was an extraordinary drought in the province. This year, especially in the period of pollination and flowering, the extreme drought and extreme temperatures of the plant has been an important reason for the decrease of oil sunflower yield in the province by 43.7% compared to the previous year (Gürkan et al., 2017).

4.4.2 Cotton

Turkey ranks ninth globally in the cotton planting area, seventh in yield amount, and second in fibre yield. Turkey is quite a convenient location for cotton farming. However, irrigation is an indispensable element in cotton farming whilst water scarcity and drought have negative effects on yield. Drought, which is more common due to climate changes, has a negative effect on cotton farming. When the cotton (bulk) product values of the 2007–2016 period were examined, the country-wide cotton yield decreased by 14.2% in 2008 compared to the previous year (Gürkan et al., 2017).

4.4.3 Rice

It has been observed that increased temperature for rice plants has resulted in increased transpiration rate, lack of vegetative growth and negative effects of grain filling period, and this will have a negative effect on rice growth and cause yield losses (Pathak and Wassmann, 2007). Rice production is anticipated to be adversely influenced by temperature increase in the Philippines. In a case where the temperature increase is 1°C, rice production is expected to decrease by 10% in the Philippines (Akalın, 2014). Floods affect paddy production areas in south and Southeast Asia from 10 to 15 million ha; that is, it causes yield losses of approximately 1 billion dollars per year (Bates et al., 2008). Drought stress greatly limits production of paddy in rain-fed systems. Drought stress impacts 10 miles of hectare area at high altitudes and 13 mil hectare area underground conditions in Asia. In Bangladesh, it is estimated that paddy production will decrease by 10% and one-third of wheat production will decrease in 2050 (Pandey et al., 2007).

4.4.4 Corn

It has been reported that corn and other major agricultural products have been significantly affected by the change in climatic conditions; for example, between 1981 and 2002, there has been a decrease in the yield of 40 megatons per year (Lobell and Field, 2007). In corn, the air temperature above 30°C will increase its sensitivity to temperatures. For example, it is determined that the yield decreases by 1% under normal irrigation conditions and 1.7% under arid conditions for every day above 30°C during the growth period (Lobell et al., 2011). If irrigation water is limited in the Çukurova region in Turkey, a decrease from 58% to 43.4% in corn yield is foreseen (Şen, 2009).

4.4.5 Wheat

A pot experiment was conducted to find out the effect of carbon dioxide content, high temperature, and limited irrigation on wheat yield under fully controlled conditions. According to the findings, the change in wheat biomass and grain yield as a result of the rise in carbon dioxide, temperature, and drought is expected in the next 70 years for the Çukurova region in Turkey. In general, increasing carbon dioxide has a positive influence on the number of siblings and the number of sibling ears. This impact increased the number of siblings by 69% and increased the number of siblings by 15%. Increasing temperature and drought decreased the number of siblings by 10% and 12%, while the number of sibling ears decreased by 10% and 18%. At maturity, plant height increased by 4% with increasing carbon dioxide. Increased temperature and limited irrigation decreased plant height by 9% and 6%, respectively. Warm conditions only cause a 7% shortening of the upper stem length; limited irrigation caused both spike 4% and upper stem length to decrease by 6% (Kapur et al., 2012).

4.4.6 Coffee

It was identified in Turkey that the total area suitable for the cultivation of coffee has reduced dramatically due to climate change and that it would decrease to less than 10% as a result of temperature increase at 2°C. In addition, it is predicted that only the high areas will remain and the other areas will become too hot to grow coffee (Maslin, 2011).

4.4.7 Cassava

The adaptation of cassava to climate change was investigated. Cassava is an important source of energy. Despite its high level of adaptability, it is affected by temperature and precipitation regimes. The climatic future suitability for cassava has been found to have noticeably reduced in regions around North India (Ceballos et al., 2011).

4.5 Report of the Influence of Climate Change on Some Ornamental Plants

4.5.1 Cyclamen

Cyclamen plant needs 15–20°C during vegetative development and 12–15°C during generative development. Higher temperatures promote leaf growth while adversely affecting flowering (Uzunoğlu et al., 2015).

4.5.2 Rose

At night 16°C temperature is suitable for many rose varieties. On sunny days, the temperature in the greenhouse can be 5–7°C higher. At higher temperatures, the development time of roses is shortened, yield increases, but quality decreases (Uzunoğlu et al., 2015).

4.5.3 Honeysuckle

It is known that global warming affects the flowering times of plants. In the western United States, honeysuckle plants bloom earlier than 3.8 days in a decade (Cayan et al., 2001).

4.5.4 Oak

The oaks normally wait for late spring temperatures to bud. In the Netherlands, oaks have been seen to open their buds earlier due to the 2°C increase in temperature since 1980 (Uzunoğlu et al., 2015).

4.5.5 Clove

Temperature, clove growth, development, flower formation, flower, leaf and stem shape, flower quality, and colour affect the vase life of flowers. Temperatures below 7°C affect flower formation and quality in a positive way but slow down the growth. In winter, 11–12°C night temperatures and 16–18°C day temperatures ensure optimal development. High temperature causes small flowers with poor growth. One of the main causes of calyx cracking in cloves is sudden temperature change. A change of more than 5.5°C within one hour period will crack the calyx. This causes loss of quality (Uzunoğlu et al., 2015).

4.5.6 Lily

The optimum temperature for lily growth is 18.5°C. High temperature and humidity cause fungal diseases in plants. In a study conducted on lilies in the Netherlands, they pollinated the lily pollen at 25°C, 30°C, 35°C and as a result, the pollen germination percentage and elongation of pollen tube at 35°C was the least (Chi et al., 2006).

4.6 Losses

Climate change causes economic losses by reducing plant productivity. Losses affect both the families who make a living from farming and the national economy. Climate change cannot be reversed in the short term, so the most important appraisal is to adapt to climate change. Therefore, adaptation activities to be done in the agricultural sector against climate changes include possible impact studies of climate change on future productivity, resistant breeding studies according to changing climatic conditions, and so on. These studies come to the forefront as important studies that can be done to prevent losses in future periods. In terms of agricultural adaptation, people will have to alter the products they produce; this may even change the sowing or planting times of crop plants (Gürkan et al., 2017). Increasing temperatures, as determined today and predicted in the future, will highlight global inequalities and will have an impact on the distribution of grain/crop regions worldwide (Denhez, 2007). In view of the fact that the existing water resources will decrease further in the future, it is necessary to develop irrigation systems that provide water increase and water usage efficiency further (Kapur et al., 2012).

4.7 Indoor Agriculture

Indoor farming is a process of cultivating crops or plants, traditionally on a large scale, exclusively within an enclosed environment. This farming practice often makes use of growing methods such as hydroponics and uses artificial lights to supply plants with the required nutrients and light levels for growth. Indoor agriculture can be practised on either small or large scales, at home and commercially. Diverse plants can be cultivated indoors, but fruits, vegetables, and herbs are most populous. However, indoor farming has distinct popularity in large cities where productive portions of land, in any size, are not readily obtainable for growing and farming. On a large scale, indoor farming is being used to strengthen local food supplies and make available fresh produce to communities in large cities. Many of these farms are vertical farms and can foster the production of many more crops covering a small land area than can be produced outdoor, soil-based farms. Some indoor farms can be erected in an area as small as an underground room and managed by a single gardener to supply fresh produce to their home. Most indoor farming uses a combination of hydroponics and man-made lights to provide plants with the nutrients and light they would have otherwise received when grown outdoors. However, some indoor farming methods, like those implemented in greenhouses, can use a blend of natural and artificial resources. In cultivating indoors, many farmers recognize the full worth of having more control over the environment than they do when they are using traditional farming procedures. Light amounts, nutrition levels, and moisture levels can all be overseen by the farmer when they are growing crops exclusively indoors. In spite of the fact that growing plants indoors can limit growing choices, gardeners and farmers have diverse plants to choose from when determining what to cultivate indoors. Some of the most sought-after plants grown indoors are generally cropping plants such as lettuce, tomatoes, peppers, and herbs (Despommier, 2010).

With the current rate of population growth and limited land for agriculture, there is the need to venture into indoor agriculture, which promises to offer more to metropolitan farming because it allows for vertical farms that grow all crops, in any place, at any time. Presently there are four methods in which indoor agriculture can be inquired into to get more food in less area. This can be further explored and heavily invested into in order to feasibly manage the food imbalance in the near future. These include hydroponics, aquaponics, aeroponics, and vertical farming.

4.7.1 Hydroponics

Hydroponics is a sub-division of hydroculture, a process of cultivating plants without soil, as opposed to traditional gardening where bacteria and fungi break down organic material to release mineral ions, and plants take their minerals via nutrient solutions in a water solvent. This cultivation approach includes growing plants in a soil-free culture with nutrient solutions, with plants suspended in a medium, such as rock wool or perlite, and provided with nutrients, or directly bathing the roots in the nutrient liquid using the nutrient-film procedure (Jones, 2016). Because of the soilless system in place, plants can grow much closer to each other. For instance, in greenhouses, hydroponic growers have complete power over the control of the indoor climate, temperature, humidity, and light intensification. Water is recirculated in this process, so that plants make use of only 10% of water as compared to field-grown plants. Nutrients are kept safe in a tank, so there are no losses or alteration of nutrients like occurs in the soil. We can easily ascertain plants pH and ensure their ideal nutrient uptake. There are better chances for improved plant growth rate as was seen in tomatoes, where the growth rate was twice as compared to field-grown ones. The non-existence of soil means less weed, fewer pest, and disease, less use of insecticides and herbicides. The cooling effect provided by air conditioning provides a constant flow of air which can be supplemented with carbon dioxide to further boost growth and development of plant. Both ambient and nutrient temperatures can be placed at specific positions that optimise the level of growth of plant. Any unutilized nutrients and water can be redistributed rather than lost to the system (Benke and Tomkins, 2017). It can be a high-risk to keep water and electricity together and if plants get any diseases or pests, these may quickly spread.

4.7.2 Aquaponics

Aquaponics consist of aquaculture and hydroponics. Aquaculture is the growing of aquatic entities in both coastal and inland areas by the process of intervening in the rearing process to promote production.

Fish waste serves as fertilizer for plants in this technique. Advantages are organic fertilization, land conservation, efficient use of water and nutrients, highest yield on a field, environmental benefits, and proximity to the markets. Disadvantages are multiple points of failure. There is more initial cost than hydroponics. Unsustainable fish food could become the most expensive factor in the entire system.

4.7.3 Aeroponics

Aeroponics is similar to the hydroponic technique that works with a water spray system that sprays plant roots with atomized nutrient solutions or mists (Christie and Nichols, 2004). It is the method of cultivating plants in an air or mist environment rather than in the soil. The aeroponic systems make use of water, liquid nutrients, and a soilless growing medium to speedily and effectively grow products. The main advantage is less water usage than hydroponics. Other advantages are higher-value crops, increased yields and growth rates, water and nutrients saving, high multiplication rate of seed, quality products, extended growing season, less incidence of pests and diseases, environment control, automatic operation, and less labour inputs. Disadvantages are dependence on the system, technical knowledge, maintenance, and high cost.

4.7.4 Vertical Farming

The vertical farming prototype is necessarily an indoor farm based on a high-rise multi-level factory designment. The vertical farm scheme aims at appreciably increasing productivity and reduce the environmental footprint within a structure of urban, indoor, climate-influenced high-rise buildings. It is an assertion that such facilities offer many prospective merits as a clean and green source of food, along with biosecurity, freedom from pests, droughts, and reduced use of transportation and fossil fuels (Benke and Tomkins, 2017). Think of the skyscraper all covered with green plants like in Singapore for beauties but in this case for agriculture. Singapore also thinks for agriculture. Sky Greens were built in the year 2017. In that project, there is a market on the ground floor, under the ground is the aquaculture area and the upper side of the building is a host for plants. Sky Greens 'A-Go-Gro' technology is based on A-shaped towers, over 6 m high, with up to 26 rows of growing levels. These rows move round at 1 mm/s to supply regular solar energy as outlined in (Krishnamurthy, 2014). The merits of this system are increased and ceaseless crop production, shelter from weather-related challenges, scaling-down or complete neglect of the use of chemical pesticides, water conservation, recycling, environment-friendly and human health-friendly energy conservation and production, sustainable urban growth, and reduced transportation costs. While vertical farming can sufficiently reduce cost of transportation, the transport industry will be disadvantaged considerably. More so, instituting vertical farming within a skyscraper would demand a substantial amount of natural daylighting for the farm to be properly fed. This will result in a rise in the inner temperature of the building in the day, and appreciably lessen the temperature after sunset (because of transpiration). Therefore more energy will be used for heating and cooling systems than that being kept by natural daylight.

4.7.4.1 Constituents of a Vertical Farm and Their Interactions

Utilizing wind turbines and storage batteries for solar panels add greater attraction value to this system. A vertical farm of several levels may take on lots of layouts, including changing warehouses no longer in use or old housing units (Benke and Tomkins, 2017).

4.8 Environmental and Economic Reasons for Shifting from Conventional Farming to Agriculture in a Regulated Environment

For several years gone by, a lot has been documented in either the scientific or non-technical press concerning the need for a secured and more authentic food and water availability. Issues of climate change, usually related to the increase of unfavourable weather conditions (floods, droughts, hurricanes), are a threat to the availability of these two basic requirements. The terrifying rate at which climate change has gained momentum in over the past 25 years has coerced us to re-evaluate some of our most valued ideas about how we conduct our daily lives. Farming is one of those human activities that has continued to exist without criticism due to its central role in aiding some 6.8 billion of us. In fact, farming is so relevantly regarded such that it makes use of most of the available freshwater, in spite of the fact that in many farming communities, water for consumption is rarely available. In industrialized countries such as the United States, the annual use of fossil fuels is for farming purposes.

Genetic engineering may be of help in the bid to boost the cultivation of crops that are compatible with regulated indoor agricultural environmental conditions. For example, it might be feasible to increase crop yield by fine-tuning the temperature, humidity, and CO_2 levels. To make maximum benefit from this system, confirming the dose-response model for light-sensitive cells would help (Benke and Benke, 2013).

REFERENCES

Adnan M, Zahir S, Fahad S, Arif M, Mukhtar A, Imtiaz AK, Ishaq AM, Abdul B, Hidayat U, Muhammad A, Inayat-Ur R, Saud S, Muhammad ZI, Yousaf J, Amanullah, Hafiz MH, Wajid N (2018) Phosphate-solubilizing bacteria nullify the antagonistic effect of soil calcification on bioavailability of phosphorus in alkaline soils. *Sci Rep* 8:4339. https://doi.org/10.1038/s41598-018-22653-7

Adnan M, Fahad S, Zamin M, Shah S, Mian IA, Danish S, Zafar-ul-Hye M, Battaglia ML, Naz RMM, Saeed B, Saud S, Ahmad I, Yue Z, Brtnicky M, Holatko J, Datta R (2020) Coupling phosphate-solubilizing bacteria with phosphorus supplements improve maize phosphorus acquisition and growth under lime induced salinity stress. *Plants* 9(900). doi: 10.3390/plants9070900

Ahmad S, Kamran M, Ding R, Meng X, Wang H, Ahmad I, Fahad S, Han Q (2019) Exogenous melatonin confers drought stress by promoting plant growth, photosynthetic capacity and antioxidant defense system of maize seedlings. *PeerJ* 7:e7793. http://doi.org/10.7717/peerj.7793

Akalın M (2014) The effects of climate change on agriculture: adaptation and mitigation strategies to eliminate these effects. *Hitit Univ J Inst Soc Sci* 7:351–377.

Akram R, Turan V, Hammad HM, Ahmad S, Hussain S, Hasnain A, Maqbool MM, Rehmani MIA, Rasool A, Masood N, Mahmood F, Mubeen M, Sultana SR, Fahad S, Amanet K, Saleem M, Abbas Y, Akhtar HM, Waseem F, Murtaza R, Amin A, Zahoor SA, ul Din MS, Nasim W (2018a) Fate of organic and inorganic pollutants in paddy soils. In: Hashmi MZ, Varma A (eds.) *Environmental pollution of paddy soils*. Springer International Publishing, Cham, Switzerland, pp. 197–214.

Akram R, Turan V, Wahid A, Ijaz M, Shahid MA, Kaleem S, Hafeez A, Maqbool MM, Chaudhary HJ, Munis, MFH, Mubeen M, Sadiq N, Murtaza R, Kazmi DH, Ali S, Khan N, Sultana SR, Fahad S, Amin A, Nasim W (2018b) Paddy land pollutants and their role in climate change. In: Hashmi MZ Varma, A (eds.) *Environmental pollution of paddy soils*. Springer International Publishing, Cham, Switzerland, pp. 113–124.

Alharby AF, Fahad S (2020) Melatonin application enhances biochar efficiency for drought tolerance in maize varieties: Modifications in physio-biochemical machinery. *Agron J:* 1–22.

Bates BC, Kundzewicz ZM, Wu S, Palutikof JP (2008) *Climate change and water, technical paper of the intergovernmental panel on climate change*. Geneva: IPCC.

Benke K, Benke K (2013) Uncertainty in health risks from artificial lighting due to disruption of circadian rhythm and melatonin secretion: A review. *Hum Ecol Risk Assess.* 19:916–929.

Benke K, Tomkins B (2017) Future food-production systems: vertical farming and controlled-environment agriculture. *Sustain Sci Pract Policy*, 13:13–26.

Cayan DR, Kammerdiener S, Dettinger MD, Caprio JM, Peterson DH (2001) Changes in the on set of spring in the Western United States. *Bull Am Meteorol Soc* 82:399–415.

Ceballos H, Ramirez J, Bellotti AC, Jarvis A, Alvarez E (2011) Adaptation of cassava to changing climates. In: Yadav SS, Redden RJ, Hatfield JL, Lotze-Campen H, Hall, AE (eds.) *Crop adaptation to climate change*, 1st edn. Wiley, New Delhi, pp. 411–425.

Chi HS, Straathorf TP, Löfler HJM, Van Tuyl JM (2006) *DLO Centre for Plant Breeding and Reproduction Research*.

Christie C, Nichols M (2004) Aeroponics: A production system and research tool. *Acta Hortic* 648:185–190.
Dellal I (2012) *Effects of climate change agriculture and food security in Turkey. Turkey's II.* Climate Change Department, Ankara, pp. 1–32.
Demir A (2009) Impact of global climate change on biodiversity and ecosystem resources. Ankara University. *J Environ Sci* 1:37–54.
Demir I (2013) TR71 Region oilseed plant cultivation and effects of climate change. *Turk J Agric Food Sci* 1:73–78.
Denhez F (2007) *Atlas of global warming.* NTV Publications, Istanbul, Turkey.
Despommier D (2010) *The vertical farm: Feeding the world in the 21st century.* Thomas Dunne Books/ St. Martin's Press, New York.
Despommier D (2011) The vertical farm: Controlled environment agriculture carried out in tall buildings would create greater food safety and security for large urban populations. *Journal für Verbraucherschutz und Lebensmittelsicherheit* 6:233–236.
Doğan S, Tüzer M (2011) Global climate change and potential impacts. *Çukurova Univ Jf Econ Adm Sci* 12:1–25.
Fahad S, Bano A (2012) Effect of salicylic acid on physiological and biochemical characterization of maize grown in saline area. *Pak J Bot* 44:1433–1438.
Fahad S, Chen Y, Saud S, Wang K, Xiong D, Chen C,Wu C, Shah F, Nie L, Huang J (2013) Ultraviolet radiation effect on photosynthetic pigments, biochemical attributes, antioxidant enzyme activity and hormonal contents of wheat. *J Food, Agric Environ* 11(3&4):1635–1641.
Fahad S, Hussain S, Bano A, Saud S, Hassan S, Shan D, Khan FA, Khan F, Chen Y, Wu C, Tabassum MA, Chun MX, Afzal M, Jan A, Jan MT, Huang J (2014a) Potential role of phytohormones and plant growth-promoting rhizobacteria in abiotic stresses: Consequences for changing environment. *Environ Sci Pollut Res* 22(7): 4907–4921. https://doi.org/10.1007/s11356-014-3754-2
Fahad S, Hussain S, Matloob A, Khan FA, Khaliq A, Saud S, Hassan S, Shan D, Khan F, Ullah N, Faiq M, Khan MR, Tareen AK, Khan A, Ullah A, Ullah N, Huang J (2014b) Phytohormones and plant responses to salinity stress: A review. *Plant Growth Regul* 75(2): 391–404. https://doi.org/10.1007/s10725-014-0013-y
Fahad S, Hussain S, Saud S, Tanveer M, Bajwa AA, Hassan S, Shah AN, Ullah A,Wu C, Khan FA, Shah F, Ullah S, Chen Y, Huang J (2015a) A biochar application protects rice pollen from high-temperature stress. *Plant Physiol Biochem* 96:281–287.
Fahad S, Nie L, Chen Y, Wu C, Xiong D, Saud S, Hongyan L, Cui K, Huang J (2015b) Crop plant hormones and environmental stress. *Sustain Agric Rev* 15:371–400.
Fahad S, Hussain S, Saud S, Hassan S, Chauhan BS, Khan F et al. (2016a) Responses of rapid viscoanalyzer profile and other rice grain qualities to exogenously applied plant growth regulators under high day and high night temperatures. *PLoS One* 11(7): e0159590. https://doi.org/10.1371/journal.pone.0159590
Fahad S, Hussain S, Saud S, Khan F, Hassan S, Jr A, Nasim W, Arif M, Wang F, Huang J (2016b) Exogenously applied plant growth regulators affect heat-stressed rice pollens. *J Agron Crop Sci* 202:139–150.
Fahad S, Hussain S, Saud S, Hassan S, Ihsan Z, Shah AN, Wu C, Yousaf M, Nasim W, Alharby H, Alghabari F, Huang J (2016c) Exogenously applied plant growth regulators enhance the morphophysiological growth and yield of rice under high temperature. *Front Plant Sci* 7:1250. https://doi.org/10.3389/fpls.2016.01250
Fahad S, Hussain S, Saud S, Hassan S, Tanveer M, Ihsan MZ, Shah AN, Ullah A, Nasrullah KF, Ullah S, AlharbyH NW, Wu C, Huang J (2016d) A combined application of biochar and phosphorus alleviates heat-induced adversities on physiological, agronomical and quality attributes of rice. *Plant Physiol Biochem* 103:191–198.
Fahad S, Bajwa AA, Nazir U, Anjum SA, Farooq A, Zohaib A, Sadia S, NasimW, Adkins S, Saud S, Ihsan MZ, Alharby H, Wu C, Wang D, Huang J (2017) Crop production under drought and heat stress: Plant responses and management options. *Front Plant Sci* 8:1147. https://doi.org/10.3389/fpls.2017.01147
Fahad S, Ishan MZ, Khaliq AK, Daur I, Saud S, Alzamanan S, Nasim W, Abdullah MA, Khan IA, Wu C, Wang D, Huang J (2018) Consequences of high temperature under changing climate optima for rice pollen characteristics-concepts and perspectives. *Archives Agron Soil Sci* DOI: 10.1080/03650340.2018.1443213

Fahad S, Rehman A, Shahzad B, Tanveer M, Saud S, Kamran M, Ihtisham M, Khan SU, Turan V, Rahman MHU (2019a) Rice responses and tolerance to metal/metalloid toxicity. In: Hasanuzzaman M, Fujita M, Nahar K, Biswas JK (eds.) *Advances in rice research for abiotic stress tolerance*. Woodhead, Cambridge, pp. 299–312.

Fahad S, Adnan M, Hassan S, Saud S, Hussain S, Wu C, Wang D, Hakeem KR, Alharby HF, Turan V, Khan MA, Huang J (2019b) Rice responses and tolerance to high temperature. In: Hasanuzzaman M, Fujita M, Nahar K, Biswas JK (eds.) *Advances in rice research for abiotic stress tolerance*. Woodhead, Cambridge, pp. 201–224.

FAO Food and Agriculture Organization (2016) Database on arable land. http://data.worldbank.org/indicator/AG.LND.ARBL.HA.PC?end=2013&start=1961&view=chart. Accessed Sept13, 2016.

Farhat A, Hafiz MH, Wajid I, Aitazaz AF, Hafiz FB, Zahida Z, Fahad S, Wajid F, Artemi C (2020) A review of soil carbon dynamics resulting from agricultural practices. *J Environ Manag* 268:110319.

Fedoroff N (2015) Food in a future of 10 billion. *Agric Food Secur* 4:1.

Gul F, Ahmed I, Ashfaq M, Jan D, Shah F, Li X, Wang D, Fahad M, Fayyaz M, Shah AS (2020) Use of crop growth model to simulate the impact of climate change on yield of various wheat cultivars under different agro-environmental conditions in Khyber Pakhtunkhwa, Pakistan. *Arabian J Geosci* 13:112. https://doi.org/10.1007/s12517-020-5118-1

Gürkan H, Bayraktar N, Bulut H (2017) Effects of increased drought on sunflower and cotton yield due to climate change. 12th Field Crops Congress, Kahramanmaraş, 12–15 October.

Habib ur Rahman M, Ahmad A, Wajid A, Hussain M, Rasul F, Ishaque W, Islam MA, Shiela V, Awais M, Ullah A, Wahid A, Sultana SR, Saud S, Khan S, Shah F, Hussain M, Hussain S, Nasim W (2017) Application of CSM-CROPGRO-Cotton model for cultivars and optimum planting dates: Evaluation in changing semi-arid climate. *Field Crops Res* http://dx.doi.org/10.1016/j.fcr.2017.07.007

Hammad HM, Farhad W, Abbas F, Fahad S, Saeed S, Nasim W, Bakhat HF (2016) Maize plant nitrogen uptake dynamics at limited irrigation water and nitrogen. *Environ Sci Pollut Res* 24(3):2549–2557. https://doi.org/10.1007/s11356-016-8031-0

Hammad HM, Abbas F, Saeed S, Shah F, Cerda A, Farhad W, Bernado CC, Wajid N, Mubeen M, Bakhat HF (2018) Offsetting land degradation through nitrogen and water management during maize cultivation under arid conditions. *Land Degrad Dev* 1–10. DOI: 10.1002/ldr.2933

Hammad HM, Ashraf M, Abbas F, Bakhat HF, Qaisrani SA, Mubeen M, Shah F, Awais M (2019) Environmental factors affecting the frequency of road traffic accidents: a case study of sub-urban area of Pakistan. *Environ Sci Pollut Res.* https://doi.org/10.1007/s11356-019-04752-8

Hammad HM, Abbas F, Ahmad A, Bakhat HF, Farhad W, Wilkerson CJ W, Shah F, Hoogenboom G (2020a) Predicting kernel growth of maize under controlled water and nitrogen applications. *Int J Plant Prod.* https://doi.org/10.1007/s42106-020-00110-8

Hammad HM, Khaliq A, Abbas F, Farhad W, Shah F, Aslam M, Shah GM, Nasim W, Mubeen M, Bakhat HF (2020b) Comparative effects of organic and inorganic fertilizers on soil organic carbon and wheat productivity under arid region. *Communications in Soil Science and Plant Analysis.* DOI: 10.1080/00103624.2020.1763385

Hussain MA, Fahad S, Sharif R, Jan MF, Mujtaba M, Ali Q, Ahmad A, Ahmad H, Amin N, Ajayo BS, Sun C, Gu L, Ahmad I, Jiang Z, Hou J (2020) Multifunctional role of brassinosteroid and its analogues in plants. *Plant Growth Regul.* https://doi.org/10.1007/s10725-020-00647-8

IFPRI International Food Policy Research Institute (2009) Impact on agriculture and costs of adaptation. *Food Policy Report.*

Ilyas M, Nisar M, Khan N, Hazrat A, Khan AH, Hayat K, Fahad S, Khan A, Ullah A (2020) Drought tolerance strategies in plants: A mechanistic approach. *J Plant Growth Regulation* https://doi.org/10.1007/s00344-020-10174-5

IPCC (2014) Climate change 2014: Synthesis report. In: Core Writing Team, Pachauri RK, Meyer LA (eds.) *Contribution of working groups I, II and III to the fifth assessment report of the Intergovernmental Panel on Climate Change*. IPCC, Geneva, Switzerland, p. 151.

Jan M, Anwar-ul-Haq M, Shah AN, Yousaf M, Iqbal J, Li X, Wang D, Shah F (2019) Modulation in growth, gas exchange, and antioxidant activities of salt-stressed rice (Oryza sativa L.) genotypes by zinc fertilization. *Arabian J Geosci* 12:775 https://doi.org/10.1007/s12517-019-4939-2

Jones J (2016) *Hydroponics: A practical guide for the soilless grower.* CRC Press, Boca Raton, FL.

Kalantari F, Tahir OM, Lahijani AM, Kalatari S (2017) A review of vertical farming technology: A guide for implementation of building integrated agriculture in cities. *Advanced Engineering Forum 24:76–91.* ISSN: 2234-991X. doi:10.4028/www.scientific.net/AEF.24.76.

Kamarn M, Wenwen C, Irshad A, Xiangping M, Xudong Z, Wennan S, Junzhi C, Shakeel A, Fahad S, Qingfang H, Tiening L (2017) Effect of paclobutrazol, a potential growth regulator on stalk mechanical strength, lignin accumulation and its relation with lodging resistance of maize. *Plant Growth Regul* 84:317–332. https://doi.org/10.1007/ s10725-017-0342-8

Kapur B, Koç M, Ozekici B (2012) Enhanced CO_2 and global climate change effects on wheat yield in Çukurova Region. *Çukurova University Journal of Science and Engineering Sciences:* 245–248.

Karaca M, Nicholls RJ (2008) Potential implications of accelerated sea-level rise for Turkey. *J Coast Res* 24 (2): 288–298.

Khan A, Tan DKY, Munsif F, Afridi MZ, Shah F, Wei F, Fahad S, Zhou R (2017a) Nitrogen nutrition in cotton and control strategies for greenhouse gas emissions: A review. *Environ Sci Pollut Res* 24:23471–23487. https://doi.org/10.1007/s11356-017-0131-y

Khan A, Kean DKY, Afridi MZ, Luo H, Tung SA, Ajab M, Fahad S (2017b) Nitrogen fertility and abiotic stresses management in cotton crop: A review. *Environ Sci Pollut Res* 24:14551–14566. https://doi.org/10.1007/s11356-017-8920-x

Krishnamurthy R (2014) *Vertical farming: Singapore's solution to feed the local urban population. Permaculture Research Institute.* http://permaculturenews.org/2014/07/25/vertical-farming-singapores-solution-feed-local-urban-population

Lobell DB, Field CB (2007) Global scale climate–crop yield relationships and the impacts of recent warming. *Environ Res Lett* 2:014002.

Lobell DB, Schlenker W, Costa-Roberts J (2011) Climate trends and global crop production since 1980, *Sci Express*, 333, (6042): 616–620.

Maslin M (2011) *Global warming*. Ankara, Turkey. Dost Publications.

Mubeen M, Ahmad A, Hammad HM, Awais M, Farid H, Saleem M, Sami ul Din M, Amin A, Ali A, Shah F, Nasim W (2020) Evaluating the climate change impact on water use efficiency of cotton-wheat in semi-arid conditions using DSSAT model. *J Water Climate Change.* doi/10.2166/wcc.2019.179/622035/jwc2019179.pdf

Muhammad B, Adnan M, Munsif F, Fahad S, Saeed M, Wahid F, Arif M, Jr. Amanullah, Wang D, Saud S, Noor M, Zamin M, Subhan F, Saeed B, Raza MA, Mian IA (2019) Substituting urea by organic wastes for improving maize yield in alkaline soil. *J Plant Nutrition*. https://doi.org/10.1080/01904167.2019.1659344

Nasim W, Ahmad A, Amin A, Tariq M, Awais M, Saqib M, Jabran K, Shah GM, Sultana SR, Hammad HM, Rehmani MIA, Hashmi MZ, Habib Ur Rahman M, Turan V, Fahad S, Suad S, Khan A, Ali S (2017) Radiation efficiency and nitrogen fertilizer impacts on sunflower crop in contrasting environments of Punjab. *Pak Environ Sci Pollut Res* 25:1822–1836. https://doi.org/10.1007/s11356-017-0592-z

Pandey S, Bhandari H, Hardy B (2007) Economic costs of drought and rice farmers' coping mechanisms: A cross-country comparative analysis. *International Rice Res Inst.* DOI: 10.22004/ag.econ.281814

Pathak H, Wassmann R (2007) Introducing greenhouse gas mitigation as a development objective in rice-based agriculture: I. Generation of technical coefficients. *Agric Syst* 94:807–825.

Rehman M, Fahad S, Saleem MH, Hafeez M, Habib ur Rahman M, Liu F, Deng G (2020) Red light optimized physiological traits and enhanced the growth of ramie (Boehmeria nivea L.). *Photosynthetica* 58 (4): 922–931.

Şahin Ü (2007) Developed a sample emergency action plan for turkey. greens climate change emergency action plan. www.yesiller.org. Accessed Feb3, 2011.

Saleem MH, Fahad S, Adnan M, Ali M, Rana MS, Kamran M, Ali Q, Hashem IA, Bhantana P, Ali M, Hussain RM (2020a) Foliar application of gibberellic acid endorsed phytoextraction of copper and alleviates oxidative stress in jute (Corchorus capsularis L.) plant grown in highly copper-contaminated soil of China. *Environ Sci Pollution Res.* https://doi.org/10.1007/s11356-020-09764-3

Saleem MH, Fahad S, Shahid UK, Mairaj D, Abid U, Ayman ELS, Akbar H, Analía L, Lijun L (2020b) Copper-induced oxidative stress, initiation of antioxidants and phytoremediation potential of flax

(Linum usitatissimum L.) seedlings grown under the mixing of two different soils of China. *Environ Sci Poll Res*. https://doi.org/10.1007/s11356-019-07264-7

Saleem MH, Rehman M, Fahad S, Tung SA, Iqbal N, Hassan A, Ayub A, Wahid MA, Shaukat S, Liu L, Deng G (2020c) Leaf gas exchange, oxidative stress, and physiological attributes of rapeseed (Brassica napus L.) grown under different light-emitting diodes. *Photosynthetica* 58(3):836–845.

Saud S, Chen Y, Long B, Fahad S, Sadiq A (2013) The different impact on the growth of cool season turf grass under the various conditions on salinity and drought stress. *Int J Agric Sci Res* 3:77–84.

Saud S, Li X, Chen Y, Zhang L, Fahad S, Hussain S, Sadiq A, Chen Y (2014) Silicon application increases drought tolerance of Kentucky bluegrass by improving plant water relations and morph physiological functions. *SciWorld J* 2014:1–10. https://doi.org/10.1155/2014/ 368694

Saud S, Chen Y, Fahad S, Hussain S, Na L, Xin L, Alhussien SA (2016) Silicate application increases the photosynthesis and its associated metabolic activities in Kentucky bluegrass under drought stress and post-drought recovery. *Environ Sci Pollut Res* 23(17): 17647–17655. https://doi.org/10.1007/s11356-016-6957-x

Saud S, Fahad S, Yajun C, Ihsan MZ, Hammad HM, Nasim W, Amanullah Jr, Arif M, Alharby H (2017) Effects of nitrogen supply on water stress and recovery mechanisms in Kentucky bluegrass plants. *Front Plant Sci* 8:983. doi: 10.3389/fpls.2017.00983

Saud S, Fahad S, Cui G, Chen Y, Anwar S (2020) Determining nitrogen isotopes discrimination under drought stress on enzymatic activities, nitrogen isotope abundance and water contents of Kentucky bluegrass. *Sci Rep* 10:6415. https://doi.org/10.1038/s41598-020-63548-w

Şen B (2009) *Determination of possible effects of climate change on 1st and 2nd crop maize yield in Çukurova Region using regional climate models*. Ph.D. Thesis, Cukurova University.

Shafer DM (2017) Climate change primer – Warm heart worlwwide environmental program https://warmheartworldwide.org/climate-change/ Accessed May13, 2019.

Shafi MI, Adnan M, Fahad S, Fazli W, Ahsan K, Zhen Y, Subhan D, Zafar-ul-Hye M, Brtnicky M, Datta R (2020) Application of single superphosphate with humic acid improves the growth, yield and phosphorus uptake of wheat (Triticum aestivum L.) in calcareous soil. *Agronomy* (10):1224. doi:10.3390/agronomy10091224

Shah F, Lixiao N, Kehui C, Tariq S, Wei W, Chang C, Liyang Z, Farhan A, Fahad S, Huang J (2013) Rice grain yield and component responses to near 2°C of warming. *Field Crop Res* 157:98–110.

Subhan D, Zafar-ul-Hye M, Fahad S, Saud S, Brtnicky M, Hammerschmiedt T, Datta R (2020) Drought Stress Alleviation by ACC Deaminase Producing Achromobacter xylosoxidans and Enterobacter cloacae, with and without timber waste biochar in maize. *Sustainability* 12(6286). doi:10.3390/su12156286

Uzunoğlu F, Bayazit S, Mavi K (2015) Effect of global climate change to ornamental plant growing. *J Agric Fac Mustafa Kemal Univ* 20 (2): 66–75.

Wahid F, Fahad S, Subhan D, Adnan M, Zhen Y, Saud S, Manzer HS, Brtnicky M, Hammerschmiedt T, Datta R (2020) Sustainable management with mycorrhizae and phosphate solubilizing bacteria for enhanced phosphorus uptake in calcareous soils. *Agriculture* 10(334). doi:10.3390/agriculture10080334

Wang D, Shah F, Shah S, Kamran M, Khan A, Khan MN, Hammad HM, Nasim W (2018) Morphological acclimation to agronomic manipulation in leaf dispersion and orientation to promote 'Ideotype' breeding: Evidence from 3D visual modeling of 'super' rice (Oryza sativa L.). *Plant Physiol Biochem*. https://doi.org/10.1016/j.plaphy.2018.11.010

Wu C, Tang S, Li G, Wang S, Fahad S, Ding Y (2019) Roles of phytohormone changes in the grain yield of rice plants exposed to heat: A review. *PeerJ* 7:e7792. DOI 10.7717/peerj.7792

Wu C, Kehui C, She T, Ganghua L, Shaohua W, Fahad S, Lixiao N,Jianliang H, Shaobing P, Yanfeng D (2020) Intensified pollination and fertilization ameliorate heat injury in rice (Oryza sativa L.) during the flowering stage. *Field Crops Res* 252:107795

Yang Z, Zhang Z, Zhang T, Fahad S, Cui K, Nie L, Peng S, Huang J (2017) The effect of season-long temperature increases on rice cultivars grown in the central and southern regions of China. *Front Plant Sci* 8:1908. https://doi.org/10.3389/fpls.2017.01908

Zafar-ul-Hye M, Naeem M, Danish S, Fahad S, Datta R, Abbas M, Rahi AA, Brtnicky M, Holatko, Jiri, Tarar ZH, Nasir M (2020a) Alleviation of cadmium adverse effects by improving nutrients uptake in bitter gourd through cadmium tolerant rhizobacteria. *Environments* 7(54). doi:10.3390/environments7080054

Zafar-ul-Hye M, Tahzeeb-ul-Hassan M, Abid M, Fahad S, Brtnicky M, Dokulilova T, Datta R D, Danish S (2020b) Potential role of compost mixed biochar with rhizobacteria in mitigating lead toxicity in spinach. *Sci Rep* 10:12159. https://doi.org/10.1038/s41598-020-69183-9.

Zahida Z, Hafiz FB, Zulfiqar AS, Ghulam MS, Fahad S, Muhammad RA, Hafiz MH, Wajid N, Muhammad S (2017) Effect of water management and silicon on germination, growth, phosphorus and arsenic uptake in rice. *Ecotoxicol Environ Saf* 144:11–18.

Zaman QU, Aslam Z, Yaseen M, Ishan MZ, Khan A, Shah F, Basir S, Ramzani PMA, Naeem M (2017) Zinc biofortification in rice: Leveraging agriculture to moderate hidden hunger in developing countries. *Arch Agron Soil Sci* 64:147–161. https://doi.org/10.1080/03650340.2017.1338343

Zamin M, Khattak AM, Salim AM, Marcum KB, Shakur M, Shah S, Jan I, Fahad S (2019) Performance of Aeluropus lagopoides (mangrove grass) ecotypes, a potential turfgrass, under high saline conditions. *Environ Sci Pollut Res* https://doi.org/10.1007/s11356-019-04838-3

Zhao C, Liu B, Piao S, Wang X, Lobell DB, Huang Y et al. (2017). Temperature increase reduces global yields of major crops in four independent estimates. *Proc Natl Acad Sci USA* 114:9326–9331. doi: 10.1073/pnas.1701762114.

5

Water Availability and Productivity under Changing Climate

Muhammad Mohsin Waqas[1], Ishfaq Ahmad[2], Yasir Niaz[3], and Shanawar Hamid[3]
[1]Center for Climate Change and Hydrological Modeling Studies, Water Management and Agricultural Mechanization Research Center, Department of Agricultural Engineering, Faculty of Engineering, Khwaja Fareed University of Engineering and Information Technology, Rahim Yar Khan, Pakistan
[2]Centre for Climate Research and Development (CCRD), COMSATS University, Islamabad, Pakistan
[3]Department of Agricultural Engineering, Faculty of Engineering, Khwaja Fareed University of Engineering and Information Technology, Rahim Yar Khan, Pakistan

5.1 Introduction

Water is the most important, crucial, scarce, and declining global resource and it is essential for living mechanisms to carry out proper functioning. Water is important resource for successful crop production (Fahad et al., 2014, 2017; Hesham and Fahad, 2020; Saud et al., 2013, 2014, 2016, 2017, 2020). The concerns arise on the effective water utilization for the crop production due to the dwindling of the water resources especially under increasing cost of water development, degradation of aquifer water, quality, quantity, and the less effective approaches of the water utilization.

The increasing water scarcity is the global issue and almost 1.2 billion and 1.6 billion pupils are under the physical and economic water scarcity, respectively. Therefore, developing countries required the new water storage infrastructure to ensure the food security, as the 17% of the global water supplies for irrigation needs to enhance for ensuring water availability. Additionally, the global crop yield needs to be enhanced by 38% to feed the increasing population.

The national's food security enhanced the importance of the water availability. Therefore, in the view of the global prospects, the expert pretends the water distribution conflicts between the nations in near future. Food security is directly coupled with the agriculture production which depends upon the timely precise application of the water. Irrigation water utilization is 80–90% of the total available water. Irrigation is vital to the agriculture-based country to ensure the food security.

The history of the water development in the continent is before the creation of Pakistan. Pakistan continues to improve the water development and management to uplift the agriculture-based country economy. The swift increasing population (1.2% per annum) of Pakistan enhances the scarcity in the scarcity-based design irrigation system to ensure the effective and sustainable utilization of water.

Poverty reduction is directly related to the effective water utilization as the 70% of the water is used for agriculture worldwide and 95% of the utilization in the south Asia. In case of Pakistan, 93% water is utilized for crop production. In Pakistan, the scarcity of water increasing at the alarming rate as the per capita of the water declines from the 5,260 m^3 per capita in 1951 to 1,038 m^3 in 2010 and 1,000 m^3 in 2017.

Generally, the water productivity is defined as the yield produced per unit amount of the water consumed. The *water productivity* term is defined differently by the different stakeholders produce the product during the season. Mostly the water productivity is distinguished in three different ways: agronomic, economic and socio-economic water productivity. Agronomic water productivity is defined as the yield per unit volume of water utilized. Economic water productivity is defined as the economic benefit per unit water utilized and socio-economics as the social benefit pert unit water utilized.

In south Asian countries, almost one-fourth of the world population is living while the water availability is only 4.5% of the total water available worldwide. As Pakistan is considered as one of the water-stressed countries in the world and the situation is becoming more verse due to the impact of the climate change and climate variability. The water resources of the country are more vulnerable to the impact of the climate change. The impact of the climate change is getting more worse on the water resources of the country from the past decade. The impacts are glaciers receding, and the levels of the reservoirs are low from the average, extreme events of the rainfall and heat waves. It causes the less water delivery for the irrigated agriculture. The water flow shrinks from the average of the 180 km^3 to the 110 km^3 under the drought conditions. It has adversely affected the water availability for irrigation and is leading towards the food insecurity.

5.2 Water Resources Availability

The water resources are divided into the three major sources: surface water, groundwater, and the rainfall. The extent and availability of these resources vary at the different spatial and temporal scale.

5.2.1 Rainfall

The impact of the climate change on the rainfall shows the potential increase in the precipitation over the 21st century from 30°N to 85°N, but in the near past three to four decades, the significant decrease in the rainfall is project for the 10°S and 30°N as shown in Figure 5.1. In the first five decades of the 20th century, precipitation increased distinctly for the 10°N and 30°N and started declining after the seventh decade of the 20th century. The variability in the precipitation is utmost uncertain, the precipitation shows the negative trends for the West Africa, South Asia and the Sahel from the start of the 20th century. But these trends show the reserve trends of the precipitation after 1979 for the tropical

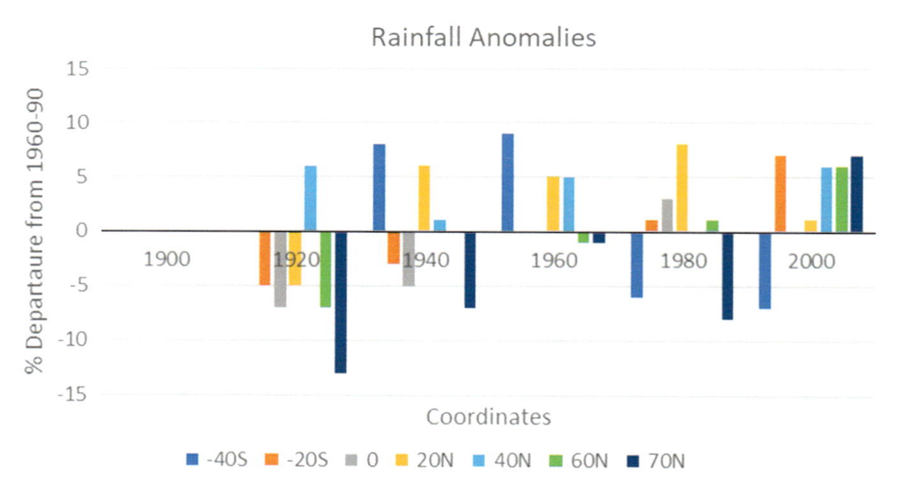

FIGURE 5.1 Mean annual anomalies for precipitation (%) over land for 1900–2005 relative to the 1961–990.

Source: Bates et al., 2008.

regions of the Africa and Sahel. An increase in the precipitation for the North-Western India was found 20% during the 1901–2005, even with the strong decreased since 1979. The projection of the multi-model ensemble approach shows the increase in the hydrological cycle components in the 21st century. The increase in the participation will be significant in the tropical region, that is, monsoon-based region and the tropical pacific. While a decrease in the subtropics will be found. The high latitude areas will receive the 20% more rainfall during the past two decades of the 21st century. While the Caribbean, Mediterranean, and Subtropical region will receive the significant decrease in the rainfall, the overall 4% and 5% rainfall is expected to increase on the land and oceans, respectively. Figure 5.1 describes the rainfall anomalies at the global scale.

Rainfall is the one of the resources that has significant variations at the different spatial and temporal scale. In case of Pakistan, the highest mean annual rainfall is about the 1,500 mm in the northern foot hills of the country, while the lowest of 100 mm is occurred in the different regions of the Baluchistan and the Sind Province. The agriculture year of Pakistan is dived into the two major cropping seasons: Rabi and Kharif. The Kharif season starts from April and ends in September. The intra seasonal variation in the Rabi season rainfall is 500 mm in some parts of the Khaber Pakhtoon Khaw province and 50 mm in some parts of the Sindh province. The variation in the Kharif season rainfall is 50 mm in some parts of Baluchistan and 800 mm in KPK and Punjab Province.

The spatial variability of the rainfall occurrence pattern in Pakistan resulted the 92% area under arid condition. The higher intensity in combination with the snow melting during the monsoon season causes the heavy floods in the Indus Basin Irrigation System (IBIS). These floods cause the huge damage to socio-economic development of the country due to the unavailability of the sufficient storage reservoirs. This higher intensity also causes the rapid runoff especially in the sloppy areas and decreases the contribution of the rainfall in the agriculture production due to less infiltration. The total contribution of the rainfall in the water availability is 13% (16.5 billion m^3) after the operation of Terbela dam.

5.2.2 Groundwater

The aquifer as a most flexible water resource provides the water at the time of the requirement and its proper management is the dire need of the farming community. The management of the groundwater quality is much more important as the climate change is bringing the variability in the availability of the water resources. Climate change will affect both chemical and thermal properties of the groundwater. The increase in the air temperature will increase the groundwater temperature having shallow watertables depth. In arid and the semi-arid regions, the increase in the evapotranspiration will increase the groundwater salinization.

Margat and Van der Gun (2013) studied the impact of the climate change on the groundwater recharge in the central America, Northern Africa (Sahara), Southern Africa, South Asia, and around the Mediterranean. The study was performed under the A2 and B2 emission scenario. The results obtained were loss of the fresh groundwater in the coastal region of the study site except the Sahara region under both A2 and B2 emission scenario. Globally, the groundwater extraction is about 982 km^3 per annum and almost 70% is used for the irrigation. The overall contribution to the irrigation is 38% worldwide. In Pakistan, the alluvial lands between the rivers started the storing of water from the seepage and percolation of the irrigation channels. As the system expands, the seepage losses created the problems of waterlogging and secondary salinization in the dominant part of the cultivated land. To overcome these problems, the Salinity Control and Reclamation projects were initiated (SCARP). The groundwater development with the introduction of the tubewell technology and the availability of the high yielding verities after the green revolution created the boom in agriculture production. The groundwater developments uplifted the living standard of the farmers community. The water storage in the aquifer of the system is providing 48 MAF for the irrigation purpose.

The groundwater contributes almost 40% in agriculture production of Pakistan. The groundwater extraction at the heads of the canals is more due to the freshwater availability (Awan and Ismaeel, 2014). The restriction for the tail user of the canal command is the poor quality of the irrigation water from the aquifer. According to the irrigation water quality standards, Pakistan has 49% of the groundwater as the good quality ($TDS < 1000$ mgl^{-1}), 36% marginal saline ($TDS < 3000$ mgl^{-1}), and 15% strongly saline ($TDS > 3000$ mgl^{-1}).

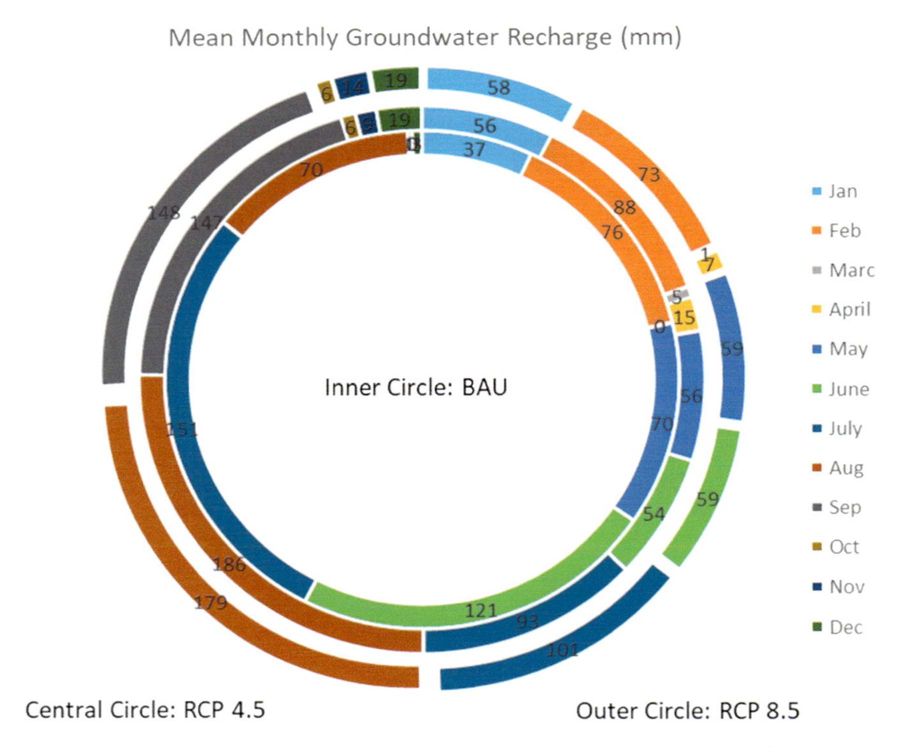

FIGURE 5.2 Mean monthly groundwater recharge under BAU, RCP4.5 and RCP 8.5 scenarios for entire Lower Chenab canal irrigation scheme.

The groundwater quality in the major provinces varies spatially in Punjab. The groundwater salinity varies from the 0.3 to the 4.6 dsm^{-1} in the Punjab and the situation is even worse in the South Punjab. While the groundwater salinity in the Sindh province varies from the 0.5 to 7.1 dsm^{-1} and the situation is much worse in the delta of the Indus basin. The longer-term groundwater availability is based on the prevailing climatic condition. The climate change is posing the impact of the regional scale hydrological cycle. The negative impacts of the climate change include the occurrence of the extreme events more rapidly and shifting of the climatic conditions. In case of too dry and too wet years of the rainfall, the availability of the groundwater will be changed. In a case study performed in the Lower Chenab Canal system, Awan and Ismaeel (2014) found the 40% increase in the rainfall under RCP 4.5 and 37% under RCP 8.5 up to 2020. While the increase in the groundwater recharge was found 40% and 37% from the business as usual scenario for the RCP 4.5 and RCP 8.5, respectively. The details of the study are presented in Figures 5.2–5.3. Figure 5.2 presents the groundwater recharge on the temporal scale of month for the Lower Chenab canal system. While Figure 5.3 presents the groundwater recharge at the spatial scale of distributary level of the largest irrigation scheme of the Indus basin irrigation system.

Groundwater potential varies spatially in Pakistan. Groundwater potential is dependent upon the continuous recharge to overcome the rapid depletion of the aquifer. The spatial variation of the groundwater quality at the provincial level shows the maximum fresh water overlain by the Punjab province. Around 79% of the Punjab has access to the fresh aquifer. The good quality water covers an area of 7.9 million acres, while the marginal saline aquifer covers an area of the 3.0 million acres. The total water potential of the Punjab province is 42.75 MAF. In case of Sindh province, only 28% of the province has access to the good-quality aquifer, while the remaining area is under the extremely saline aquifer. The groundwater depth is shallow in the major part of the Sindh province. The total groundwater potential of the Sindh province is 18 MAF. In case of KPK, the groundwater recharge is less as compared to the extraction especially in Dera Ismail Khan, Banu, and Karak districts. The total groundwater potential of the KPK province is 3.11 MAF. In terms of area, Baluchistan is the largest

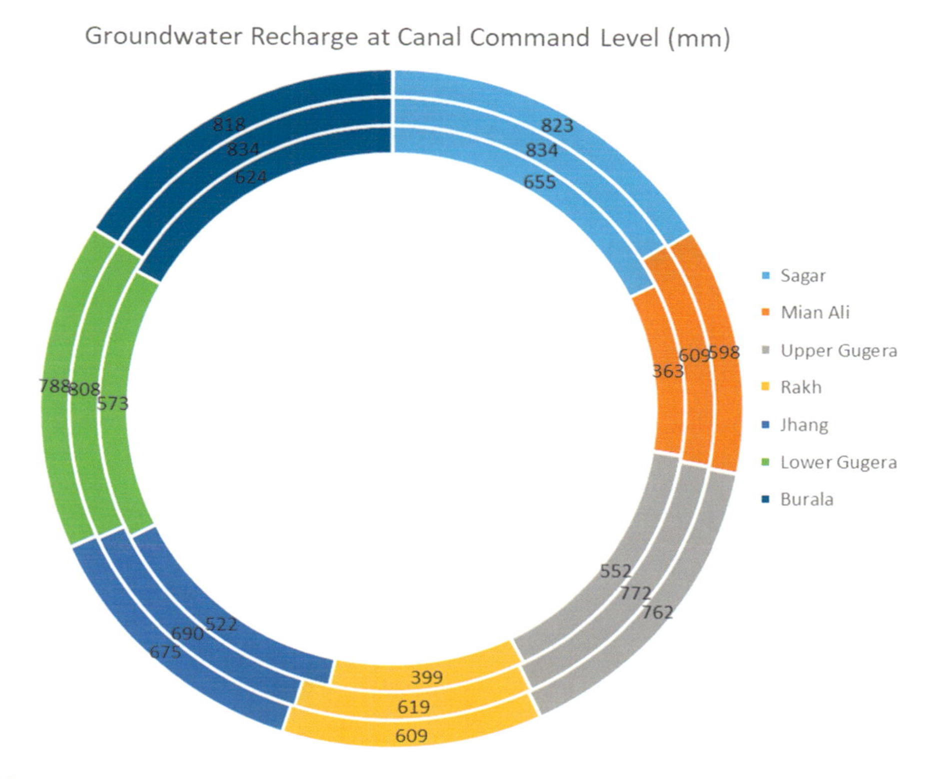

FIGURE 5.3 Mean annual groundwater recharge for BAU, RCP 4.5 and RCP 8.5 scenarios for different canal command areas of Lower Chenab canal irrigation scheme.

province of Pakistan, but it is at the lowest in case of the total water potential of only 2.13 MAF. The Baluchistan province is under the worse physical scarcity of the water in Pakistan, where people drink the water with the TDS of more than 3,000 mgl^{-1}.

5.2.3 Surface Water

Global surface water resources under changing climate were analysed from the catchment to the global scale during the 20th century. Studies showed the direct relation of participation and temperature with the rivers flow under changing climate. However, the studies were also conducted to describe the impact of rainfall and temperature on the runoff as well as the human activities in the catchment. The change in the annual discharge is observed at the global scale. The higher latitudes areas, that is, the USA are considered under the high runoff areas, while the other parts of the world shows decreased in the runoff, that is, South America, West Africa, and Southern Europe. The main driving factor of the runoff variations in the different areas of the world are the El Niño-Southern Oscillation (ENSO) and North Atlantic oscillation (NAO) patterns.

Figure 5.4 describes the SRES emission scenarios-based water stress calculation with the increasing population. The highest stress is found under the A2 emission scenario, and major causes of the water stress under this scenario are due to higher projection of the population. In case of A1 and B1 emission scenario, the population of the most of the countries will decline after 2050 and this resulted the decrease in the water stress between 2050 and 2100 agreeing A1B, B1, and A1FI emissions scenarios.

The total runoff of 4% increases with the increases in 1°C at the global scale during the 20th century. There is no consistency in the relation of the runoff and the rainfall due to the human activities in the catchment. Rivers flows were delayed in some parts due to the occurrence of the snowfall instead of the rainfall. Similarly, the flow in the rivers occurs early due to the increase in the temperature that causes

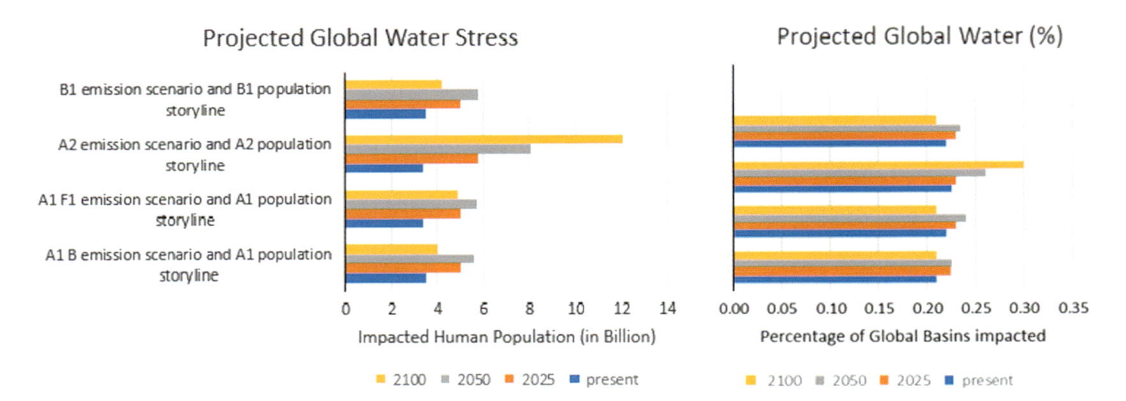

FIGURE 5.4 Global projected water stress (based on per capita water availability).

the occurrence of the rainfall instead of the snowfall and snow melting due to increase in temperature in some parts of the world of which the best example is the New England where the snowmelt shifted one to two weeks late between 1936 and 2000. The studies are less in Asia that describe the impacts of the climate change on the river's flows. On the other hand, several studies are available on the impact of climate change on the river flow in Australia, Europe, and North America. Arnell (1999) presented the change in regional average annual runoff by the 2050s, by major river basin.

The treaty was signed on the water rights between Pakistan and India as Known as Indus Water Treaty. Under this treaty, the both countries have the rights to discharge the floods in the natural stream of the any river and the none of them will discharge the industrial effluent into the any natural stream of the river without the proper treatment. The impact of the climate change shows the increase in the flood up to 2050 and then low discharge in the natural stream of the rivers. This will result in the less water availability under both extreme conditions of flood and drought to Pakistan (Table 5.1).

One of the world's largest contagious irrigation system of Pakistan was developed to irrigate all cultivated area command by the five rivers. The annual average water flow in the system is 157 MAF, out of which 106 MAF is diverted to the main canals from the barrages. While the unattended water is 8–92 MAF, at an average of 39 MAF, that is falling annually into the Arabian Sea. The required amount of water is 5–10 MAF to stop the salt intrusion and to protect the delta mangroves. The escape of the water from the Kotri barrage towards seas is mostly in the 70–100 summer days. The regular discharge pattern is not available to protect the Indus delta from the salt's intrusion from the sea. The other example of the climate change impacts includes the salt intrusion from the sea water due to the increase in the sea level under the impact of the increase in the sea surface temperature. The intrusion of the salt will be triggered more due to the over pumping in the coastal areas of the less and unregulated discharge to the sea from the river's deltas. Table 5.2 describes the surface water availability in the Indus basin irrigation system.

As the major aim of the dam in the agriculture-based country is water saving for irrigation, Pakistan constructed the two major dams under IWT, that is, Terbela and Mangla. Both of these reservoirs are

TABLE 5.1

Rivers flows and % contribution

Sr No	River	Mean Annual Volume (billion m3)	Contribution (%)
1	Indus	74.1	42.1
2	Jhelum	28.3	16.1
3	Chenab	7.14	4.1
4	Ravi Sutlej	7.14	4.1
5	Kabul	27.43	15.6

TABLE 5.2

Surface water availability (million acre feet)

Period	Kharif	Rabi	Total	% Increase and Decrease over the Avg
Average system usage	67.1	36.4	103.5	-
2009–2010	67.3	25.0	92.3	−10.8
2010–2011	53.4	34.6	88.0	−15.0
2011–2012	60.4	29.4	89.8	13.2
2012–2013	57.7	31.9	89.6	−13.4
2013–2014	65.5	32.5	98.0	−5.3
2014–2015	69.3	31.1	102.4	−1.1
2015–2016	65.5	32.9	98.4	−4.9
2016–2017	71.4	29.7	101.1	−2.3
2017–2018	70.0	24.2	94.2	−9.0
2018–2019	59.6	24.8	84.4	−18.5

Source: Indus River System Authority.

TABLE 5.3

Existing dams on the Indus and its tributaries

River	Project	Storage Capacity (BCM)
Indus	Terbela	11.38
	Chashma	0.75
	Manchar Lake	0.92
	Kinjhar Lake	0.39
	Chotiari Lake	0.95
Jhelum	Mangla	5.9
Kabul	Warsak	0.05
Kurram	Baran Dam	0.04
Hub	Hub	0.93
Haro	Khanpur	0.11
Kohat Toi	Tanda	0.07
Kurang	Rawal	0.05
Soan	Simply Dam	0.02
Pishin	BKD Dam	0.05
	Hamal Lake	0.09

Source: WAPDA.

losing their capacity due to the sedimentation. Both reservoirs lost their capacity of 5.8 MAF. The completed projects and their storage capacity are given in Table 5.3.

In case of Pakistan, Rashid studied the impact of the climate change on the water availability of the Jhelum river basin. Study showed in the Tables 5.4–5.5 described the projected changes in the annual stream flow at the different gauges of the Jhelum river basin under A2 and B2 scenario of the HadCM3 model with respect to the base line period. The flow is increased under the 2020s and 2050s period of the study with respect to the 1961–1990 under the A2 scenario as shown in Table 5.4. The maximum discharge passes at the Azad-Pattan gauge where 87% of the flow passes. The increase in the flow is projected by 30% and 50% for the 2020s and 2050s period, respectively. The increase in the discharge for 2080s is projected 34% at the same gauge. Similarly, under the B2 scenario, increased in the flow was found for the 2080s as it was projected for the A2 scenario but for the other period, the flow was less increased under the B2 scenario as compared to the A2 scenario as shown in Table 5.4.

TABLE 5.4

Future changes (%) in annual stream flow under A2 scenario

Under A2 Emission Scenario							
Stream Flow \ % Changes Under Climate Change	**Naran**	**Gari-Habibulla**	**Muzaffarabad**	**Domel**	**Kohala**	**Azad-Pattan**	**Kotli**
Stream flow (m^3/s), 1961–1990	47.9	104.7	356.2	362.4	811.0	846.6	133.8
Future changes (%) in 2020s	2.9	29.3	20.2	3.2	29.5	30.5	37.8
Future changes (%) in 2050s	4.6	23.8	18.2	7.2	26.5	25.6	4.1
Future changes (%) in 2080s	7.9	42.1	25.0	8.0	35.5	36.5	74.3
Under B2 Emission Scenario							
Stream Flow \ % Changes Under Climate Change	**Naran**	**Gari-Habibulla**	**Muzaffarabad**	**Domel**	**Kohala**	**Azad-Pattan**	**Kotli**
Stream flow (m^3/s), 1961–1990	47.9	104.7	356.2	362.4	811.0	846.6	133.8
Future changes (%) in 2020s	7.5	33.3	22.4	1.0	30.2	30.9	22.7
Future changes (%) in 2050s	6.4	30.4	23.3	5.7	28.5	26.5	11.3
Future changes (%) in 2080s	15.5	42.9	24.1	5.5	33.3	34.3	56.2

TABLE 5.5

Majors crop production (thousand tonnes)

	Cotton (000 bales)	Sugarcane	Rice	Maize	Wheat
2012–2013	13,031	36,750	5,536	4,220	24,211
2013–2014	12,769	67,460	6,798	4,944	25,979
2014–2015	13,960	62,826	7,003	4,937	25,086
2015–2016	9,917	65,482	6,801	5,271	25,633
2016–2017	10,671	75,482	6,849	6,134	26,674
2017–2018	11,946	83,333	7,450	5,902	25,076
2018–2019	9,861	67,174	7,202	6,309	25,195

P: Provisional (July–March).
Source: Pakistan Bureau of Statistics.

5.2.4 Hill Torrents

Hill torrents are the flows of water from the mountain areas toward the low land areas. Hill torrents are the rainfall waters that move with high speed and huge sedimentation with varying discharge depending upon the rainfall intensity in the catchment. It is locally known as the Rod Kohi Irrigation system, which is acombination of two words: *Rod* (a natural stream) and *Kohi* (originating from *Koh*, meaning hill). Internationally, thissy stem of irrigation is known as the spate irrigation.

The hill torrent system in Pakistan has great potential to contribute to food security. Sufi et al. reported on 23 BCM in the major 14 hill torrents of Pakistan. The efficiency of this system is less than 50%. The spate irrigation is applied a single time through the construction of the bunds using earth work. The water is stored in the rootzone and farmers conserve the moisture using the Sugha. The crop production depends upon the conserved moisture and on the direct rainfall that is less than the 15% of the total crop water requirement.

5.3 Water Allowance

Cheema et al. (2016) described the average water allowance in the canals as 3.49 cusec per 1,000 acres. The water allowance is lower as compared to other agriculture-based countries, e.g., 4.8, 5.31, and 10 cusecs is the average water allowance for India, Egypt, and Mexico, respectively. The water allowance of the canal commands is given in Figure 5.5.

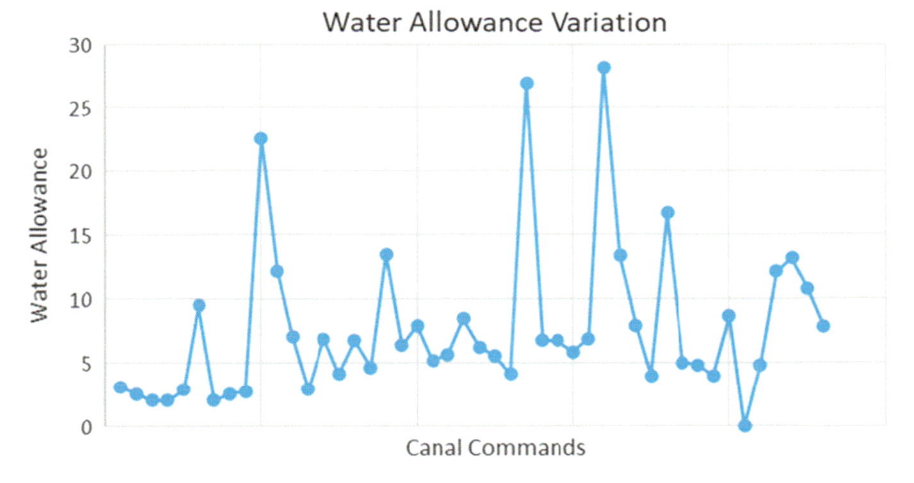

FIGURE 5.5 Water allowance (cusec)/1000 acres of CCA.

Source: IWMI, Lahore, Pakistan.

5.4 Water Distribution Pattern

Warabandi is the pattern of water availability to the end users of an irrigation system. *Warbandi* is the local term of the Saraiki language which is the combination of the two words *Wara* and *Bandi*. *Wara* means turn and *Bandi* means to divert. Thus, *warabandi* means diverting the water toward the field channels for irrigation at the time of one's turn. The current warabandi system is seven or ten days turn system. The dominant warabandi system in the Punjab is the seven days and it starts from the Monday 6.00 am to the farmers at the first inlet of the watercourse. The rotation of this turn is changed each year, the night irrigators receive the water at the morning time. As the warabnadi system is under the provincial irrigation department, the water allocation is the 25 minutes per acre and a single watercourse commands the area of the 400–500 acres based on the discharge of the outlet. The climate change will impact the water supply under dry and the wet period. The supply-based irrigation system of Pakistan and India will move towards the further water scarcity. During the flooding periods, the system is unable to convey water towards the fields from the rivers according to demand. In case of the drought conditions, the canals closure periods will be increased.

5.5 Water Productivity

The global water productivity benchmark is defined as wheat ranges 0.6–1.7 kg m^{-3}, rice ranges 0.6–1.6 kg m^{-3}, cottonseed ranges 0.41–0.95 kg m^{-3}, cotton lint ranges 0.14–0.33 kg m^{-3}, and maize ranges 1.1–2.7 kg m^{-3}. These variations were found based on the variability of the climate, soil nutrients, and irrigation water management. Saud et al., (2013) described the global-level variation of 20–49% within the agriculture year to year for rice, wheat, maize, and soy. The variation of 18–43% of the crop yield was witnessed due to extreme climate events such as heat and cold waves, drought, and high-intensity rainfall.

Pakistan is blessed with the four climatic seasons, that is, a cold winter, an extensive, hot summer, autumn, and a pleasant spring. These seasons provide opportunities to the farmers to grow a diversity of crops on their farms. The major crops of Pakistan include te cotton, wheat, rice, maize and sugarcane. Pakistan is the fourth-largest producer of cotton in the world. Wheat is the main staple and Pakistan is exporting and donating extra wheat to neighbouring countries. The details of the major crop production in Pakistan is given in Table 5.5.

The crop growth reduced due to the increase in the input cost and the other climatic extremities. The cotton growth is expected to decrease 17.5% during 2018–2019 as compared to the considered base period of 2012–2013. While the growth was 11.9% higher in the last year from the base period of 2012–2013. The highest reduction of 19.4% in the sugarcane growth is also expected this year from the base period. The rice crop shows the 3.3% growth reduction, while maize and the wheat show the increase in the growth rate of 6.9% and 0.5%, respectively. The details and the yearly updates on the crop growths in Pakistan can be accessed in the economic survey of Pakistan annual report.

5.6 Water Productivity: A Case Study of Pakistan

Water productivity not only varies spatially in the Indus basin but also within the canal command areas. Water productivity was found higher in the canal commands of the low water allowance due to the effective utilization of the groundwater. Figure 5.6 describes the highest wheat water productivity of the 1.2 kgm^{-3} in the upper and lower Jhelum canal command, while the lowest water productivity of 0.10 kgm^{-3} was found in the rice canal command. Figure 5.6 describes the highest rice water productivity of 1.1 kgm^{-3} at the Upper Depalpur canal and the lowest of 0.3 kgm^{-3} at the Begari canal command.

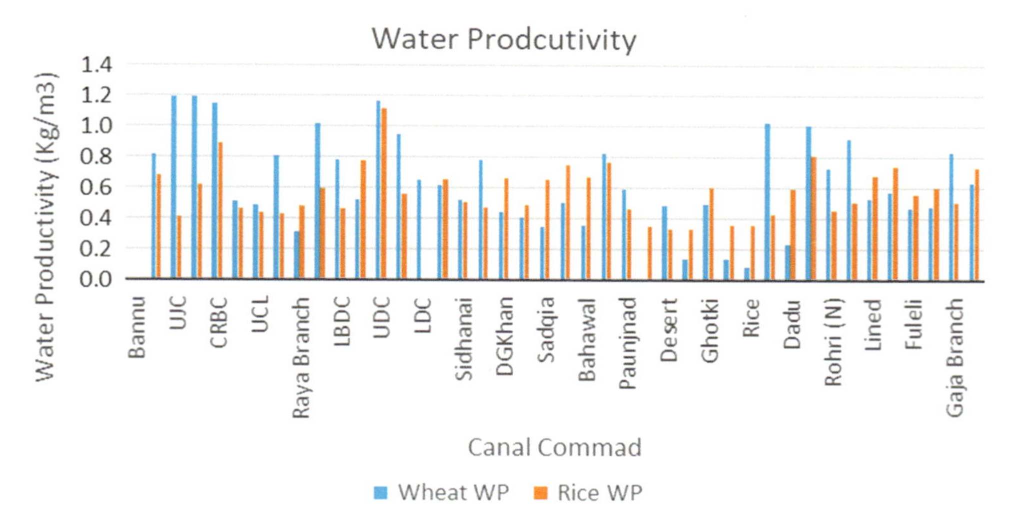

FIGURE 5.6 Water productivity of the wheat and rice and different canal commands.

The impact of climate change varies regionally in the different river basins. The variation includes the increase in the frequency of the floods and droughts under changing climatic condition. Efficient application of irrigation will be very important for the enhanced water productivity. Although the different water management techniques have been applied in the different region of the world to mitigate the climate change impacts on the crop production, there is dire need to explore and apply the mitigation techniques in the Indus river basin. This should be directed towards the estimation of water availability under fold and drought condition based on the frequency of the rainfall and the snow melt under changing climatic condition. The impact of the climate change on the crop production varies under different latitude and irrigation application techniques. The crop production for the water loving crops can be increased under increased precipitation, while the production will be decreased for the sensitive water crops. The higher rainfall intensity will reduce the crop production of the cotton crop, while the higher rainfall intensity and magnitude will increase the crop production of the rice crop. The crops are more sensitive to the precipitation as compared to the temperature. If the frequency or the magnitude of the rainfall will be decreased in the arid (Barani) and spate irrigation-based crop production areas, the crop production can be rescued by increasing the water holding capacity of the soil.

The variation in temperature due to climate change impacts will cause changes to the cropping pattern on a regional scale. The crop duration will get shortened due to increase in the temperature that will leads toward the early maturity of the crops. The higher crop production will require the altered sowing and harvesting time of the major crops to maximize the crop production. This impact will change the input parameters of the crops, that is, the short duration of the crops required less interval of the fertilizer application to get the maximum production by increasing the vegetative, flowering, and maturity stage of the crops. The timely and precise application can be used to increase the crop production. The challenge of the 21st-century agriculture is to ensure the food security of the swift increasing population under scare water condition. The climate change impacted further increases the risks of food insecurity due to increasing in the drought and floods based on the unexpected pattern of the rainfall. The food security is directly dependent upon the enhanced water productivity. The increased in the water productivity means increase in the crop production with the efficient water utilization. The water productivity is not only depending upon the water availability or the efficiency of the water application, it also includes the other external factors under the changing climatic condition. The external factors include the GCMs prediction especially regarding the climate variability, soil water storage and fertility condition of the soil under changing climate, levels of the CO_2 concentration, climatic variables, and uncertainty in the agro-hydrological models.

In Pakistan, the water availability from the surface and rainfall is not uniform. Therefore, farmers require additional storage to store the water at the time of surplus water supply for effective application to fields for better crop production. The stored water will help farmers to apply the water according to better irrigation scheduling. The on-farm water storage will help to practice the high-efficiency irrigation system, that is, drip irrigation and sprinkler irrigation. The sprinkler system efficiency is almost 70–80% and the drip irrigation system has 80–90%. The adoptability and application of the system varies under specific soil, water, terrain, wind, and its crop-specific nature.

Skimming wells are a technology in which the freshwater is collected from the fresh aquifer layers with the help of multiples bores in this layer. In this technique, the farmers will install a bore hole to store the surplus water into the groundwater bank. Similarly, the rainwater harvesting is the technique that combats the water scarcity in the rainfed areas through proper planning and installation of the rainwater harvesting tanks. This practice will be more adoptive in mega cities of developing countries, that is, Karachi is the city where proper rainwater harvesting will help to secure the water availability for domestic water use. Check dams or mini dams are a good solution to improve the water availability in the rainfed areas.

5.7 Conclusion

The climate change impacts on the surface water availability and productivity are variable. In case of surface water availability, the glaciers melting in the early period will bring the heavy floods, while later on drought will occurr due to less snow melting. Similarly, the variation in the rainfall is changing its position of occurrence and the intensity and frequency is also changing. The extreme events of the rainfall will reduce the water availability for the better crop production due to low infiltration and higher runoff. While in irrigated areas, the groundwater recharge is expected to increase due to increase in the rainfall. It highly depends upon the site-specific crop and soil characteristics. In most of the coastal areas, the salt intrusion will be due more to the increase in the surface temperature and increase in the sea-level rise. The water productivity will be affected due to occurrence of the extreme events and the huge losses of water as compared to the water utilization. It will be better to conserve the soil moisture using the on-farm water storage, aquifer recovery, use of skimming wells, and adoption of high efficiency irrigation systems.

REFERENCES

Arnell NW (1999) Climate change and global water resources. *Glob Environ Chang* 9:S31–S49.

Awan UK, Ismaeel A (2014) A new technique to map groundwater recharge in irrigated areas using a SWAT model under changing climate. *J Hydrol [Internet]* 519(PB):1368–1382. http://dx.doi.org/10.1016/j.jhydrol.2014.08.049

Babel MS, Wahid SM (2008) Freshwater under threat: Vulnerability assessment of freshwater resources to environmental change – Ganges-Brahmaputra – Meghna river basin Helmand river basin Indus river basin. *United Nations Environment Programme (UNEP), Nairobi, Kenya.* http://wwwuneporg/pdf/southasia_reportpdf, accessed August 28, 2015.

Boesen J, Munk Ravnborg H (2004) From waterwars to waterriots? Lessons from transboundary water management. Proceedings of the international conference, December 2003, DIIS, Copenhagen. DIIS Working Paper.

Cheema MJM, Bakhsh A, Mahmood T, Liaqat MU (2016) Assessment of water allocations using remote sensing and GIS modeling for Indus basin, *Pakistan. PSSP Work Pap* 36:48.

Dragoni W, Sukhija BS (2008) Climate change and groundwater: A short review. *Geol Soc London, Spec Publ* 288(1):1–12.

Fahad S, Hussain S, Bano A, Saud S, Hassan S, Shan D, Khan FA, Khan F, Chen Y, Wu C, Tabassum MA, Chun MX, Afzal M, Jan A, Jan MT, Huang J (2014) Potential role of phytohormones and plant growth-promoting rhizobacteria in abiotic stresses: Consequences for changing environment. *Environ Sci Pollut Res* 22(7):4907–4921. https://doi.org/10.1007/s11356-014-3754-2

Fahad S, Bajwa AA, Nazir U, Anjum SA, Farooq A, Zohaib A, Sadia S, Nasim W, Adkins S, Saud S, Ihsan MZ, Alharby H, Wu C, Wang D, Huang J (2017) Crop production under drought and heat stress: Plant responses and management options. *Front Plant Sci* 8:1147. https://doi.org/10.3389/fpls.2017.01147

Gillani NA (2001) Supplement to the framework for action for achieving the Pakistan Water Vision 2025. *Pakistan Water Partnersh.*
Hesham FA, Fahad S (2020) Melatonin application enhances biochar efficiency for drought tolerance in maize varieties: Modifications in physio-biochemical machinery. *Agron J* 112(4):1–22.
Hunter PR, MacDonald AM, Carter RC (2010) Water supply and health. *PLoS Med* 7(11):e1000361.
Irfan M, Qadir A, Ali H, Jamil N, Ahmad SR (2019) Vulnerability of environmental resources in Indus basin after the development of irrigation system. In: Irrigation-Addressing Past Claims and New Challenges. *IntechOpen.*
Kang Y, Khan S, Ma X (2009) Climate change impacts on crop yield, crop water productivity and food security – A review. *Prog Nat Sci* 19(12):1665–1674.
Mahmood R, Jia S (2016) Assessment of impacts of climate change on the water resources of the trans-boundary Jhelum river basin of Pakistan and India. *Water* 8 (6):246.
Margat J, Van der Gun J (2013) *Groundwater around the world: A geographic synopsis.* CRC Press.
Monheit AC (2011) Running on empty. Inq J Heal Care Organ Provision. *Financ* 48(3):177–182.
Parish ES, Kodra E, Steinhaeuser K, Ganguly AR (2012) Estimating future global per capita water availability based on changes in climate and population. *Comput Geosci* 42:79–86.
Ranjan P, Kazama S, Sawamoto M (2006) Effects of climate change on coastal fresh groundwater resources. *Glob Environ Chang* 16(4):388–399.
Sathar Z, Royan R, Bongaarts J (2013) Capturing the demographic dividend in Pakistan. Islamabad: Population Council.
Saud S, Chen Y, Long B, Fahad S, Sadiq A (2013) The different impact on the growth of cool season turf grass under the various conditions on salinity and drought stress. *Int J Agric Sci Res* 3:77–84.
Saud S, Li X, Chen Y, Zhang L, Fahad S, Hussain S, Sadiq A, Chen Y (2014) Silicon application increases drought tolerance of Kentucky bluegrass by improving plant water relations and morph physiological functions. *SciWorld J* 2014:1–10. https://doi.org/10.1155/2014/368694
Saud S, Chen Y, Fahad S, Hussain S, Na L, Xin L, Alhussien SA (2016) Silicate application increases the photosynthesis and its associated metabolic activities in Kentucky bluegrass under drought stress and post-drought recovery. *Environ Sci Pollut Res* 23(17):17647–17655. https://doi.org/10.1007/s11356-016-6957-x
Saud S, Fahad S, Yajun C, Ihsan MZ, Hammad HM, Nasim W, Amanullah Jr, Arif M, Alharby H (2017) Effects of nitrogen supply on water stress and recovery mechanisms in Kentucky bluegrass plants. *Front Plant Sci* 8:983. doi: 10.3389/fpls.2017.00983
Saud S, Fahad S, Cui G, Chen Y, Anwar S (2020) Determining nitrogen isotopes discrimination under drought stress on enzymatic activities, nitrogen isotope abundance and water contents of Kentucky bluegrass. *Sci Rep* 10:6415. https://doi.org/10.1038/s41598-020-63548-w
Siebert S, Burke J, Faures J-M, Frenken K, Hoogeveen J, Döll P et al. (2010) Groundwater use for irrigation – A global inventory. *Hydrol Earth Syst Sci* 14(10):1863–1880.
Tarar RN (1997) Pakistan's surface water scenario in 21st century and needed actions. In: Tariq AR, Latif M (eds.) *Water for the 21st century: Demand, supply, development and socio-environmental issues. Proceedings of an International Symposium, Centre of Excellence in Water Resources Engineering Publications* 110, Lahore, pp. 49–60.
Vogel E, Donat MG, Alexander LV, Meinshausen M, Ray DK, Karoly D et al. (2019) The effects of climate extremes on global agricultural yields. *Environ Res Lett* 14(5):54010.
Zwart SJ, Bastiaanssen WGM (2004) Review of measured crop water productivity values for irrigated wheat, rice, cotton and maize. *Agric Water Manag* 69(2):115–133.

6

Impact of Climate Change on Biodiversity of Insect Pests

Abdel Rahman Al-Tawaha[1], Syed Kamran Ahmad[2], Huma Naz[3], and Abdelrazzaq Al-Tawaha[4]
[1]Department of Biological Sciences, Al Hussein Bin Talal University, Ma'an, Jordan
[2]Department of Entomology, School of Agriculture, ITM University Gwalior, M.P., India
[3]Mohammad Ali Nazeer Fatima Degree College, Hardoi, U.P., India
[4]Department of Crop Science, Faculty of Agriculture, Universiti Putra Malaysia, Selangor, Malaysia

6.1 Introduction

Climate change is a highly debated issue in vogue and recently it raised a significant stockpile of literature (Adnan et al., 2018, 2020; Ahmad et al., 2019; Akram et al., 2018a,b; Bezemer and Jones, 1998; Cornelissen, 2011; Wang et al., 2018; Farhat et al., 2020; Gul et al., 2020; Habib ur Rahman et al., 2017; Hammad, 2016, 2018, 2019, 2020a,b; Hussain et al., 2020; Ilyas et al., 2020; Jan et al., 2019; Kamaran et al., 2017; Khan et al., 2017a,b; Lindroth, 1996; Mubeen et al., 2020; Muhammad et al., 2019; Naseem et al., 2017; Parmesan, 2006; Rehman et al., 2020; Sajjad et al., 2019; Saleem et al., 2020a,b,c; Saud et al., 2013, 2014, 2016, 2017, 2020; Shah et al., 2013; Stiling and Cornelissen, 2007; Subhan et al., 2020; Tylianakis et al., 2008; Wahid et al., 2017, 2020; Watt et al., 1995; Wu et al., 2011, 2019, 2020; Yang et al., 2017; Zafar-ul-Hye et al., 2020a,b; Zahida et al., 2017; Zamin et al., 2017; Zvereva and Kozlov, 2006). Authors from every section of the academic world have addressed the issue and have connected their areas of expertise. Climate change is a global event as proved by enhancements in the worldwide mean temperature of air and oceans, resulting in the melting down of ice glaciers and the rising of mean sea levels (IPCC, 2007; Fahad and Bano, 2012; Fahad et al., 2013, 2014a,b, 2015a,b, 2016a,b,c,d, 2017, 2018, 2019a,b; Alharby and Fahad, 2020). The regular universal air temperature of the planet atmosphere has been enhanced 0.74 ± 0.18°C for the period of the past century ending in (IPCC, 2007). As a fact of nature, the energy of incoming solar radiation is stable by outgoing long wave radiation in the atmosphere (Sable and Rana, 2016).

Apart from the aforementioned physical changes, the climate change also alters the balanced ecosystems and life attributes of associated species (Parmesan, 2006; Root et al., 2003; Walther et al., 2002).

It is expected that due to the extreme past and current human commerce, the fauna and flora will face a new climate in the near future in the form of elevated greenhouse gases concentrations, altered ultraviolet radiations, temperature, changes in rainfall patterns, and so forth (Cornelissen, 2011). Insects, the most abundant creatures representing almost half of the total biodiversity reported to date (Speight et al., 1999), are an important driving force of ecosystem structure and function (Crawley, 1983). Insects in nature are dependent on various ecosystem members, namely, plants, mammals, debris, soils, and water bodies. An alteration in climate can affect the insects through the changes experienced by their associated living beings, namely, host plants and mammals. More precisely, the climatic change is believed to affect the insect-plant interactions in the following ways:

1. **Direct effect:** alterations in interaction pattern, behaviour, and life attributes (Ju et al., 2017; Cornelissen, 2011).
2. **Indirect effect:** alterations in morphology, biochemistry, physiology, species richness, diversity and abundance of host plants, associated competitors, or natural enemies (Cornelissen, 2011; Gifford et al., 1996; Gutierrez and Ponti, 2014; Jamieson et al., 2012; Kazakis et al., 2007; Thomson et al., 2010; Thuiller et al., 2005; Yadugiri, 2010; Yuan et al., 2009).

The impact of climate has also been documented in the case of aquatic insects affecting their diversity, abundance, and feeding behaviour due to alterations in aquatic resources such as temporary rivers, lakes, and ponds (Brown et al., 2007, Hering et al., 2009; Woodward et al., 2010). Universally, the insects interact with the ecosystem in the form of predators and parasitoids, pollinators, herbivores, and so forth, and changes in their abundance and diversity have the potential to modify the services they provide (Hillstrom and Lindroth, 2008). Therefore, any transformation in insect-associated entities will equally spawn the changes in insect behaviour and diversity which thereby force the insects to look for survival option. One of the survival options may be the migration to other places of suitable climatic conditions and resources. Such shifts and changes in insect niches and habitats may occur in two probable ways:

1. Self/adaptive shift ('Acta-shift' in Latin)
2. Forced/dependant shift ('Coactus-shift' in Latin)

Self or adaptive shifting: When insects are not affected by host or food and suffer from climate change directly, they are required to shift in order to protect themselves. This shifting is irrespective of food or host availability.

Forced or dependent shifting: Insects are dependant creatures for their survival, e.g., plants, mammals, soil, and water. If the host is affected by the climate change (survives or dies), the dependant insects will be forced to migrate or shift in order to survive.

In general, individual factors can influence one or more associated factors for insect survival, namely, degree days requirement, host plants of insects, and parasitism. However the phenomenon of climate change takes place in a cumulative and complex manner affecting numerous dependent factors (abiotic factors, host plants, and parasitism) that ultimately force insects to shift either adaptively or forced one. This shifting fabricates multiple future events related to insects, e.g., (1) invasion of non-native insects in to new area (Hill et al., 2017; Ju et al., 2017; Wan and Yang, 2016). According to Ju et al., (2015, 2017), climate warming may enhance the occurrence of alien invasive insects. (2) Changes in defined pest complexity of plants and animals after invasion are considered (Logan et al., 2003; Regniere et al., 2009). (3) Development of new insect biotypes shows adaptive changes in feeding and other behaviour patterns in a similar species (Kambrekar et al., 2015; Sharma, 2014). (4) Sudden pest outbreaks of a particular insect species invade to places where that species is already thriving well (Volney and Fleming, 2000). (5) Shifting may result in changes in status of a particular insect as pests may or may not force both host and parasite to migrate at a time in absence of natural enemies because of the changes in climatic factors (Figure 6.1).

Due to this shift, insects are subjected to sharp extinction. Moreover, notwithstanding the change taking place in the climate, failure to migrate may result in species extinction as well. On an average, almost 45 to 275 species are flattering died out each day globally (Kharouba et al., 2015; Sangle et al., 2015; van-Asch and Visser, 2007).

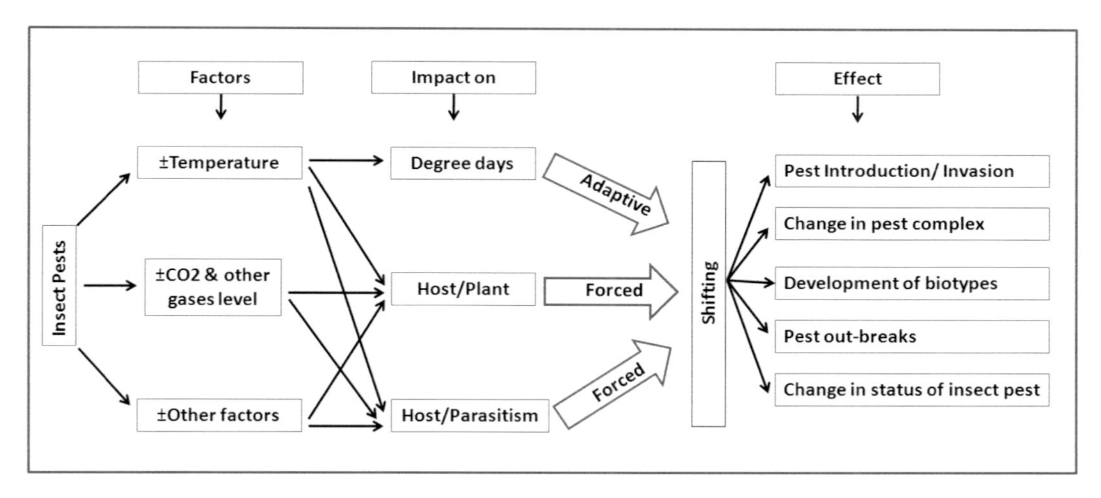

FIGURE 6.1 Systematic process of climate change affecting insects.

6.2 Effect of Increased Level of CO_2 on Biodiversity of Insect Pests

Atmospheric CO_2 concentration (CO_2) is nearly doubling during this century and is at its highest in the past 26 million years. Because of such variations in atmospheric CO_2 points, temperatures, and rainfall, great effects will occur in those natural enemies of insects. These changes affect the relationship among plants, insect herbivores, and natural enemies in such a way that it is not easy to comprehend and anticipate. To study the direct effects of climate change on species of plants at the first trophic level, some studies were done (Long et al., 2004).

From the literature, to evoke climate change effects, two advances are noticeable. In the first advance, to anticipate the reaction of certain groups of natural enemies to climate change, mathematical models were applied (Guitierrez et al., 2008; Stireman et al., 2005). In the second advance, empirical proofs were given in all three trophic levels: the plant-herbivore-natural enemy, which came from experiments showing the reactions to either elevated CO_2 or temperature, that is, individual climate change parameters (Bezemer et al., 1998). The outcome of climate alteration on parasitoids and all types of natural enemies of agricultural pests has been summed up in a new review. The reaction to increased CO_2, temperature, and changes in precipitation is highlighted in this review.

6.3 Climate Change Impacts on the Agriculture Sector

In former IPCC reports, important findings on the evolution of the climate as well as its important physical effects were there, e.g., effects for land and ocean temperature change, sea-level rise, and ocean acidification, and the latest IPCC report affirms it. Improved reasoning of possible spatial modification in intensity, precipitation, and seasonal distribution has also been brought. Furthermore, at a much more localized scope, better projections on a medium-term perspective are being made because of advancements in modelling and data collection. These advancements are of the utmost significance to improve interpretation and possible properties in agricultural schemes (Figure 6.1).

6.4 Insects and Environment

Insects are considered to be economically and environmentally important organisms. Insects are distinguished from other living organisms in terms of their environmental adherence as they are cold-blooded and the temperature of these insects is equal to the temperature of the external environment. Therefore, many scientists consider that the temperature is one of the most important environmental factors affecting the behaviour of many types of insects. Also, many scientists worldwide agree that heat is the dominant factor affecting insect distribution, development, survival, and reproduction (Yamamura and Kiritani, 1998; Bale et al., 2002). It is possible that climate change can increase the temperature at the level of the universe, and that the increase in temperature affects agricultural pests in very complicated ways. Many scientists point out that warm temperatures in temperate regions may be considered one of the most important factors that lead to more types and numbers of insects.

6.5 General Impact on Insects

It is known that insects have a thermal effect that depends on the temperature of the surrounding environment, and that there is no specific mechanism for regulating the temperature of insects in order to adapt them to the external environment. The effect of temperature affects directly or indirectly the insect physiology (Thomson et al., 2010).

Many studies indicate that due to the increase in temperatures, migration increased among insects toward the north (Das et al., 2011; Hamilton et al., 2005) (Figure 6.2).

The effects of CO_2 on insects are believed to be mainly indirect: the focus on insect damage comes from changes in host culture. Few researchers have observed that a cause of the existence of the CO_2 level, negative effects on pest problems, can be observed. Ultimately, the technology of enrichment of the concentration of gas in the open air (FACE) was used to create the opportunity with concentrations of CO_2 and O_2 similar to that foreseen by the change models for the global environment in the 21st century. FACE provides testing of the cultivation conditions in the field with less attention than those present in confined spaces. At the beginning of the season, soybeans grown in an accumulated CO_2 atmosphere suffered 57% more damage from insects (mainly the Japanese cockroach, potato grasshopper, corn root, and Mexican bean beetle) than those grown in the current atmosphere. To continue the experiment, treatment with insecticides was also necessary. Whiteflies and aphids can cause greater harm in vegetables, fruit crops, cereals, and grain legumes by moving to temperate parts (Sharma, 2010).

Some changes that are evident are changes in the state of pests, the growing threat of invasive alien species, and large variations in the enemy's natural activity. Before normal Helicoverpaarmigera (Hub.), infestations are common in northern India due to temperatures (Sharma, 2010), with a consequent increase in peas. In recent years, threats to Indian agriculture include invasion of exotic insect species, which suck on pests, namely papaya mealybug, woolly afid of sugar cane, coconut mite, and biocco, on many plantations and spiral orchards. Examples of native species are the accumulated outbreaks of parasites that suck like jassidae and white flies in cotton and thrips as virus carriers in peanuts, sunflowers, and tomatoes. We need to understand that somehow these changes in the distribution of pests and epidemics are related to climate change and climate variability. In New York, in the last ten years and longer, a network of pheromone traps in sweet corn fields has been used to monitor the ear of corn (Helicoverpazea) throughout the central and western part of the state (personal communication of sailors). The cob worm, a migratory parasite of sweet corn, does not overwinter normally in the north of New York State and which is mainly a late season insect, and capture begins in mid-July (Figure 6.3).

6.6 Effect of Precipitation on Biodiversity of Insect Pests

Many researchers in the fields of biology agree that climate change is a difference in the average state of the climate or the variability of its components, which remains for a long time. Many scientists also point out that there is a close relationship between biological diversity and climate change, and that they affect each other. Many studies also indicate that sustainable resources for biological diversity can reduce the impacts of climate change on people and ecosystems. (Coviella and Trumble, 1999).

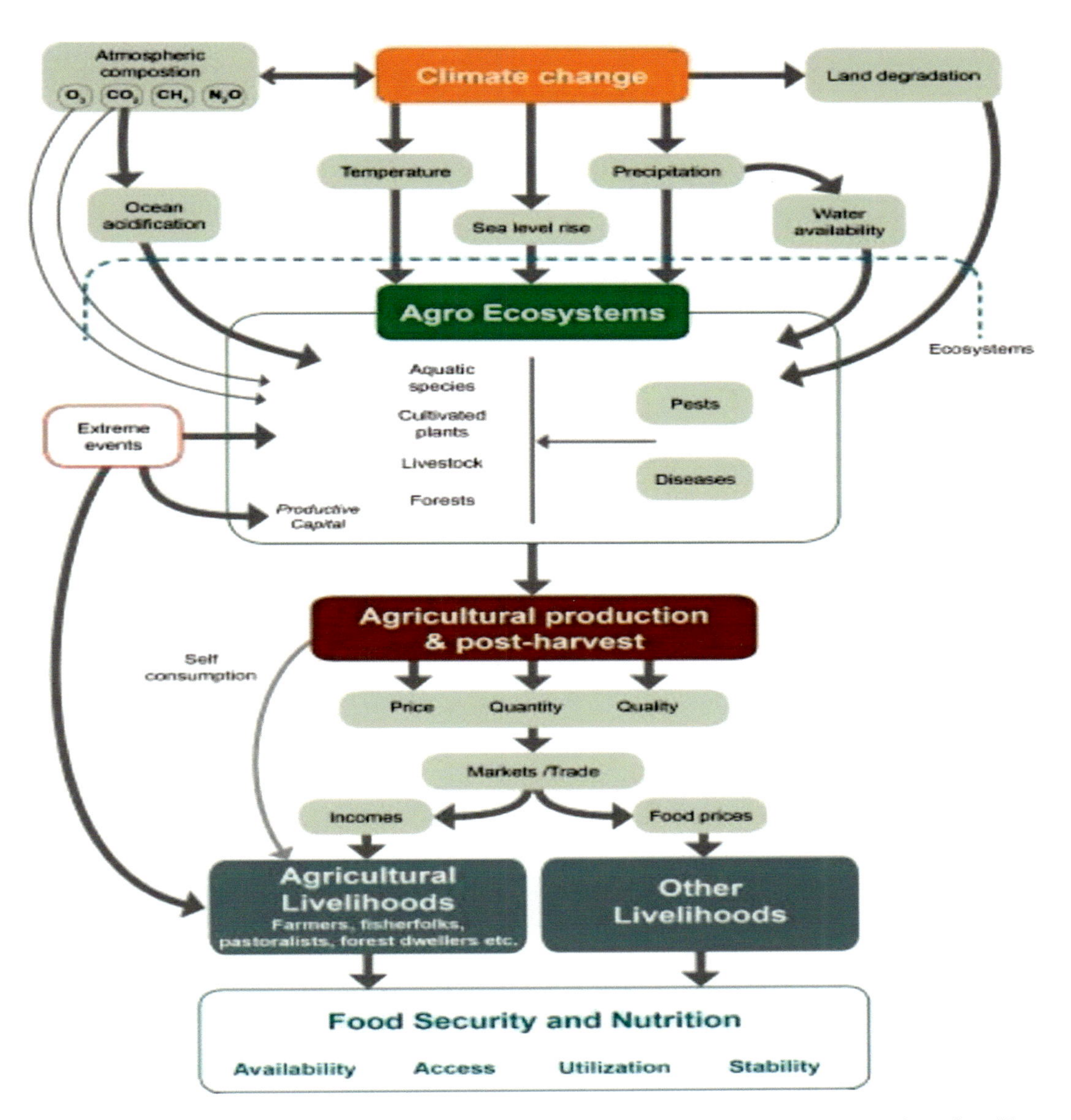

FIGURE 6.2 Schematic representation of the cascading effects of climate change impacts on food security and nutrition.

Source: Surender (2017).

Biodiversity includes many aspects of changes in nature, ranging from genes to species and ecosystems to biomes (Gaston and Spicer, 1998). Most research has traditionally been done on species and populations (Angermeier and Schlosser, 1995). Termites are found in a wide range of terrestrial environment and distributed throughout the tropical, subtropical, and temperate regions of the world. They are polyphagous pests and one of the most destructive creatures of agricultural, horticultural, and plantation crops, including forest trees and buildings. They feed on cellulose rich roots of crop plants and such plants can easily be uprooted from the soil. The damage by termites is greater in rainfed than irrigated crops, that is, 20–25% infestation in rainfed crops and up to 10% in irrigated crops (Figure 6.4).

FIGURE 6.3 (Left) corn earworm adult (from Seaman); (right) corn earworm larva.

Source: Curtis Petzoldt and Abby Seaman (2012).

FIGURE 6.4 Anthills in a spruce forest.

Source: Bjorn Kristersson, Shutterstock.

6.7 Conclusion

The concentration of atmospheric CO_2 is high in this century and is the highest its ever been in the last 26 million years. On the other hand, studies show that changes in atmospheres, such as changes in carbon dioxide levels, temperatures, and temperature variations, will affect the natural enemies of insects, damage to vegetables, fruit crops, grains, and legumes. Insects are considered to be economically and environmentally important organisms. Insects are distinguished from other living organisms by their adhesion to the environment because they are considered cold-blooded organisms and the temperature of these insects is equal to the temperature of the external environment. There are many insects that do not pose any threat to agricultural and fruit production, and they can be a benefit to the ecosystem and farmers, such as insects used for biological control and pollinators.

REFERENCES

Adnan M, Zahir S, Fahad S, Arif M, Mukhtar A, Imtiaz AK, Ishaq AM, Abdul B, Hidayat U, Muhammad A, Inayat-Ur R, Saud S, Muhammad ZI, Yousaf J, Amanullah, Hafiz MH, Wajid N (2018) Phosphate-solubilizing bacteria nullify the antagonistic effect of soil calcification on bioavailability of phosphorus in alkaline soils. *Sci Rep* 8:4339. https://doi.org/10.1038/s41598-018-22653-7

Adnan M, Fahad S, Zamin M, Shah S, Mian IA, Danish S, Zafar-ul-Hye M, Battaglia ML, Naz RMM, Saeed B, Saud S, Ahmad I, Yue Z, Brtnicky M, Holatko J, Datta R (2020) Coupling phosphate-solubilizing bacteria with phosphorus supplements improve maize phosphorus acquisition and growth under lime induced salinity stress. *Plants* 9(900). doi: 10.3390/plants9070900

Ahmad S, Kamran M, Ding R, Meng X, Wang H, Ahmad I, Fahad S, Han Q (2019) Exogenous melatonin confers drought stress by promoting plant growth, photosynthetic capacity and antioxidant defense system of maize seedlings. *PeerJ* 7:e7793 http://doi.org/10.7717/peerj.7793

Akram R, Turan V, Hammad HM, Ahmad S, Hussain S, Hasnain A, Maqbool MM, Rehmani MIA, Rasool A, Masood N, Mahmood F, Mubeen M, Sultana SR, Fahad S, Amanet K, Saleem M, Abbas Y, Akhtar HM, Waseem F, Murtaza R, Amin A, Zahoor SA, ul Din MS, Nasim W (2018a) Fate of organic and inorganic pollutants in paddy soils. In Hashmi MZ, Varma A (eds.) *Environmental pollution of paddy soils*. Springer, Cham, Switzerland, pp. 197–214.

Akram R, Turan V, Wahid A, Ijaz M, Shahid MA, Kaleem S, Hafeez A, Maqbool MM, Chaudhary HJ, Munis, MFH, Mubeen M, Sadiq N, Murtaza R, Kazmi DH, Ali S, Khan N, Sultana SR, Fahad S, Amin A, Nasim W (2018b) Paddy land pollutants and their role in climate change. In Hashmi MZ, Varma A (eds.) *Environmental pollution of paddy soils*. Springer, Cham, Switzerland, pp. 113–124.

Alharby AF, Fahad S (2020) Melatonin application enhances biochar efficiency for drought tolerance in maize varieties: Modifications in physio-biochemical machinery. *Agron J:* 1–22.

Angermeier PL, Schlosser IJ (1995) Conserving aquatic biodiversity: Beyond species and populations. *American Fisheries Society Symposium* 17:402–414.

Arora R, Dhawan AK (2011) Climate change and insect pest management, recent trends in integrated pest management, 3rd Insect Science Congress, 20th April 2011, PAU, Ludhiana, India, pp. 77–88a.

Aziz K, Daniel KYT, Fazal M, Muhammad ZA, Farooq S, FanW, Fahad S, Ruiyang Z (2017a) Nitrogen nutrition in cotton and control strategies for greenhouse gas emissions: a review. *Environ Sci Pollut Res* 24:23471–23487. https://doi.org/10.1007/s11356-017-0131-y

Aziz K, Daniel KYT, Muhammad ZA, Honghai L, Shahbaz AT, Mir A, Fahad S (2017b) Nitrogen fertility and abiotic stresses management in cotton crop: A review. *Environ Sci Pollut Res* 24:14551–14566. https://doi.org/10.1007/s11356-017-8920-x

Bale JS, Masters GJ, Hodkinson ID et al. (2002) Herbivory in global climate change research: direct effects of rising temperature on insect herbivores. *Global Change Biol* 8:1–16.

Bezemer TM, Jones TM (1998) Plant-insect herbivore interactions in elevated atmospheric CO_2: quantitative analysis and guild effects. *Oikos* 82:212–222.

Bezemer TM, Jones TH, Knight KJ (1998) Long-term effects of elevated CO_2 and temperature on populations of the peach potato aphid Myzuspersicae and its parasitoid Aphidiusmatricariae. *Oecologia* 116:128–135.

Brown LE, Hannah DM, Milner AM (2007) Vulnerability of alpine stream biodiversity to shrinking glaciers and snowpacks. *Glob Change Biol* 13:958–966.

Cicero K (1993) Making a home for beneficial insects. The New Farm, February, 28–33. ISSN: 0163-0369

Colley MR, Luna JM (2000) Relative attractiveness of potential beneficial insectary plants to aphidophagous hoverflies (Diptera: Syrphidae). *Environ Entomol* 29(5):1054–1059.

Connor DJ (2008) Organic agriculture cannot feed the world. *Field Crops Res* 106:187–190.

Cornelissen T (2011) Climate change and its effect on terrestrial insects and herbivory patterns. *Neotropical Entomol* 40(2):155–163.

Coviella C, Trumble J (1999) Effects of elevated atmospheric carbon dioxide on insect plant interactions. *Conserv Biol* 13:700–712.

Crawley MJ (1983) *Herbivory: The dynamics of animal-plant interactions*. Oxford:Blackwell Scientific Publications, p. 437.

Curtis P, Abby S (2012) *Climate change effects on insects and pathogens, climate change and agriculture: Promoting practical and profitable responses*. New York State Agricultural Extension Station, Geneva.

Das DK, Singh J, Vennila S (2011) Emerging crop pest scenario under the impact of climate change. *J Agric Phys* 11:13–20.

Fahad S, Bano A (2012) Effect of salicylic acid on physiological and biochemical characterization of maize grown in saline area. *Pak J Bot* 44:1433–1438.

Fahad S, Chen Y, Saud S,Wang K, Xiong D, Chen C, Wu C, Shah F, Nie L, Huang J (2013) Ultraviolet radiation effect on photosynthetic pigments, biochemical attributes, antioxidant enzyme activity and hormonal contents of wheat. *J Food, Agric Environ* 11(3&4):1635–1641.

Fahad S, Hussain S, Bano A, Saud S, Hassan S, Shan D, Khan FA, Khan F, Chen Y, Wu C, Tabassum MA, Chun MX, Afzal M, Jan A, Jan MT, Huang J (2014a) Potential role of phytohormones and plant growth-promoting rhizobacteria in abiotic stresses: Consequences for changing environment. *Environ Sci Pollut Res* 22(7):4907–4921. https://doi.org/10.1007/s11356-014-3754-2

Fahad S, Hussain S, Matloob A, Khan FA, Khaliq A, Saud S, Hassan S, Shan D, Khan F, Ullah N, Faiq M, Khan MR, Tareen AK, Khan A, Ullah A, Ullah N, Huang J (2014b) Phytohormones and plant responses to salinity stress: A review. *Plant Growth Regul* 75(2):391–404. https://doi.org/10.1007/s10725-014-0013-y

Fahad S, Hussain S, Saud S, Tanveer M, Bajwa AA, Hassan S, Shah AN, Ullah A,Wu C, Khan FA, Shah F, Ullah S, Chen Y, Huang J (2015a) A biochar application protects rice pollen from high-temperature stress. *Plant Physiol Biochem* 96:281–287.

Fahad S, Nie L, Chen Y, Wu C, Xiong D, Saud S, Hongyan L, Cui K, Huang J (2015b) Crop plant hormones and environmental stress. *Sustain Agric Rev* 15:371–400.

Fahad S, Hussain S, Saud S, Hassan S, Chauhan BS, Khan F et al. (2016a) Responses of rapid viscoanalyzer profile and other rice grain qualities to exogenously applied plant growth regulators under high day and high night temperatures. *PLoS One* 11(7):e0159590. https://doi.org/10.1371/journal.pone.0159590

Fahad S, Hussain S, Saud S, Khan F, Hassan S, Jr A, Nasim W, Arif M, Wang F, Huang J (2016b) Exogenously applied plant growth regulators affect heat-stressed rice pollens. *J Agron Crop Sci* 202:139–150.

Fahad S, Hussain S, Saud S, Hassan S, Ihsan Z, Shah AN, Wu C, Yousaf M, Nasim W, Alharby H, Alghabari F, Huang J (2016c) Exogenously applied plant growth regulators enhance the morphophysiological growth and yield of rice under high temperature. *Front Plant Sci* 7:1250. https://doi.org/10.3389/fpls.2016.01250

Fahad S, Hussain S, Saud S, Hassan S, Tanveer M, Ihsan MZ, Shah AN, Ullah A, Nasrullah KF, Ullah S, AlharbyH NW, Wu C, Huang J (2016d) A combined application of biochar and phosphorus alleviates heat-induced adversities on physiological, agronomical and quality attributes of rice. *Plant Physiol Biochem* 103:191–198.

Fahad S, Bajwa AA, Nazir U, Anjum SA, Farooq A, Zohaib A, Sadia S, NasimW, Adkins S, Saud S, Ihsan MZ, Alharby H, Wu C, Wang D, Huang J (2017) Crop production under drought and heat stress: Plant responses and management options. *Front Plant Sci* 8:1147. https://doi.org/10.3389/fpls.2017.01147

Fahad S, Ishan MZ, Khaliq AK, Daur I, Saud S, Alzamanan S, Nasim W, Abdullah MA, Khan IA, Wu C, Wang D, Huang J (2018) Consequences of high temperature under changing climate optima for rice pollen characteristics-concepts and perspectives. *Archives Agron Soil Sci.* DOI: 10.1080/03650340.2018.1443213

Fahad S, Rehman A, Shahzad B, Tanveer M, Saud S, Kamran M, Ihtisham M, Khan SU, Turan V, Rahman MHU (2019a) Rice responses and tolerance to metal/metalloid toxicity. In: Hasanuzzaman M, Fujita M, Nahar K, Biswas JK (eds.) *Advances in rice research for abiotic stress tolerance*. Woodhead, Cambridge. pp. 299–312.

Fahad S, Adnan M, Hassan S, Saud S, Hussain S, Wu C, Wang D, Hakeem KR, Alharby HF, Turan V, Khan MA, Huang J (2019b). Rice responses and tolerance to high temperature. In: Hasanuzzaman M, Fujita M, Nahar K, Biswas JK (eds.) *Advances in rice research for abiotic stress tolerance*. Woodhead, Cambridge, pp. 201–224.

Farhat A, Hafiz MH, Wajid I, Aitazaz AF, Hafiz FB, Zahida Z, Fahad S, Wajid F, Artemi C (2020) A review of soil carbon dynamics resulting from agricultural practices. *J Environ Manag* 268(2020):110319.

Gaston KJ, Spicer JI (1998) *Biodiversity: An introduction*. Oxford: Blackwell Science.

Gifford RM, Barrett DJ, Lutze FL et al. (1996) Agriculture and global change: Scaling direct carbon dioxide impacts and feedbacks through time. In: Walker B, Steffen W (eds.) *Global change and terrestrial ecosystems*. Cambridge University Press, Cambridge, pp. 229–259.
Guitierrez AP, Ponti L, d'Oultremont T, Ellis CK (2008) Climate change effects on poikilothermtritrophic interactions. *Climatic Change* 87:167–192.
Gul F, Ahmed I, Ashfaq M, Jan D, Shah F, Li X, Wang D, Fahad M, Fayyaz M, Shah AS (2020) Use of crop growth model to simulate the impact of climate change on yield of various wheat cultivars under different agro-environmental conditions in Khyber Pakhtunkhwa, Pakistan. *Arabian J Geosci* 13:112 https://doi.org/10.1007/s12517-020-5118-1
Gutierrez AP, Ponti L (2014) Analysis of invasive insects: Links to climate change. In: Ziska LH, Dukes JS (eds.) *Invasive species and global climate change* (Vol. 4). CABI, 45–61.
Habib ur Rahman M, Ahmad A, Wajid A, Hussain M, Rasul F, Ishaque W, Islam MA, Shiela V, Awais M, Ullah A, Wahid A, Sultana SR, Saud S, Khan S, Shah F, Hussain M, Hussain S, Nasim W (2017) Application of CSM-CROPGRO-Cotton model for cultivars and optimum planting dates: Evaluation in changing semi-arid climate. *Field Crops Res.* http://dx.doi.org/10.1016/j.fcr.2017.07.007
Hamilton JG, Dermody O, Aldea M, Zangerl AR, Rogers A, Berenbaum MR, Delucia E (2005) Anthropogenic changes in tropospheric composition increase susceptibility of soybean to insect herbivory. *Environ Entomol* 34(2):479–485.
Hammad HM, Farhad W, Abbas F, Fahad S, Saeed S, Nasim W, Bakhat HF (2016) Maize plant nitrogen uptake dynamics at limited irrigation water and nitrogen. *Environ Sci Pollut Res* 24(3):2549–2557. https://doi.org/10.1007/s11356-016-8031-0
Hammad HM, Abbas F, Saeed S, Shah F, Cerda A, Farhad W, Bernado CC, Wajid N, Mubeen M, Bakhat HF (2018) Offsetting land degradation through nitrogen and water management during maize cultivation under arid conditions. *Land Degrad Dev* 1–10. DOI: 10.1002/ldr.2933
Hammad HM, Ashraf M, Abbas F, Bakhat HF, Qaisrani SA, Mubeen M, Shah F, Awais M (2019) Environmental factors affecting the frequency of road traffic accidents: a case study of sub-urban area of Pakistan. *Environ Sci Pollut Res.* https://doi.org/10.1007/s11356-019-04752-8
Hammad HM, Abbas F, Ahmad A, Bakhat HF, Farhad W, Wilkerson CJ W, Shah F, Hoogenboom G (2020a) Predicting kernel growth of maize under controlled water and nitrogen applications. *Int J Plant Prod.* https://doi.org/10.1007/s42106-020-00110-8
Hammad HM, Khaliq A, Abbas F, Farhad W, Shah F, Aslam M, Shah GM, Nasim W, Mubeen M, Bakhat HF (2020b) Comparative effects of organic and inorganic fertilizers on soil organic carbon and wheat productivity under arid region. *Communications in Soil Science and Plant Analysis.* DOI: 10.1080/00103624.2020.1763385
Hering D, Schmidt-Kloiber A, Murphy J et al. (2009) Potential impacts of climate change on aquatic insects: a sensitivity analysis for European caddisflies (Trichoptera) based on distribution patterns and ecological preferences. *Aquat Sci* 71:3–14.
Hill MP, Gallardo B, Terblanche JS (2017) A global assessment of climatic niche shifts and human influence in insect invasions. *Glob Ecol Biogeogr* 26:679–689.
Hillstrom ML, Lindroth, RL (2008) Elevated atmospheric carbon dioxide and ozone alter forest insect abundance and community composition. *Insect Conser Divers* 1:233–241.
Hussain S, Mubeen M, Ahmad A, Akram W, Hammad HM, Ali M, Masood N, Amin A, Farid HU, Sultana SR, Shah F, Wang D, Nasim W (2019) Using GIS tools to detect the land use/land cover changes during forty years in Lodhran district of Pakistan. *Environ Sci Pollut Res* https://doi.org/10.1007/s11356-019-06072-3
Hussain MA, Fahad S, Sharif R, Jan MF, Mujtaba M, Ali Q, Ahmad A, Ahmad H, Amin N, Ajayo BS, Sun C, Gu L, Ahmad I, Jiang Z, Hou J (2020) Multifunctional role of brassinosteroid and its analogues in plants. *Plant Growth Regul* https://doi.org/10.1007/s10725-020-00647-8
Ilyas M, Mohammad N, Nadeem K, Ali H, Aamir HK, Kashif H, Fahad S, Aziz K, Abid U, (2020) Drought tolerance strategies in plants: A mechanistic approach. *J Plant Growth Regul.* https://doi.org/10.1007/s00344-020-10174-5
IPCC (2007) Climate Change 2007 – The Physical Science Basis Contribution of Working Group I to the Fourth Assessment Report of the IPCC.
Jamieson MA, Trowbridge AM, Raffa KF et al. (2012) Consequences of climate warming and altered precipitation patterns for plant-insect and multitrophic interactions. *Plant Physiol* 160:1719–1727.

Jan M, Anwar-ul-Haq M, Shah AN, Yousaf M, Iqbal J, Li X, Wang D, Shah F (2019) Modulation in growth, gas exchange, and antioxidant activities of salt-stressed rice (Oryza sativa L.) genotypes by zinc fertilization. *Arabian J Geosci* 12:775 https://doi.org/10.1007/s12517-019-4939-2

Jones GA, Gillett JL (2005) Intercropping with sunflowers to attract beneficial insects in organic agriculture. *Florida Entomol* 88(1):91–96.

Ju RT, Zhu HY, Gao L et al. (2015) Increases in both temperature means and extremes likely facilitate invasive herbivore outbreaks. *Sci Rep* 5(15715):1–10.

Ju RT, Gao L, Wei SJ et al. (2017) Spring warming increases the abundance of an invasive specialist insect: links to phenology and life history. *Sci Rep* 7(14805):1–12.

Kamarn M, Wenwen C, Irshad A, Xiangping M, Xudong Z, Wennan S, Junzhi C, Shakeel A, Fahad S, Qingfang H, Tiening L (2017) Effect of paclobutrazol, a potential growth regulator on stalk mechanical strength, lignin accumulation and its relation with lodging resistance of maize. *Plant Growth Regul* 84:317–332. https://doi.org/10.1007/ s10725-017-0342-8

Kambrekar DN, Guledgudda SS, Katti A et al. (2015) Impact of climate change on insect pests and their natural enemies. *Karnataka J Agricul Sci* 28(5):814–816.

Kazakis G, Ghosn D, Vogiatzakis N et al. (2007) Vascular plant diversity and climate change in the alpine zone of the Lefka Ori, Crete. *Biodivers Conserv* 16:1603–1615.

Kharouba HM, Vellend M, Sarfraz RM et al. (2015) The effects of experimental warming on the timing of a plant-insect herbivore interaction. *J Anim Ecol* 84:785–796.

Lewis T (1997) Major crops infested by thrips with main symptoms and predominant injurious species (Appendix II). In: Lewis T (ed.) *Thrips as crop pests*. CAB International, New York, pp. 675–709.

Lindroth RL (1996) Consequences of elevated atmospheric CO_2 for forest insects. In: Koch GW, Mooney HA (Eds.), *Carbon dioxide and terrestrial ecosystems*. Academic Press, San Diego, pp. 105–120.

Logan JA, Regniere J, Powell JA (2003) Assessing the impacts of global warming on forest pest dynamics. *Front Ecol Environ* 1:130–137.

Long SP, Ainsworth EA, Rogers A et al. (2004) Rising atmospheric carbon dioxide: plants FACE the future. *Annu Rev Plant Biol* 55:591–628.

McCarthy JJ, Canziani OF, Leary NA et al. (2001) Climate change 2001: Impacts, adaptation, and vulnerability. United Nations Inter governmental Panel on Climate Change. www.grida.no/climate/ipcctar/wg2/001htm.

Menéndez R (2007) How are insects responding to global warming. *Tijdschrift voor Entomologie* 150(2):355.

Mubeen M, Ahmad A, Hammad HM, Awais M, Farid H, Saleem M, Sami ul Din M, Amin A, Ali A, Shah F, Nasim W (2020) Evaluating the climate change impact on water use efficiency of cotton-wheat in semi-arid conditions using DSSAT model. *J Water Climate Change*. doi/10.2166/wcc.2019.179/622035/jwc2019179.pdf

Muhammad B, Adnan M, Munsif F, Fahad S, Saeed M, Wahid F, Arif M, Jr. Amanullah, Wang D, Saud S, Noor M, Zamin M, Subhan F, Saeed B, Raza MA, Mian IA (2019a) Substituting urea by organic wastes for improving maize yield in alkaline soil. *J Plant Nutr*. https://doi.org/10.1080/01904167.2019.1659344

Nasim W, Ahmad A, Amin A, Tariq M, Awais M, Saqib M, Jabran K, Shah GM, Sultana SR, Hammad HM, Rehmani MIA, Hashmi MZ, Habib Ur Rahman M, Turan V, Fahad S, Suad S, Khan A, Ali S (2017) Radiation efficiency and nitrogen fertilizer impacts on sunflower crop in contrasting environments of Punjab. *Pakistan Environ Sci Pollut Res* 25:1822–1836. https://doi.org/10.1007/s11356-017-0592-z

Gavkare O, Anil, Devi N, Thakur SK, Devi KK (2014) Termites: An economic polyphagous pest. Popular Kheti 2(2):121–123.

Parmesan C (2006) Ecological and evolutionary responses to recent climate change. *Annu Rev Ecol Evol System* 37:637–669.

Regniere J, Nealis V, Porter K (2009). Climate suitability and management of the gypsy moth invasion into Canada. *Biol Invasions* 11:135–148.

Rehman M, Fahad S, Saleem MH, Hafeez M, Muhammad Habib ur Rahman, Liu F, Deng G (2020) Red light optimized physiological traits and enhanced the growth of ramie (Boehmeria nivea L.). *Photosynthetica* 58(4):922–931.

Reiners S, Petzoldt C (2005) Integrated crop and pest management guidelines for commercial vegetable production. Cornell Cooperative Extension publication #124VG http://www.nysaes.cornell.edu/recommends/.

Root TL, Price JT, Hall KR et al. (2003) Fingerprints of global warming on wild animals and plants. *Nature* 421:57–60.

Sable MG, Rana DK (2016) Impact of global warming on insect behavior – A review. *Agric Rev* 37(1):81–84.

Saleem MH, Fahad S, Adnan M, Ali M, Rana MS, Kamran M, Ali Q, Hashem IA, Bhantana P, Ali M, Hussain RM (2020a) Foliar application of gibberellic acid endorsed phytoextraction of copper and alleviates oxidative stress in jute (Corchorus capsularis L.) plant grown in highly copper-contaminated soil of China. *Environ Sci Pollution Res* https://doi.org/10.1007/s11356-020-09764-3

Saleem MH, Rehman M, Fahad S, Tung SA, Iqbal N, Hassan A, Ayub A, Wahid MA, Shaukat S, Liu L, Deng G (2020b) Leaf gas exchange, oxidative stress, and physiological attributes of rapeseed (Brassica napus L.) grown under different light-emitting diodes. *Photosynthetica* 58(3):836–845.

Saleem MH, Fahad S, Shahid UK, Mairaj D, Abid U, Ayman ELS, Akbar H, Analía L, Lijun L (2020c) Copper-induced oxidative stress, initiation of antioxidants and phytoremediation potential of flax (Linum usitatissimum L.) seedlings grown under the mixing of two different soils of China. *Environ Sci Poll Res* https://doi.org/10.1007/s11356-019-07264-7

Sangle PM, Satpute SB, Khan FS et al. (2015) Impact of climate change on insects. *Trends in Biosci* 8(14):3579–3582.

Saud S, Chen Y, Long B, Fahad S, Sadiq A (2013) The different impact on the growth of cool season turf grass under the various conditions on salinity and drought stress. *Int J Agric Sci Res* 3:77–84.

Saud S, Li X, Chen Y, Zhang L, Fahad S, Hussain S, Sadiq A, Chen Y (2014) Silicon application increases drought tolerance of Kentucky bluegrass by improving plant water relations and morph physiological functions. *SciWorld J* 2014:1–10. https://doi.org/10.1155/2014/ 368694

Saud S, Chen Y, Fahad S, Hussain S, Na L, Xin L, Alhussien SA (2016) Silicate application increases the photosynthesis and its associated metabolic activities in Kentucky bluegrass under drought stress and post-drought recovery. *Environ Sci Pollut Res* 23(17):17647–17655. https://doi.org/10.1007/s11356-016-6957-x

Saud S, Fahad S, Yajun C, Ihsan MZ, Hammad HM, Nasim W, Amanullah Jr, Arif M and Alharby H (2017) Effects of nitrogen supply on water stress and recovery mechanisms in Kentucky bluegrass plants. *Front Plant Sci* 8:983. doi: 10.3389/fpls.2017.00983

Saud S, Fahad S, Cui G, Chen Y, Anwar S (2020) Determining nitrogen isotopes discrimination under drought stress on enzymatic activities, nitrogen isotope abundance and water contents of Kentucky bluegrass. *Sci Rep* 10:6415. https://doi.org/10.1038/s41598-020-63548-w

Shafi MI, Adnan M, Fahad S, Fazli W, Ahsan K, Zhen Y, Subhan D, Zafar-ul-Hye M, Brtnicky M, Datta R (2020) Application of single superphosphate with humic acid improves the growth, yield and phosphorus uptake of wheat (Triticum aestivum L.) in Calcareous Soil. *Agronomy* (10):1224. doi:10.3390/agronomy10091224

Shah F, Lixiao N, Kehui C, Tariq S, Wei W, Chang C, Liyang Z, Farhan A, Fahad S, Huang J (2013) Rice grain yield and component responses to near 2°C of warming. *Field Crop Res* 157:98–110.

Sharma HC (2010) Effect of climate change on IPM in grain legumes. 5th International Food Legumes Research Conference (IFLRC V), and the 7th European Conference on Grain Legumes (AEP VII), 26–30th April 2010, Anatalaya, Turkey.

Sharma HC (2010) Global warming and climate change: Impact on arthropod biodiversity. Pest Management, and Food Security. In: Thakur RK, Gupta, PR, Verma AK (eds.) *Souvenir, National Symposium on Perspectives and Challenges of Integrated Pest Management for Sustainable Agriculture, 19–21 Nov 2010.* Solan, Himachal Pradesh, India, pp. 1–14.

Sharma HC (2014) Climate change effects on insects: Implications for crop protection and food security. *J Crop Improv* 28(2):229–259.

Speight MR, Hunter MD, Watt AD (1999) *Ecology of insects: Concepts and applications*. Blackwell Science Oxford, p. 340.

Stiling P, Cornelissen T (2007) How does elevated carbon dioxide (CO_2) affect plant-herbivore interactions? A field experiment and a meta-analysis of CO_2-mediated changes on plant chemistry and herbivore performance. *Glob Change Biol* 13:1823–1842.

Stireman JO, Dyer LA, Janzen DH et al. (2005). Climatic unpredictability and parasitism of caterpillars: implications of global warming. *Proc Natl Acad Sci* 102:17384–17387.

Subhan D, Zafar-ul-Hye M, Fahad S, Saud S, Martin B, Tereza H, Rahul D (2020) Drought stress alleviation by ACC deaminase producing Achromobacter xylosoxidans and Enterobacter cloacae, with and without timber waste biochar in maize. *Sustainability* 12(6286). doi:10.3390/su12156286

Surender S (2017) Diverse effects of climate change on agricultural pests. DOI: 10.13140/RG.2.2.15264.48646.

Thomson LJ, Macfadyen S, Hoffmann AA (2010) Predicting the effects of climate change on natural enemies of agricultural pests. *Biol Control* 52:296–306.

Thuiller W, Lavorel S, Araujo MB et al. (2005) Climate change threats to plant diversity in Europe. *PNAS* 102:8245–8250.

Tylianakis JM, Didham RK, Bascompte J et al. (2008) Global change and species interactions in terrestrial ecosystems. *Ecol Lett* 11:1351–1363.

van-Asch M, Visser ME (2007) Phenology of forest caterpillars and their host trees: The importance of synchrony. *Annu Rev Entomol* 52:37–55.

Vincent C, Hallman G, Panneton B et al. (2003). Management of agricultural insects with physical control methods. *Annu Rev Entomol* 48:261–281.

Volney WJA, Fleming RA (2000) Climate change and impacts of boreal forest insects. *Agricu Ecosyst Environ* 82:283–294.

Wahid F, Fahad S, Subhan D, Adnan M, Zhen Y, Saud S, Manzer HS, Brtnicky M, Hammerschmiedt T, Datta R (2020) Sustainable management with mycorrhizae and phosphate solubilizing bacteria for enhanced phosphorus uptake in calcareous soils. *Agriculture* 10(334). doi:10.3390/agriculture10080334

Walther GR, Post E, Convey P (2002) Ecological responses to recent climate change. *Nature* 416(6879):389–395.

Wan FH, Yang NW (2016) Invasion and management of agricultural alien insects in China. *Annu Rev Entomol* 61:77–98.

Wang D, Shah F, Shah S, Kamran M, Khan A, Khan MN, Hammad HM, Nasim W (2018) Morphological acclimation to agronomic manipulation in leaf dispersion and orientation to promote 'Ideotype' breeding: Evidence from 3D visual modeling of 'super' rice (Oryza sativa L.). *Plant Physiol Biochem*. https://doi.org/10.1016/j.plaphy.2018.11.010

Watt AD, Whittaker JB, Docherty M et al. (1995) The Impact of Elevated Atmospheric CO_2 on Insect Herbivores, in: Harrington R, Stork NE (Eds.), *Insects in a changing environment*. Academic Press, London, pp. 198–217.

Woodward G, Perkins DM, Brown LE (2010) Climate change and freshwater ecosystems: impacts across multiple levels of organization. *Philosophical Transactions of the Royal Society B: Biological Sciences* 365:2093–2106.

Wu C, Tang S, Li G, Wang S, Fahad S, Ding Y (2019) Roles of phytohormone changes in the grain yield of rice plants exposed to heat: A review. *PeerJ* 7:e7792 DOI 10.7717/peerj.7792

Wu C, Kehui C, She T, Ganghua L, Shaohua W, Fahad S, Lixiao N, Jianliang H, Shaobing P, Yanfeng D (2020) Intensified pollination and fertilization ameliorate heat injury in rice (Oryza sativa L.) during the flowering stage. *Field Crops Res* 252:107795.

Wu Z, Dijkstra P, Koch GW et al. (2011). Responses of terrestrial ecosystems to temperature and precipitation change: a meta-analysis of experimental manipulation. *Global Change Biol* 17:927–942.

Yadugiri VT (2010) Climate change: The role of plant physiology. *Curr Sci* 99:423–425.

Yamamura K, Kiritani K (1998) A simple method to estimate the potential increase in the number of generations under global warming in temperate zones. *Appl Entomol Zool* 33:289–298.

Yang Z, Zhang Z, Zhang T, Fahad S, Cui K, Nie L, Peng S, Huang J (2017) The effect of season-long temperature increases on rice cultivars grown in the central and southern regions of China. *Front Plant Sci* 8:1908. https://doi.org/10.3389/fpls.2017.01908

Yuan JS, Himanen SJ, Holopainen JK et al. (2009) Smelling global climate change: mitigation of function for plant volatile organic compounds. *Trends Ecol Evol* 24:323–331.

Zafar-ul-Hye M, Naeem M, Danish S, Fahad S, Datta R, Abbas M, Rahi AA, Brtnicky M, Holatko, Jiri, Tarar ZH, Nasir M (2020a) Alleviation of cadmium adverse effects by improving nutrients uptake in bitter gourd through cadmium tolerant rhizobacteria. *Environments* 7(54). doi:10.3390/environments7080054

Zafar-ul-Hye M, Tahzeeb-ul-Hassan M, Abid M, Fahad S, Brtnicky M, Dokulilova T, Datta R D, Danish S (2020b) Potential role of compost mixed biochar with rhizobacteria in mitigating lead toxicity in spinach. *Sci Rep* 10:12159. https://doi.org/10.1038/s41598-020-69183-9.

Zahida Z, Hafiz FB, Zulfiqar AS, Ghulam MS, Fahad S, Muhammad RA, Hafiz MH, Wajid N, Muhammad S (2017) Effect of water management and silicon on germination, growth, phosphorus and arsenic uptake in rice. *Ecotoxicol Environ Saf* 144:11–18.

Zamin M, Khattak AM, Salim AM, Marcum KB, Shakur M, Shah S, Jan I, Fahad S (2019) Performance of Aeluropus lagopoides (mangrove grass) ecotypes, a potential turfgrass, under high saline conditions. *Environ Sci Pollut Res.* https://doi.org/10.1007/s11356-019-04838-3

Zvereva EL, Kozlov MV (2006) Consequences of simultaneous elevation of carbon dioxide and temperature for plant-herbivore interactions: A meta-analysis. *Glob Change Biol* 12:27–41.

7

Pheromonal and Microbial-Symbiotic-Associated Insect Behaviour

Abid Ali, Ismail Zeb, and Hafsa Zahid
Department of Zoology, Abdul Wali Khan University, Mardan, Khyber Pakhtunkhwa, Pakistan

7.1 Introduction

'What animals do' is the fundamental definition of animal behaviour. More precisely, it is the way in which organisms adjust and interact with its total environment incorporating the association of same or other species and with the physical environment. Such behaviours of insects including movements, orientations, disperses, feeding, mating, protection, exclusion, and social behaviours are mediated by the nervous and endocrine system (White and Ewer, 2014). Other perspectives of communications also include visual interaction, mechanoreceptor, rubbing body structure, and clicking calls through sensory receptors.

In the case of mating behaviour, both male and female have different approaches for copulation (Silva et al., 2012), specifically the pheromones are produced by males involved in male-female encounters, courtship, and matting (Hanks and Millar, 2013). The future mating of a male can be modified according to previous sexual experiences (Iglesias et al., 2019), and it has been suggested that males can learn from previous experiences for adjusting the behaviours of mating (Pérez et al., 2010). The migration of insects is multidimensional and diverse, ranges from a few metres to hundreds of kilometres to exploit temporary breeding sites (Chapman et al., 2010). Insect migration is greatly related with the meteorological conditions such as a change in the climatic perspectives affecting migration habitats of insects in many directions that involve the understanding of insect ecological perspectives along with knowledge of meteorology, aerodynamics, remote sensing, and climatology (Adnan et al., 2016; Su et al., 2013). Insects spatially distribute their offspring to a wide range of environmental settings (Holland et al., 2006). Among these migrations, chemical approaches are the useful traits used to prevent migrants from scattering in a restricted zone (Pérez et al., 2010). Such highly variable and inheritable migratory traits of insects are evolutionarily conserved in insects (Bilton et al., 2001). But the behavioural adaptations that facilitate the movements of insects remain largely unknown (Chapman et al., 2010).

The rhythmicity has been greatly recognized in insects, particularly ubiquitous circadian rhythms are well known (Sandrelli et al., 2008). It has been assumed that organisms originated in the tropical regions have developed circadian clocks that predict the 24 hour environmental cycles characterized by similar patterns of irradiance and temperature throughout the seasons (Helfrich et al., 2018). In social insects such as honey bees and ants, the insect colonies face daily problems such as sizes of the colony, environmental conditions, food sources, and pressure of predation (Krittika and Yadav, 2019). Other insects such as beetles, mites, flies, thrips, and locusts use aggregation behaviour against antagonists

(Yew and Chung, 2015). Pheromones of ticks comprised purines, substituted phenols, or cholesteryl esters used for various purposes such as assemblage, attachment for feeding, and sexual behaviour (Sonenshine, 2006). The aggregation pheromones of ticks were first time discovered in the genus *Argas*, then reported in *Ixodid* (Sonenshine, 2004). Alarm pheromones are less specialized in assessing the general communication clue to other species (Vandermoten et al., 2012). All the trail marking, alarm producing, aggregation, mating, and other pheromones are best known in social insects such as ants, termites, bees, and wasps (Czaczkes et al., 2015). The characteristics of eusocial insect societies are shown by their contributions in brood care and labour division between a reproductive queen and particularly sterile workers. The reproductive queen of the colony releases pheromones such as conserved saturated linear and branched hydrocarbons for maintaining social hierarchy as well as advertising its fecundity and suppressing breeding capacity in workers (Van et al., 2014).

7.2 Pheromone-Mediated Insect Behaviour

In 1959, Karlson Luscher originally proposed the term *pheromone* (derived from Greek words, pherin [to transfer] and hormone []to excite]) as a substance which is secreted by an individual to the environment as a message of specific activity and received by other individuals (Butenandt et al., 1959; Packianather et al., 2009). Bombykol was the first chemically identified pheromone from *Bombyx mori* (Silk moth) (Butenandt et al., 1959), and the second was 9-oxo-2(E)-decenoic acid (9-ODA) 1, secreted by the mandibular gland of honey bee queen's (Gary, 1962). These chemical substances bring communication between members of the same species by controlling different functions like attraction, aggression, and sexual desires (Yew and Chung, 2015). The entomologists noticed the excitement of male moths in the absence of female moth because of some fragrance that is responsible for sexual activities (Gomez and Benton, 2013). These small fragrances of pheromones were extracted and their activities were identified which revealed that the small volume of pheromones can stimulate millions of moths. Pheromones are categorized in to four distinct classes: (1) releaser pheromones that initiate the instant behavioural responses in the receiver; (2) the primer pheromones that elicit the physiological alterations and ultimately result in a behaviour response; (3) the modulator pheromones are chemosensory signals that modify the effect on other species; and (4) the signaler pheromones provide a variety of information in the form of reproductive status, sex of the sender, age, and dominance to the receiver (Semwal et al., 2013).

Various alarming pheromones of different insects have been identified, among these (E)-beta-farnesene is best studied which is released by most of the aphids for signalling member of the same family to stop feeding, predation, and dispersion (Joachim et al., 2013). In termites, the sternal gland secreted trail pheromone for foraging which coordinated the activities of hundreds or thousands of isopterans (Leonardo et al., 2009). Using 3.7 kg of dried of *Atta texana*, methyl 4-methylpyrrole-2-carboxylate was identified as the first ant trail pheromone found in the venom gland (Tumlinson, 1971). Ant's species show varieties in trail pheromone producing glands (Billen et al., 2005); for instance, postpygidial gland in genus *Aenictus* (army ants) (Billen et al., 1999); Dufour gland in *Pogonomyrmex, Monomorium pharaonis*, and *Solenopsis invicta* (Hölldobler et al., 2004), pre-tarsal gland in *Amblyopone reclinate* (Billen et al., 2005); paired tibial glands in case of the genus *Crematogaster* (Inwood, 2008); venom gland as the source of trail pheromone in some *Myrmecine* and *Ponerine* ants (Mashaly et al., 2011); and pavan's gland in *Dolichoderus thoracicus* (Choe et al., 2012). But the several compounds contributing to the trail pheromone are still at infancy (Jarau et al., 2010). A substantial amount of 2,6-dichlorophenol of the *Amblyomma americanum* was the first sex pheromone mediating the recruitment, selection, and gametes fusion between the mating partners (Berger, 1972). In *Argas* ticks, the assemblage pheromone was discovered used for the recruitment of the members of the same species to the host for feeding (Leahy et al., 1973). The assemblage of various ticks species including soft and hard including was reported due to pheromone, *Ixodes holocyclus* and *Aponnomca concolor* (Treverrow et al., 1977), *Ixodes ricinus* (Graff, 1978), *Hyalomma dromedarii* (Leahy et al., 1981), *Rhipicephalus appendiculatus* and *Amblyomma cohaerans* (Otieno et al., 1985), *Rhipicephalus evertsi* (Gothe and Neitz, 1985), and and *Ixodes scapularis* (*I. scapularis*) (Allan and Sonenshine, 2002; Fahad et al., 2019). On the other hand, limited is known about the biosynthetic pathways of pheromone which is used in the aggregation of other ticks to the host body.

The first identified trail pheromone was the Hexyl Decanoate from the labial gland secretions of a stingless bee (*Trigona recursa*), which is the main foraging component (Jarau et al., 2006). The majority of the trail pheromones are secreted from labial gland as in *Geotrigona mombuca*; however, the mandibular glands of foragers' secretes citral chemical, which disproves its supposed role as a trail pheromone (Stangler et al., 2009). Variations in quality and quantity of trail pheromones between species and castes are important for creating specific, nest-marking, nestmate recognition, and egg-marking pheromones (Abdalla and Cruz, 2001). Pheromones play an important role in the proper screening and management of targeted pests which rely on basic knowledge about the production and biosynthesis pathways. For the production of synthetic pheromones, the researcher must know the actual chemical and physical nature of the pheromone (Olszewski et al., 2011) (Table 7.1).

7.3 Microbial-Symbionts and Insect Host Interface

Microbial-symbionts influence the host gene expression directly, by secreting neuroactive compounds, sharing metabolic processes or the modulation of signal perceptions (Engl and Kaltenpoth, 2018). These microbial communities may considerably affect many aspects of their host's ecology, behaviour, physiology, and responses to climate changes (Oliver et al., 2010). However, the relative contribution of the microbial-symbionts with insect host is less explicitly acknowledged which are the emerging area of research.

The endosymbionts act as a metabolic unit for the *Cicadas* host which are extremely dependent on each other and cannot survive without one another (Fisher et al., 2017). Higher mating rates was observed in the *Drossophila simulans* and *Drossophila melanogaster* infected with *Wolbachia* through the exchange in pheromone signalling pathways between microbial-symbionts and host species (Engl and Kaltenpoth, 2018). The innate fear of rats to cat urine was noticed into attraction due to *Toxoplasma gondii* which epigenetically re-programmes the brain of infected rats hence indirectly increases the risk of cats predation and parasite transmission from the intermediate host to its definitive host (Engl and Kaltenpoth, 2018). The microbial-symbionts in the gut of lower termites play a crucial role by digesting lignocellulose and produce high levels of acetate for their host (Engl and Moran, 2013). The *Wigglesworthia* provides the biosynthesis of vitamins and cofactors, while biosynthesis of essential amino acids devoted to *Sulcia muelleri* reveals the significance of both microbial-symbionts to the insect host (McCutcheon and Moran, 2007). Computational screening, phylogenetic and experimental analyses of the entire *Acyrthosiphon pisum* genome provided strong support for the transfer of twelve genes or fragments from bacteria to the aphid genome used for the production of certain amino acids (Nikoh et al., 2010). Furthermore, a transcriptome and proteome level should be planned to understand the phenomenon of how microbial-symbionts interfere with the host metabolic pathways.

In the field of chemical ecology, the characterization of biosynthetic enzymes and their precursors influenced through microbial-symbionts will be applicable to control pest and therapeutic potentials, and assess the evolutionary perspectives of host and associated microbiomes (Gu et al., 2013). The implementation allows manipulation of insect behaviour in a specific manner as compared to general pesticides, which are highly toxic (Witzgal et al., 2008). Thus, the manipulation and the exploitation of the microbial-symbionts could result in important practical applications for the management of insect related problems (Crotti et al., 2012). The products originated from the insect microbial-symbionts have a vast biochemical diversity which is a powerful resource for drug discovery. As well as tee vectorial capacity of the insect vector is also influenced by the symbiotic microorganisms and thus provide targets for potential disease control (Engl and Moran, 2013). Therefore, it is

TABLE 7.1

Represents a summary of various pheromones of insect species

References	Year	Insect Species	Chemical Compounds	Behavoral Response	Extraction/ Identification	Origin	Sex of Reproductives
Tentschert et al.	2000	*Eutetramorium mocquerysi*	2,3-Dimethyl-5-(2-methylpropyl) (pyrazine)	Foraging and nest emigration	No information	Poison gland	No information
Peppuy et al.	2001	*Macrotermes annandalei, M. barneyi, Odontotermes hainanensis, O. maesodensis*	(Z)-dodec-3-en-1-ol (dodecenol)	Foraging	LPME and SPME, GC and GC-MS	Sternal gland	Both male and female workers
Hölldobler et al.	2001	*Pogonomyrmex barbatus, P. maricopa, P. occidentalis P. rugosus*	Pyrazines [2,5-dimethylpyrazine, trimethylpyrazine and 3-ethyl-2,5-dimethylpyrazine (EDMP	Recruitment	GC-MS	Poison gland	Worker females
		Macrotermes annandalei	(Z)–dodec-3-en-1-ol (dodecenol)	Recruitment	LPME and SPME	Sternal gland	Both male and female workers
Kohl et al.	2001	*Camponotus socius*	Formic acid (2 S,4 R,5 S)-(2 S,4 R,5 S)-2,4-dimethyl-5-hexanolide and 2,3-dihydro-3,5-dihydroxy-6-methylpyran-4-one.	Recruitment to food sources and nest emigrations	GC-MS, GC and EI	Poison gland of hind gut	Female worker
Blatrix et al.	2002	*Gnamptogeny striatula*	Methylgeraniol (2E)-3,4,7-trimethyl-2,6-octadien-1-ol	Recruitment	GC-MS	Dufour's gland	Workers
Bordereau et al.	2002	*Cornitermes bequaerti*	Dodecatrienol (3Z, 6Z, 8E)-dodecatrien-1-ol	Induced attraction of male termite	LPME & SPME/ GC-MS	Abdominal tergal glands	Female
Steinmetz et al.	2003	*Vespula vulgaris*	4-methyl-branched or monounsaturated Hydrocarbons	Chemical terrestrial trails functions	GC-MS	Cuticular lipid	Forager females workers
Robert et al.	2003	*Ancistrotermespakistanicus*	DodecadienolC12 alcohol (Z,Z)-dodeca-3,6-dien-1-ol	Recruitment	Immersion in hexane GC-MS and FTIR, SPME	Surface of sternal gland	Female alates, male and workers
Kohl et al.	2003	*Camponotus castaneus, C. balzani, C. sericeiventris*	3,5-dimethyl-6-(1'-ethylpropyl)-tetrahydro-2H-pyran-2-one,3,4-dihydro-8-hydroxy-3,5,7-trimethylisocoumarin.	Recruit nest mates to new target areas.	GC, GC-MS and behavioural analysis	Rectal bladder Poison and Dufour gland	Workers
Morgan et al.	2004	*Crematogaster castanea*	(R)-2-dodecanol	Recruitment	GC-MS	Tibial glands	Workers

(continued)

TABLE 7.1 (Continued)

References	Year	Insect Species	Chemical Compounds	Behavoral Response	Extraction/ Identification	Origin	Sex of Reproductives
Jarau et al.	2004	*Melipona seminigra*	41 new compounds of groups Alkanes, Alkenes, Methyl alkanes, Aldehyde, Esters	Foraging, Marking food sources	GC-MS,	Claw, retractor tendons	Foragers
Sillam-Dussès et al.	2005	*Prorhinotermes canalifrons, P. simplex*	Neocembrene A, (1E,5E,9E,12 R)-1,5,9-trimethy l-12-(1-methylethenyl)-1,5,9-cyclotetradecatriene	Recruitment	SPME-GC/MS	Sternal gland and abdominal tergal integument	Female alates
Granero et al.	2005	*Bombus terrestris*	Eucalyptol	Foraging	GC-MS with Electronic Flow Control (EFC) ion-trap	Tergal glands	Workers
Jarau et al.	2006	*Trigona recursa*	Decyl octanoate/octyl	Foraging and nest recruitment	SPME	Labial gland	Female
Schorkopf et al.	2006	*Trigona spinipes*	Octyl octanoate	Guiding nest mates to a profitable food source	GC-MS	Cephalic labial glands	No information
Jarau et al.	2006	*Trigona corvina*	Octyl octanoate, decyl hexanoate and decyl octanoate/octyll decanoate	Nestmate recruitment,	GC-EAD,	Labial gland	Foragers
Jarau et al.	2006	*Coptotermes gestroi*	(Z,Z,E)-dodeca-3,6,8-trien-1-ol	Recruitment	GC-MS and GC-FTIR	No information	Workers
		Mastotermes darwiniensis Termopsidae Porotermitinae Stolotermitinae	(E)-2,6,10-trimethylundeca-5,9-dien-1-ol	Recruitment	GC-MS and GC-FTIR	No information	Workers
		Termopsidae family	4,6-dimethyl-dodecanal	Recruitment	GC-MS and GC-FTIR	No information	Workers
		Kalotermitida Cryptotermes, K. Procryptotermes,	(Z)-dodec-3-en-1-ol (dodecenol)	Recruitment	GC-MS and GC-FTIR	No information	Workers
		Rhinotermitidae	(Dodecatrienol) (3Z,6Z,8E)-dodeca-3,6,8-trien-1-ol)	Recruitment	GC-MS and GC-FTIR	No information	Workers
Stangler et al.	2008	*Geotrigona mombuca*	Farnesyl butanoate, neryl, geranyl octanoate, octyl hexanoate, octyl octanoate, and octyl decanoate	Recruited nestmates to a rich food source	GC/MS analyses on a HP GC	Labial gland	Workers foragers

(continued)

TABLE 7.1 (Continued)

References	Year	Insect Species	Chemical Compounds	Behavoral Response	Extraction/ Identification	Origin	Sex of Reproductives
Sillam-Dussès et al.	2007	*Mastotermes darwiniensis, Porotermes adamsoni Stolotermes Victoriensis*	(E)-2,6,10-trimethyl-5,9-undecadien-1-ol	Recruitment and orientation	SPME, GC and GC-MS	Sternal glands	Workers
Sillam-Dusses et al.	2009	*Prorhinotermes simplex, P. canalifrons, P. inopinatus*	(3Z,6Z,8E)-Dodeca-3,6,8-trien-1-ol (dodecatrienol) 1E,5E,9E,12 R)-1,5,9-Trimethyl-12-(1-methylethenyl)-1,5,9-cyclotetradecatrie ne (neocembrene)	Recruitment and orientation	GC/MS,EAG and GC-EAD, SPME	Sternal glands	Female alates
Sillam-Dussès et al.	2009	*Kalotermes flavicollis, Cryptotermes brevis, Cryptotermes pallidus, Cryptotermes darlingtonae*	(Z)-Dodec-3-en-1-ol	Recruitment	SPME, GC and EI	Sternal gland	No information
Bordereau et al.	2010	*Zootermopsis nevadensis Z. angusticollis*	4,6-Dimethyldodecanal	Sex attraction and orientation	SPME, Positive chemical ionization, GC-MS	Sternal gland	Male and female alates
Kotoklo et al.	2010	*Amitermes evuncifer*	(3Z,6Z,8*E*) dodeca-3,6,8-trien-1-ol (dodecatrienol), neocembrene, Dodecatrienol	Recriutment	GC-MS and SPME	Sternal glands	Worker
Wen & Sillam-Dussès	2014	*Odontotermes formosanus (Shiraki)*	(3Z)-dodec-3-en-1-ol and (3Z,6Z)-dodeca-3,6-dien-1-ol	Orientation, recruitment effect	Video analysis, SPME, GC-MS	Sternal glands	Foraging workers
Lubes	2018	*Microcerotermes exiguous*	(3Z, 6Z, 8E) -dodeca-3,6,8-trien-1-	Trail following pheromone	GC,(SPME)	Sternal glands	Workers
Nakamura	2019	*Tetramorium tsushimae*	*Methyl 6-methylsalicylate*		Liquid–liquid extraction, GC–MS	Poison gland, hindgut, Dufour's gland	Workers

GC: Gas Chromatography, MS: Mass Spectrophotometry, FID: Flame Ionization Detector, DART-MS: Direct Analysis in Real Time Mass Spectrometry, ESI-MS: Electrospray Ionization Mass Spectrometry, NMRS: Nuclear Magnetic Resonance, SPME: Solid Phase Micro Extraction, LPME: Liquid Phase Microextraction, EAG: Electro antennography, GC-EAD: Gas Chromatography-Electro Antennographic Detection, EI Electron Impact.

deemed intractable to study and explore their greatest value in scientific ways which are not yet realized.

7.4 Microbial-Symbiotic-Mediated Behavioural Manipulation of Insect Host

Various insects' species have been identified as harbouring communities of symbiotic microbes. Other than harmful effects, beneficial services are also performed by these microbes for their host, such as synthesis of missing nutrients, protection against some pathogenic agents, adaptations to environment, and enhancement of insect–insect and insect–host interactions (Gibson and Hunter, 2010). The exact location of the microbes may vary according to its genotype and the host (Oliver et al., 2009). They may be primary symbionts (obligatory and mutualistic) or secondary symbionts (reside in specialized tissue of the host) (Su et al., 2013). A projected 10–20% of insect species such as hemipterans, beetles, lice, some flies, and ants bear intracellular symbionts in specialized cells and approximately 70% of *Wolbachia* alone could infect insect species manipulating the host reproductive status (Douglas, 2011; Ferrari and Vavre, 2011). These symbiotic microbes can also manipulate the host behaviour for their successful transmission to another host (Adamo and Webster, 2013), using several strategies including perpetuating their own genetic system (Libersat et al., 2009). The *Ophiocordyceps unilateralis* species of fungi (also called zombie ant fungi) are known to manipulate the behaviour of ants host which then leave the nest and start biting on vegetation (Araújo et al., 2018). The *Dicrocoelium dendriticum* (Lancet liver fluke) change the behaviour of *Formica fusca* navigational skills which climb to the tip of grasses for transmission into its final host (Libersat et al., 2018). A transcriptomic study revealed that protists in the hindgut of termites possess potentially important cellulases contributing to cellulolytic activity and are important in hydrogen cycling (Peterson and Scharf, 2016). The pea aphids are susceptible to *Aphidius ervi* predation due to its facultative endosymbionts (Su et al., 2013). Many terrestrial insects infest hairworms which ultimately commit suicide in water and exit the parasite into an aquatic environment (Libersat et al., 2018). A novel host can be infected by mosquitoes containing malarial parasite which influences the blood feeding behaviour of the host and increases the probability of its transmission to the other host (Cator et al., 2012). The *Coxiella* like endosymbionts are effectively transmitted to >99% of tick progeny and has been reported as essential microbes for tick survival and reproduction specifically in *Amblyomma americanum* (Bonnet et al., 2017). The indirect enhancement of tick borne diseases is related to the greater motility of *Dermacentor variabilis* infected with rickettsia as compared to uninfected ticks (Kagemann and Clay, 2013). One of the more disturbing interaction is induced by *Entomophthora muscae* infecting female houseflies which alter the attraction behaviour of male to infected females than uninfected females. Insects are the major models for the rapid screening of microbial-symbionts and its evolution (Scully and Bidochka, 2006). Such microbes can play a crucial role in the control of diseases, economic losses, and other pathogenic associated problems. But the factors (biotic and abiotic) driving the diversity of microbiome are largely uncertain. Therefore, we argue that interactions between microbial-symbionts and insect host are of great importance for the control of future insect related problems.

7.5 Microbial-Symbionts Modulating Insect Pheromones

A microbiome is a community of commensal, symbiotic, and pathogenic agents (Pathobiome) associated with the host. Currently, these microbiomes are the major concern of researchers due to their crucial role in insect biology. Most of the studies have shown the vital roles of the microbial volatiles in the maintenance of their host (Schmidt et al., 2015), ranging from bacteria, viruses, fungi, and another wide array of microorganisms adapted to live in a close association with several hosts (Cristaldo et al., 2014; Kanchiswamy et al., 2015). These symbionts influence the insect fitness by contributing in the nutrition, digestion, detoxification, reproduction, protection as well as the components of chemical communication responsible for mate recognition, choice, attraction, localization, aggregation, nestmate, trail markings, and kin recognition in insects, by modulating the host chemical profiles through direct biosynthesis of pheromone components and precursors, or through general changes in the metabolic pathways of the host (Engl and Kaltenpoth, 2018). The microbial-symbionts and their releasing volatile chemical components, genes involved in the volatile biosynthesis, and their interference with the regulatory pathways of the host are largely unknown (Schmidt et al., 2015; Schmidt et al., 2017).

An ant (*Atta sexdens rubropilosa*)-associated bacteria (*Serratia marcescens*) produce a family of pyrazines upon the induction of L-threonine concentration in the host. The bacterial production of the pyrazines is common trail pheromone for the host *Eutetramorium mocquerysi* and alarm pheromone for *Solenopsis invicta* (Silva et al., 2018). The fecal pellets of locust (*Schistocerca gregaria*)-released volatile compounds (Guaiacol (2-methoxy and phenol) are synthesized by gut bacteria which was previously known to produce by insect host itself (Dillon et al., 2000). The cultivable gut bacteria of *Chrysolina herbacea* metabolize the monoterpenes of *Mentha aquatica* into its own pheromone activity (Pizzolante et al., 2017). Similarly, endosymbionts (*Rickettsia*) such as *Rickettsia buchneri* isolated from *Ixodes pacificus* and *I. scapularis* have been found to encode genes for folate biosynthesis (Hunter et al., 2015). The gut bacteria of specific *Ades aegypti* are associated with the carboxylic acids and methyl esters preparation for stimulating oviposition (Ponnusamy et al., 2008). The insect microbial-symbionts (*Streptomyces*) are associated with the growth inhibition of other pathogens by secreting certain chemicals (Chevrette et al., 2019). It was demonstrated that highly volatile carboxylic acids (VCAs) are found in the faeces of German cockroaches (Family Ectobiidae) containing a large number of gut bacteria than normal cockroaches relatively free of VCAs (Wada et al., 2015). Insects contain large-scale microbiome for their own purposes which help the researcher to select specific microbial-symbionts for the production of antibiotics. Microbes manipulating the behaviour of insect host by interfering in pheromone production and modulating the production pathways is a challenging phenomenon for the biologists. The limited knowledge about these microbial-symbionts requires co-operative potentials of both researchers studying host symbiotic relationships and chemical ecologists to obtain more inclusive understandings in this regard.

7.6 Extraction and Identification of Insect Pheromones

The bioassays for an insect's pheromones are relatively easy to perform due to its feasible implementation in the laboratory. The essential execution in performing the experiments is to avoid contamination. Currently, various techniques are developed for quick and precise extraction and identification of pheromones (Antony et al., 2015). The recent advances in genome engineering technologies based on the clustered regularly interspaced short palindromic repeats associated with RNA-guided endonuclease Cas9 can easily modify the structure and even manipulate the production of biosynthetic chemicals in almost any organism of choice (Hsu et al., 2014). The identification, detection, quantification, and structural characterization of volatile chemicals are predominantly performed by gas chromatography (GS). In a combination with mass spectrometry (MS), various branched and linear alkane constituents of pheromones can be structurally identified (Kalinová et al., 2006). Combination of GC and flame ionization detector (FID) analysis is used for structural confirmation; however, solid-phase microextraction with GC/MS method detects and quantifies semichemicals in a very small amount (Levi et al., 2012). Direct analysis in real-time mass spectrometry (DART-MS) is useful for the rapid identification of organic compounds without any extractions, derivatizations, and chromatographic procedures (Manfredi et al., 2016). Triacylglycerides, long chain fatty alcohols, sterols, and other solid or liquid bulk materials can be detected through DART-MS (Gross, 2014). The matrix-assisted laser desorption/ionization mass spectrometry also called ionization mass spectrometry are helpful in the analysis of a wide range of biomolecules especially more applicable for the analysis of intact insects or crude extract and non-volatile pheromones (Kusano et al., 2014). Rapid analysis of insect's required small amount for sample preparation subjected to DART in combination with MS followed by laser interrogation, but alkanes are not detected, and no stereochemistry is obtained in this technique (Yew and Chung, 2015).

Under atmospheric conditions, the electrospray ionization mass spectrometry (ESI-MS) generates ions from liquid samples, allowing the ionization of a broad range of biomolecules (Murphy and Axelsen, 2011). Furthermore, liquid chromatography with ESI-MS can significantly enhance the detection and increases the analytical resolution of low abundance molecules even stereoisomers can also be identified using special chromatographic columns (Awad and El-Aneed, 2013). Such a technique can be used to identify putative nematode lipid pheromones from the extract. It is obvious that the technique has few applications on ESI-MS for pheromone analysis. The nuclear magnetic resonance spectroscopy (NMRS) has become a universal method in lipid analysis and is considered as a powerful tool for the structure determination of compounds. In comparison to common chromatographic analysis, NMRS is a non-destructive advanced method which provides a detail description of chemical composition within the same analysis (Xu et al., 2015). Gas chromatography using a time-of-flight mass spectrometric detector is another analytical method used specifically for chemical identification. GC with electrospray ionization-time of flight mass spectrometry detection operates with a high precision independent of concentration range and is more advanced than GC coupled with electroantennogram detection (EAD). Because no compounds can be detected by FID detection in combination with GC-EAD. This Gas chromotography technique showed the presence of several chemicals (Kalinová et al., 2006).

7.7 Omic Era

For many decades, the chemical communication system in insects and insect symbionts was an exciting challenge for researchers. Several approaches were applied to characterize specific enzymes encoded by host genes or associated microbial-symbionts for the synthesis of pheromones. Identification of pheromones-encoding genes through in-situ hybridization at cellular site and the transcriptomics studies using next-generation sequencing (NGS) (Caballero et al., 2013), RNA interference (RNAi), DNA microarrays, clustered regularly interspaced short palindromic repeats (CRISPR), computational and bioinformatics analysis have advanced the previous approaches (Mahato et al., 2017). With the rise of these novel omics technologies, biological systems are further investigated for generating large-scale and heterogeneous datasets (Gomez et al., 2014). The first transcriptomic analysis revealed the catalogue of genes expressed in pheromone gland tissues (Caballero et al., 2013). The transcriptomes obtained from pheromone producing gland of *Agrotis ipsilon* revealed the presence of transcripts that encode pheromone biosynthetic enzymes among them fifteen transcripts were chiefly expressed in production sites of pheromone (Gu et al., 2013). The transgenic Arabidopsis (genus of family Brassicaceae) with terpene synthase gene resulted in the production of (E)-b-farnenese which is an alarm pheromone of aphid simultaneously protecting the plant species from aphids. In *Lutzomyia longipalpis*, the protein-encoding transcripts were identified, encoding four enzymes responsible for the terpenoid production (Caballero et al., 2013). The identification of bacterial operational taxonomic units (OTU) showed 97% sequence identity by16S rRNA gene analysis which revealed that the gut of most insects bears <20–30 taxa including proteobacteria, firmicutes, and protists (Wilson et al., 2006). Through NGs, many tick-borne pathogens of both medical and veterinary importance are identified including Crimean-Congo haemorrhagic fever virus, Kyasanur Forest disease virus, and bacterial species *Anaplasma, Borrelia, Coxiella, Ehrlichia, Francisella, Acinetobacter, Rickettsia, Babesia*, and the parasitic *Theileria* (Greay et al., 2018), suggesting that insects are the ideal model to study various aspects of interactions between the host and microorganisms (Douglas, 2011), which can be helpful for the novel drug targets by understanding the chemical profiles of microbial-symbionts used for the behavioural manipulation of the insect host. These microbial-symbionts can open new horizons for researchers to study their interference in insect host metabolic pathways.

7.8 Conclusion and Future Directions

Pheromones enhance the chemical communication between insects modulating insects behaviour. Microbial-symbionts interfere in various biochemical pathways of insect host provide chemicals that can be used directly by insect host or interfere with the metabolic pathways of insect host. Understanding the genomic levels of insect pheromone production and the biochemical nature of microbial-symbionts interaction with the insect host warrant further investigation. Therefore, it is deemed to apply the novel omic techniques on a wide range of insects to control insect pests, pathogens transmitted by insects to other organisms, and to protect beneficial insects from diseases and stresses.

Acknowledgement

We would like to thank the Higher Education Commission and Pakistan Science Foundation for their financial support.

REFERENCES

Abdalla FC, Cruz CD (2001) Dufour glands in the hymenopterans (Apidae, Formicidae, Vespidae): A review. *Rev Bras Biol* 61:95–106. doi:10.1590/s0034-71082001000100013.

Adamo SA, Webster JP (2013) Neural parasitology: How parasites manipulate host behaviour. *J Exp Biol* 216:1–2. doi:10.1242/jeb.082511.

Adnan, M, Shah Z, Arif M, Khan MJ, Mian IA, Sharif M, Alam M, Basir A, Ullah H, Rahman I, Saleem N (2016) Impact of rhizobial inoculum and inorganic fertilizers on nutrients (NPK) availability and uptake in wheat crop. *Can J Soil Sci* 96:169–176.

Allan SA, Sonenshine DE (2002) Evidence of an assembly pheromone in the black-legged deer tick, Ixodes scapularis. *J Chem Ecol* 28:15–27. DOI: 10.1023/A:1013554517148.

Andersson MN, Newcomb RD (2017) Pest control compounds targeting insect chemoreceptors: Another silent spring? *Front Ecol Evol* 5:1–5. https://doi.org/10.3389/fevo.2017.00005.

Antony B, Soffan A, Jakše, Alfaifi S et al. (2015) Genes involved in sex pheromone biosynthesis of *Ephestia cautella*, an important food storage pest, are determined by transcriptome sequencing. *BMC Genom* 16:1–26. doi:10.1186/s12864-015-1710-2.

Araújo JP, Evans HC, Kepler R et al. (2018) Zombie-ant fungi across continents: 15 new species and new combinations within Ophiocordyceps Myrmecophilous hirsutelloid species. *Stud Mycol* 90:119–160. doi:10.1016/j.simyco.2017.12.002.

Atri C, Kumar B, Kumar H et al. (2012) Development and characterization of Brassica juncea–fruticulosa introgression lines exhibiting resistance to mustard aphid (Lipaphis erysimi Kalt). *BMC Genet* 13:1–9. doi:10.1186/1471-2156-13-104.

Awad H, El-Aneed A (2013) Enantioselectivity of mass spectrometry: Challenges and promises. *Mass Spectrom Rev* 32:466–483. doi:10.1002/mas.21379.

Bahrndorff S, Alemu T, Alemneh T et al. (2016) The microbiome of animals: Implications for conservation biology. *Int J Genom* 2016:1–16. http://dx.doi.org/10.1155/2016/5304028.

Berger RS (1972) 2,6-dichlorophenol, sex pheromone of the lone star tick. *Science* 177:704–705. doi:10.1126/science.177.4050.704.

Billen J, Gobin B, Ito F (1999) Fine structure of the postpygidial gland in Aenictus army ants. *Acta Zool.* 80:307–310. https://doi.org/10.1046/j.1463-6395.1999.00026.x.

Billen J, Thys B, Ito F et al. (2005) The pretarsal footprint gland of the ant Amblyopone reclinata (Hymenoptera, Formicidae) and its role in nestmate recruitment. *Arthropod Struct Dev* 34:111–116. https://doi.org/10.1016/j.asd.2004.11003.

Bilton DT, Freeland JR, Okamura B (2001) Dispersal in freshwater invertebrates. *Annu Rev Ecol Syst* 32:159–181. https://doi.org/10.1146/annurev.ecolsys.32.081501.114016.

Blatrix R, Schulz C, Jaisson P et al. (2002) Trail pheromone of ponerine ant Gnamptogenys striatula: 4-methylgeranyl esters from Dufour's gland. *J Chem Ecol* 28:2557–2567. doi:10.1023/a:1021444321238.

Bonnet SI, Binetruy F, Hernández JAM et al. (2017) The tick microbiome: why non-pathogenic micro-organisms matter in tick biology and pathogen transmission. *Front Cell Infect Microbiol* 7:229–236. doi:10.3389/fcimb.2017.00236.

Bordereau C, Cancello EM, Sémon E et al. (2002) Sex pheromone identified after solid phase microextraction from tergal glands of female alates in Cornitermes bequaerti (Isoptera, Nasutitermitinae). *Insectes Sociaux* 49:209–215. https://doi.org/10.1007/s00040-002-8303-1.

Bordereau C, Lacey MJ, Semon E et al. (2010) Sex pheromones and trail-following pheromone in the basal termites *Zootermopsis nevadensis* (Hagen) and *Zootermopsis angusticollis* (Hagen)(Isoptera:

Termopsidae: Termopsinae). *Biol J Linn Soc* 100:519–530. https://doi.org/10.1111/j.1095-8312.2010.01446.x.

Butenandt A, Beckmann R, Stamm D et al. (1959) Über den Sexuallockstoff des Seidenspinners Bombyx mori – Reindarstellung und Konstitution. *Z Naturforschung Part B – Chem Biochem Biophys Biol Verwandten Geb* 14:283–284.

Caballero GN, Valenzuela JG, Ribeiro JM et al. (2013) Transcriptome exploration of the sex pheromone gland of Lutzomyia longipalpis (Diptera: Psychodidae: Phlebotominae). *Parasites Vectors* 6:1–16. doi:10.1186/1756-3305-6-56.

Cator LJ, Lynch PA, Read AF et al. (2012) Do malaria parasites manipulate mosquitoes?. *Trends Parasitol* 28:466–470. doi:10.1016/j.pt.2012.08.004.

Chapman JW, Nesbit RL, Burgin LE et al. (2010) Flight orientation behaviors promote optimal migration trajectories in high-flying insects. *Science* 327:682–685. doi:10.1126/science.1182990.

Chevrette MG, Carlson CM, Ortega HE et al. (2019) The antimicrobial potential of Streptomyces from insect microbiomes. *Nat Commun* 10:1–11. doi:10.1038/s41467-019-08438-0.

Choe DH, Villafuerte DB, Tsutsui ND (2012) Trail pheromone of the argentine ant, Linepithema humile (Mayr) (Hymenoptera: Formicidae). *PLoS One* 7:1–7.doi:10.1371/journal.pone.0045016.

Cristaldo PF, DeSouza O, Krasulová J et al. (2014) Mutual use of trail-following chemical cues by a termite host and its inquiline. *PLoS One* 9:1–9. doi:10.1371/journal.pone.0085315.

Crotti E, Balloi A, Hamdi C et al. (2012) Microbial symbionts: A resource for the management of insect-related problems. *Microbial Biotechnol* 5:307–317. doi:10.1111/j.1751-7915.2011.00312.x.

Czaczkes TJ, Grüter C, Ratnieks FL (2015) Trail pheromones: An integrative view of their role in social insect colony organization. *Annu Rev Entomol* 60:581–599. https://doi.org/10.1146/annurev-ento-010814-020627.

Dillon RJ, Vennard CT, Charnley AK (2000) Pheromones: Exploitation of gut bacteria in the locust. *Nature* 403:851–852. doi:10.1038/35002669.

Douglas AE (2011) Lessons from studying insect symbioses. *Cell Host Microbe* 10:359–367. doi:10.1016/j.chom.2011.09.001.

Engl P, Moran NA (2013) The gut microbiota of insects–diversity in structure and function. *FEMS Microbiol Rev* 37:699–735. doi:10.1111/1574-6976.12025.

Engl T, Kaltenpoth, M (2018) Influence of microbial symbionts on insect pheromones. *Nat Prod Rep* 35:386–397. doi:10.1039/C7NP00068E.

Engl T, Michalkova V, Weiss BL et al. (2018) Effect of antibiotic treatment and gamma-irradiation on cuticular hydrocarbon profiles and mate choice in tsetse flies (Glossina m. morsitans). *BMC Microbiol* 18:159–292. doi:10.1186/s12866-018-1292-7.

Fahad, S, Khan FA, Hussain S, Khan IA, Saeed M, Saud S, Hassan S, Adnan M, Amanullah MA, Alam M, Ullah H (2019) Suppressing photorespiration for the improvement in photosynthesis and crop yields: A review on the role of S-allantoin as a nitrogen source. *J Environ Manag* 237:644–651.

Ferrari J, Vavre F (2011) Bacterial symbionts in insects or the story of communities affecting communities. *Philos Trans R Soc B: Biol Sci* 366:1389–1400. doi:10.1098/rstb.2010.0226.

Fisher RM, Henry LM, Cornwallis CK et al. (2017) The evolution of host-symbiont dependence. *Nature Commun* 8:2955–2969. doi:10.1038/ncomms15973.

Gary NE (1962) Chemical mating attractants in the queen honey bee. *Science* 136:773–774. doi:10.1126/science.136.3518.773.

Gibson CM, Hunter MS (2010) Extraordinarily widespread and fantastically complex: Comparative biology of endosymbiotic bacterial and fungal mutualists of insects. *Ecol Lett* 13:223–234. doi:10.1111/j.1461-0248.2009.01416.x.

Gomez C, Benton R (2013) The joy of sex pheromones. *EMBO Rep* 14:874–883. doi: 10.1038/embor.2013.140.

Gomez D, Abugessaisa I, Maier D et al. (2014) Data integration in the era of omics: Current and future challenges. *BMC Syst Biol* 8:1–11. doi:10.1186/1752-0509-8-S2-I1.

Gothe R, Neitz AWH (1985) Investigation into the participation of male pheromones of Rhipicephalus evertsi evertsi during infestation. *Onderstepoort J Vet Res* 52:6–70.

Graff JF (1976) Ecologie and ethology d'Ixodes ricinus L. en Suisse (Ixodoidea: Ixodidae). 2 e partie. *Bull Soc Entomol Suisse* 51:241–253.

Granero AM, Sanz JMG, Gonzalez FJE et al. (2005) Chemical compounds of the foraging recruitment pheromone in bumblebees. *Naturwissenschaften* 92:371–374. doi:10.1007/s00114-005-0002-0.

Greay TL, Gofton AW, Paparini A et al. (2018) Recent insights into the tick microbiome gained through next-generation sequencing. *Parasites Vectors* 11:1–12. Doi:10.1186/s13071-017-2550-5.

Gross JH (2014) Direct analysis in real time – A critical review on DART-MS. *Anal Bioanal Chem* 406:63–80. doi:10.1007/s00216-013-7316-0.

Gu SH, Wu KM, Guo YY et al. (2013) Identification of genes expressed in the sex pheromone gland of the black cutworm *Agrotis ipsilon* with putative roles in sex pheromone biosynthesis and transport. *BMC Genom* 14:1–21. doi:10.1186/1471-2164-14-636.

Hanks LM, Millar JG (2013) Field bioassays of cerambycid pheromones reveal widespread parsimony of pheromone structures, enhancement by host plant volatiles, and antagonism by components from heterospecifics. *Chemoecology* 23:21–44. doi:10.1007/s00049-012-0116-8.

Haque, SZ, Haque M (2017) The ecological community of commensal, symbiotic, and pathogenic gastro-intestinal microorganisms–an appraisal. *Clin Exp Gastroenterol* 10:91–103. doi:10.2147/CEG.S126243.

Helfrich C, Bertolini E, Menegazzi P (2018) Flies as models for circadian clock adaptation to environmental challenges. *Eur J Neurosci*:1–16. doi:10.1111/ejn.14180.

Holland RA, Wikelski M, Wilcove DS (2006) How and why do insects migrate? *Science* 313:794–796. doi:10.1126/science.1127272.

Hsu PD, Lander ES, Zhang F (2014) Development and applications of CRISPR-Cas9 for genome engineering. *Cell* 157:1262–1278. doi:10.1016/j.cell.2014.05.010.

Hunter DJ, Torkelson JL, Bodnar J et al. (2015) The Rickettsia endosymbiont of *Ixodes pacificus* contains all the genes of de novo folate biosynthesis. *PLoS One* 10:1–15. doi:10.1371/journal.pone.0144552.

Hölldobler, B, Morgan ED, Oldham NJ et al. (2001) Recruitment pheromone in the harvester ant genus Pogonomyrmex. *J Insect Physiol* 47:369–374. doi:10.1016/s0022-1910(00)00143-8.

Hölldobler B, Morgan ED, Oldham NJ et al. (2004) Dufour gland secretion in the harvester ant genus Pogonomyrmex. *Chemoecology* 14:101–106.

Iglesias CM, Fox RJ, Vincent A et al. (2019) No evidence that male sexual experience increases mating success in a coercive mating system. *Anim Behav* 150:201–208. https://doi.org/10.1016/j.anbehav.2019.02.012.

Inwood M (2008) Trail pheromones of ants. *Physiol Entomol* 34:1–17. doi:10.1111/j.1365-3032.2008.00658.x.

Jarau S, Hrncir M, Ayasse M et al. (2004) A stingless bee (Melipona seminigra) marks food sources with a pheromone from its claw retractor tendons. *J Chem Ecol* 30:793–804. doi:10.1023/b:joec.0000028432.29759.ed.

Jarau S, Schulz CM, Hrncir M et al. (2006) Hexyl decanoate, the first trail pheromone compound identified in a stingless bee, Trigona recursa. *J Chem Ecol* 32:1555–1564. doi:10.1007/s10886-006-9069-0.

Jarau S, Dambacher J, Twele R et al. (2010) The trail pheromone of a stingless bee, Trigona corvina (Hymenoptera, Apidae, Meliponini), varies between populations. *Chem Senses* 35:593–601. doi:10.1093/chemse/bjq057.

Jeanson R, Ratnieks FL, Deneubourg JL (2003) Pheromone trail decay rates on different substrates in the Pharaoh's ant, Monomorium pharaonis. *Physiol Entomol* 28:192–198. doi:10.1046/j.1365-3032.2003.00332.x.

Joachim C, Hatano E, David A et al. (2013) Modulation of aphid alarm pheromone emission of pea aphid prey by predators. *J Chem Ecol* 39:773–782. doi:10.1007/s10886-013-0288-x.

Kagemann J, Clay K (2013) Effects of infection by Arsenophonus and Rickettsia bacteria on the locomotive ability of the ticks Amblyomma americanum, Dermacentor variabilis, and Ixodes scapularis. *J Med Entomol* 50:155–162. doi:10.1603/me12086.

Kalinová B, Jiroš P, Žd'árek J et al. (2006) GC× GC/TOF MS technique—A new tool in identification of insect pheromones: Analysis of the persimmon bark borer sex pheromone gland. *Talanta* 69:542–547. https://doi.org/10.1016/j.talanta.2005.10.045.

Kanchiswamy CN, Malnoy M, Maffei ME (2015) Chemical diversity of microbial volatiles and their potential for plant growth and productivity. *Front Plant Sci* 6:1–23. doi:10.3389/fpls.2015.00151.

Karlson P, Lüscher M (1999) 'Pheromones': A new term for a class of biologically active substances. *Nature* 183:55–56. doi:10.1038/183055a0.

Kohl E, Hölldobler B, Bestmann HJ (2001) Trail and recruitment pheromones in Camponotus socius (Hymenoptera: Formicidae). *Chemoecology* 11:67–73. doi:10.1007/PL00001834.

Kohl E, Hölldobler B, Bestmann HJ (2003) Trail pheromones and Dufour gland contents in three Camponotus species (C. castaneus, C. balzani, C. sericeiventris: Formicidae, Hymenoptera). *Chemoecology* 13:113–122. doi:10.1007/s00049-003-0237-1.

Kotoklo EA, Sillam-Dussès D, Kétoh G et al. (2010) Identification of the trail-following pheromone of the pest termite Amitermes evuncifer (Isoptera: Termitidae). *Sociobiology* 55:579–588.

Krittika S, Yadav P (2019) Circadian clocks: An overview on its adaptive significance. *Biol Rhythm Res* 8:1–24. https://doi.org/10.1080/09291016.2019.1581480.

Kusano M, Kawabata SI, Tamura Y et al. (2014) Laser desorption/ionization mass spectrometry (LDI-MS) of lipids with iron oxide nanoparticle-coated targets. *Mass Spectrom* 3:1–8. https://doi:10.5702/massspectrometry.A0026.

Leahy MG, Hajkova Z, Bourchalova J (1981) Two females pheromones in the metastriate ticks, Hyalomma dromedarii (Acarina: Ixodidae). *Acta Entomol Bohemoslavia* 78:224–230. doi: 10.1007/bf00132316.

Leonardo CAM, Casarin FE, Lima JT (2009) Chemical communication in Isoptera. *Neotropical Entomol* 38:1–6. doi: 10.1590/s1519-566x2009000100001.

Levi A, Nestel D, Fefer D et al. (2012) Analyzing diurnal and age-related pheromone emission of the olive fruit fly, Bactrocera oleae by sequential SPME-GCMS analysis. *J Chem Ecol* 38:1036–1041. doi:10.1007/s10886-012-0167-x.

Libersat F, Delago A, Gal R (2009) Manipulation of Host Behavior by Parasitic Insects and Insect Parasites. *Annu Rev Entomol* 54:189–207. doi:10.1146/annurev.ento.54.110807.090556.

Libersat F, Kaiser M, Emanuel S (2018) Mind control: How parasites manipulate cognitive functions in their insect hosts. *Front Psychol* 9:566–572. doi:10.3389/fpsyg.2018.00572.

Llan SA, Sonenshine DE, Burridge MJ (2006) Ticks pheromones and uses thereof. *United States Patent Office* 6:297–331.

Lu M, Hulcr J, Sun J (2006) The role of symbiotic microbes in insect invasions. *Annual Rev Ecol Evol Syst* 47:487–505. https://doi.org/10.1146/annurev-ecolsys-121415-032050.

Lubes G, Cabrera A (2018) Identification and evaluation of (3Z,6Z,8E)-dodeca-3,6,8-trien-1-ol as the trail following pheromone on Microcerotermes exiguus (Isoptera: Termitidae). *Revista Biol Tropical* 66:303–311. doi:10.15517/rbt.v66i1.27111.

Manfredi M, Robotti E, Bearman G et al. (2016). Direct analysis in real time mass spectrometry for the nondestructive investigation of conservation treatments of cultural heritage. *J Anal Methods Chem* 2016:1–12. http://dx.doi.org/10.1155/2016/6853591.

Mashaly AMA, Ahmed AM, Al-Abdullah MA et al. (2011) The trail pheromone of the venomous samsum ant, Pachycondyla sennaarensis. *J Insect Sci* 11:1–12. doi: 10.1673/031.011.0131.

McCutcheon JP, Moran NA (2007) Parallel genomic evolution and metabolic interdependence in an ancient symbiosis. *Proc Natl Acad Sci* 104:19392–19397. doi:10.1073/pnas.0708855104.

Morgan ED, Brand JM, Mori K et al. (2004) The trail pheromone of the ant Crematogaster castanea. *Chemoecology* 14:119–120. doi:10.1007/s00049-003-0262-0.

Murphy RC, Axelsen PH (2011) Mass spectrometric analysis of long-chain lipids. *Mass Spectrom Rev* 30:579–599. doi:10.1002/mas.20284.

Nakamura T, Harada K, Akino T (2019) Identification of methyl 6-methylsalicylate as the trail pheromone of the Japanese pavement ant Tetramorium tsushimae (Hymenoptera: Formicidae). *Appl Entomol Zoology* 54:297–305. doi:10.1007/s13355-019-00626-0.

Nikoh N, McCutcheon JP, Kudo T et al. (2010) Bacterial genes in the aphid genome: Absence of functional gene transfer from Buchnera to its host. *PLoS Genet* 6:1–21. https://doi.org/10.1371/journal.pgen.1000827.

Oldham NJ, Morgan ED, Gobin B et al. (1994) First identification of a trail pheromone of an army ant (Aenictus species). *Naturwissenschaften* 81:313–316. doi:10.1007/BF01919378.

Oliver KM, Degnan PH, Hunter MS et al. (2009) Bacteriophages encode factors required for protection in a symbiotic mutualism. *Science* 325:992–994. doi:10.1126/science.1174463.

Oliver KM, Degnan PH, Burke GR, Moran NA (2010) Facultative symbionts in aphids and the horizontal transfer of ecologically important traits. *Annu Rev Entomol* 55:247–266. doi:10.1146/annurev-ento-112408-085305.

Olszewski CG, Bomont C, Coutrot P (2011) Synthesis of insect pheromones: Improved method for the preparation of queen substance and royal jelly of honeybees Apis mellifera. *In Annales de la Société entomologique de France* 47:45–54. doi:10.1080/00379271.2011.10697695.

Otieno DA, Hassanali A, Obenchain FD et al. (1985) Identification of guanine as an assembly pheromones of ticks. *Insect Sci Appl* 6:667–670. doi:10.1007/bf01202877.

Packianather MS, Landy M, Pham DT (2009) Enhancing the speed of the Bees Algorithm using Pheromone-based Recruitment. In 2009 7th IEEE International Conference on Industrial Informatics 89–794.

Peppuy A, Robert A, Sémon E et al. (2001) Species specificity of trail pheromones of fungus-growing termites from northern Vietnam. *Insectes Sociaux* 48:245–250. https://doi.org/10.1007/PL00001773.

Peterson BF, Scharf ME (2016) Lower termite associations with microbes: synergy, protection, and interplay. *Front Microbiol* 7:1–7. doi:10.3389/fmicb.2016.00422.

Pizzolante G, Cordero C, Tredici SM et al. (2017) Cultivable gut bacteria provide a pathway for adaptation of Chrysolina herbacea to Mentha aquatica volatiles. *BMC Plant Biol* 17:1–20. https://doi.org/10.1186/s12870-017-0986-6.

Ponnusamy L, Xu N, Nojima S et al. (2008) Identification of bacteria and bacteria-associated chemical cues that mediate oviposition site preferences by Aedes aegypti. *Proc Natl Acad Sci* 105:9262–9267. doi:10.1073/pnas.0802505105.

Pérez SD, Martínez HMG, Aluja M (2010) Male age and experience increases mating success but not female fitness in the Mexican fruit fly. *Ethology* 116:778–786. doi:10.1016/j.jinsphys.2012.11.010.

Robert A, Peppuy A, Sémon E et al. (2004) A new C12 alcohol identified as a sex pheromone and a trail-following pheromone in termites: The diene (Z, Z)-dodeca-3, 6-dien-1-ol. *Naturwissenschaften* 91:34–39. doi:10.1007/s00114-003-0481-9.

Sandrelli F, Costa R, Kyriacou CP et al. (2008) Comparative analysis of circadian clock genes in insects. *Insect Mol Biol* 17:447–463. https://doi.org/10.1111/j.1365-2583.2008.00832.x.

Saran RK, Millar JG, Rust MK (2007) Role of (3Z, 6Z, 8E)-dodecatrien-1-ol in trail following, feeding, and mating behavior of Reticulitermes hesperus. *J Chem Ecol* 33:369–389. doi:10.1007/s10886-006-9229-2.

Schmidt R, Cordovez V, De Boer W et al. (2015) Volatile affairs in microbial interactions. *ISME J* 9:23–29. doi:10.1038/ismej.2015.42.

Schmidt R, Dejager V, Zühlke D et al. (2017) Fungal volatile compounds induce production of the secondary metabolite Sodorifen in Serratia plymuthica PRI-2C. *Sci Rep* 7:1–14. doi:10.1038/s41598-017-00893-3.

Scully LR, Bidochka MJ (2006) Developing insectmodels for the study of current and emerging human pathogens. *FEMS Microbiol Lett* 263:1–9. https://doi.org/10.1111/j.1574-6968.2006.00388.x.

Semwal A, Kumar R, Teotia UVS et al. (2013) Pheromones and their role as aphrodisiacs: A review. *J Acute Dis* 2:253–261. https://doi.org/10.1016/S2221-6189(13)60140-7.

Sillam-Dusses D, Kalinova B, Jiroš P et al. (2009) Identification by GC-EAD of the two-component trail-following pheromone of Prorhinotermes simplex (Isoptera, Rhinotermitidae, Prorhinotermitinae). *J Insect Physiol* 55:751–757. doi:10.1016/j.jinsphys.2009.04.007.

Sillam-Dussès D, Sémon E, Lacey MJ et al. (2007) Trail-following pheromones in basal termites, with special reference to Mastotermes darwiniensis. *J Chem Ecol* 33:1960–1977. doi:10.1007/s10886-007-9363-5.

Silva CCA, Laumann RA, Ferreira JBC, Moraes MCB, Borges M, Cokl A (2012) Reproductive biology, mating behavior, and vibratory communication of the brown-winged stink bug, Edessa meditabunda (Fabr.) (Heteroptera: Pentatomidae). *Psyche: J Entomol* 2012:1–9. http://dx.doi.org/10.1155/2012/598086.

Silva JEA, Ruzzini AC, Paludo CR et al. (2018) Pyrazines from bacteria and ants: Convergent chemistry within an ecological niche. *Sci Rep* 8:25–95. doi:10.1038/s41598-018-20953-6.

Snyder AK, Rio RV (2015) 'Wigglesworthia morsitans' folate (vitamin B9) biosynthesis contributes to tsetse host fitness. *Appl Environ Microbiol* 81(16):5375–5386. doi: 10.1128/AEM.00553-15.

Sonenshine DE (2004) Pheromones and other semiochemicals of ticks and their use in tick control. *Parasitology* 129:405–425. https://doi.org/10.1017/S003118200400486X.

Sonenshine DE (2006) Tick pheromones and their use in tick control. *Annu Rev Entomol* 51:557–580. doi:10.1146/annurev.ento.51.110104.151150.

Stangler ES, Jarau S, Hrncir M et al. (2009) Identification of trail pheromone compounds from the labial glands of the stingless bee Geotrigona mombuca. *Chemoecology* 19:13–19. doi:10.1007/s00049-009-0003-0.

Steinmetz I, Schmolz E, Ruther J (2003) Cuticular lipids as trail pheromone in a social wasp. Proceedings of the Royal Society of London. *Series B: Biol Sci* 270:385–391. doi:10.1098/rspb.2002.2256.

Su Q, Zhou X, Zhang Y (2013) Symbiont-mediated functions in insect hosts. *Commun Integr Biol* 6:1–7. https://doi.org/10.4161/cib.23804.

Tentschert J, Bestmann HJ, Hölldobler B et al. (2000) 2, 3-dimethyl-5-(2-methylpropyl) pyrazine, a trail pheromone component of Eutetramorium mocquerysi Emery (1899)(Hymenoptera: Formicidae). *Naturwissenschaften* 87:377–380. doi:10.1007/s001140050745.

Tumlinson JH, Silverstein RM, Moser JC et al. (1971) Identification of the trail pheromone of a leaf-cutting ant, Atta texana. *Nature* 234:344–348. doi:10.1038/234348b0.

Van OA, Oliveira RC, Holman L et al. (2014) Conserved class of queen pheromones stops social insect workers from reproducing. *Science* 343:287–290. doi: 10.1126/science.1244899.

Vandermoten S, Mescher MC, Francis F et al. (2012) Aphid alarm pheromone: An overview of current knowledge on biosynthesis and functions. *Insect Biochem Mol Biol* 42:155–163. doi: 10.1016/j.ibmb.2011.11.008.

Wen P, Ji BZ, Sillam-Dussès D (2014) Trail communication regulated by two trail pheromone components in the fungus-growing termite Odontotermes formosanus (Shiraki). *PloS One* 9:1–12. doi:10.1371/journal.pone.0090906.

White BH, Ewer J (2014) Neural and hormonal control of postecdysial behaviors in insects. *Annu Rev Entomol* 59:363–381. doi:10.1146/annurev-ento-011613-162028.

Wilson AC, Dunbar HE, Davis GK et al. (2006) A dual-genome microarray for the pea aphid, Acyrthosiphon pisum, and its obligate bacterial symbiont, Buchnera aphidicola. *BMC Genom* 7:44–50. doi:10.1186/1471-2164-7-50.

Witzgal P, Stelinski L, Gut L et al. (2008) Codling moth management and chemical ecology. *Annu Rev Entomol* 53:503–522. doi:10.1146/annurev.ento.53.103106.093323.

Witzgall P, Kirsch P, Cork A (2010) Sex pheromones and their impact on pest management. *J Chem Ecol* 36:80–100. https://doi.org/10.1007/s10886-009-9737-y.

Xu JL, Riccioli C, Sun DW (2015) An overview on nondestructive spectroscopic techniques for lipid and lipid oxidation analysis in fish and fish products. *Compr Rev Food Sci Food Saf* 14:466–477. https://doi.org/10.1111/1541-4337.12138.

Yew JY, Chung H (2015) Insect pheromones: An overview of function, form, and discovery. *Prog. Lipid Res* 59:88–105. https://doi.org/10.1016/j.plipres.2015.06.001.

Zurek L, Watson DW, Krasnoff SB et al. (2002) Effect of the entomopathogenic fungus, Entomophthora muscae (Zygomycetes: Entomophthoraceae), on sex pheromone and other cuticular hydrocarbons of the house fly, Musca domestica. *J Invertebr Pathol* 80:171–176. doi:10.1016/s0022-2011(02)00109-x.

8

The Chemistry of Atmosphere

Nayab Gul*
Climate Change Centre, The University of Agriculture, Peshawar

8.1 Introduction

Atmospheric chemistry is an exciting field of science that deals with the study of earth's atmosphere and other planets (Jacob, 1999; Seinfeld and Pandis, 2016). It studies global chemistry, the regions close to earth, remote regions, and polluted and clean environments. This branch of science is focused on the chemical processes occurring in the earth's atmosphere, including biogeochemical cycles, formation of clouds, emission, photochemistry, transport and diffusion of aerosol and gas particles, and so forth (Finlayson-Pitts and Pitts Jr, 1986; Andreae and Crutzen, 1997).

The atmosphere is the outermost layer of the earth and it is the mixture of different gases in different proportions surrounding the earth. The density of gases varies with altitude. The atmosphere has no precise limit but 99% of gaseous mixtures exist up to 32 km due to the earth's gravitational pull (Vallero, 2014).

The chemical composition of earth's atmosphere is important for many reasons but mostly for the interaction of living organisms with their environment. The earth's atmosphere is composed of different gases existing in different percentages. The major constituents of atmosphere are nitrogen (N_2) gas which is 78% by volume and O_2 gas which is 21% (Seinfeld and Pandis, 2016). Other gases like nitrous oxide, argon, hydrogen, methane, and ozone are the trace gases present in less than 1% (Jacob, 1999). The volumewise presence of gases in the atmosphere is shown in Table 8.1. The presence of a high percentage of N_2 is responsible for the growth of all vegetation and O_2 is vital for the presence of life on earth. The water vapours of about 2–3% and CO_2 absorb solar energy which keep the average temperature of the atmosphere at about 15 °C (Andrady et al., 2017). CO_2 is used by green plants for photosynthesis, for water vapours for the precipitation and cloud formation, and for replenishing of the moisture of the soil. Water vapours also protect living tissues from drying and dehydration. The upper layers of atmosphere protect the living organisms on earth from the solar radiations that could damage living tissue. Besides protection from solar radiation, the upper atmosphere protects the earth from the bombardment of meteorites and charged particles.

Atmospheric chemistry is a multidisciplinary field of research focusing on meteorology, environmental chemistry, physics, computer modelling, climatology, and so forth. The composition of atmosphere was first studied by Antoine Lavoisier, Joseph Priestley, and Henry Cavendish in the 18th century. They regarded air as one of the four elements of earth. The trace constituents of air were discovered in the late 19th and early 20th centuries. The most important discovery in 1840 was done by Christian Friedrich Schoenbein who discovered ozone (Rubin, 2001). In the 20th century, atmospheric science moved toward the understanding of how the concentration of trace gases changes with time and how different compounds in the air are created and destroyed by chemical reactions. Two important phenomena, the ozone layer and photochemical smog, were explained by Sydney Chapman and Gordon Dobson, and Haagen-Smit (Haagen-Smit, 1952), respectively. In the 21st century, atmospheric chemistry was studied as an important component of the earth's system. The focus of this century is to study the atmosphere, biosphere, geosphere, the link between climate and atmospheric chemistry, the impact of climate change on ozone and their solution, and so forth (Warneck, 2000).

TABLE 8.1
The percentage distribution of atmospheric gases

S.No	Name	Symbol	Percentage
1	Nitrogen	N_2	78
2	Oxygen	O_2	21
3	Argon	Ar	0.94
4	Hydrogen	H_2	0.00005
5	Carbon dioxide	CO_2	0.03
6	Neon	Ne	0.002
7	Methane	CH_4	0.0002
8	Krypton	Kr	0.0001
9	Nitrous oxide	N_2O	0.00005

The natural and anthropogenic activities change the composition of earth's atmosphere which ultimately affects human health, ecosystem, and crops. The problems occurring in the atmosphere due to changes in atmospheric composition are *acid rain*, *global warming*, *photochemical smog*, *greenhouse gases*, and *ozone layer depletion* (Hegglin et al., 2014). The researchers seek to understand the sources, formation, and possible solution of these problems. They evaluate the government policies of environment to help in problem solution.

8.2 Acid Rain

Acid rain is a form of rainfall with acidic composition such as nitric acid (HNO_3) and sulphuric acid (H_2SO_4) (Dove, 1996). The acidic rain would fall on the ground in the form of hail, fog, rain, snow, as well as dust.

The acid rain is caused by the emission of oxides of sulphur (SO_x) and nitrogen (NO_x) gases emitted in the atmosphere by natural and anthropogenic activities, including volcanic eruption and burning of fossil fuel in the industries, transportation, and generation of electricity (Lal, 2016). The SO_x and NO_x are transported via long distances through wind and air currents. These oxides react with either rainfall water or other chemicals and form HNO_3 and H_2SO_4 before falling on the ground (Shukla et al., 2013). The acid rain is not the problem of those who live near the sources but it is a problem for everyone because the winds blow the air to distant areas across the border (Kara, 2015; Wei et al., 2017).

The composition of acid rain would either be wet or dry depending on the presence of moisture in the atmosphere. The acid rain is caused by the mixing of SO_x and NO_x with snow, rain, hail, and fog in the atmosphere as the wet deposition of acid rain while the deposition of acidic particles and dust in the absence of moisture result in dry deposition. The deposition may occur over vegetation, buildings, and water bodies that can be harmful to human health, infrastructure, plants, and animals (Lal, 2016; Wei, Liu et al., 2017). The next rain washes the accumulated acids on the infrastructure and plant flow in the acidic water resulting in the lowering of waterbody pH (Shukla et al., 2013). The acid rain greatly affects the agriculture productivity by diminishing the mineral content of the soil carrying minerals to the lower layers of soil (McCormick, 2013; Kara, 2015).

8.2.1 Measurement of Acid Rain

The alkalinity and acidity of a substance can be measured through a pH scale range of 0 to 14; a pH of 7.0 on the scale is considered neutral (Wei et al., 2017). The lower the pH of a substance, the more acidic the substance, and the greater the pH above 7.0, the more alkaline the substance. The normal pH of rainwater is 5.6 but may vary between 4.2 and 4.4 (McCormick, 2013; Wei et al., 2017). The slight

TABLE 8.2
The pH of different substances

S.No.	Name	pH	Fate
1	Battery acid	0	Acidic
2	Lemon juice	2	Acidic
3	Soda	2.5	Acidic
4	Acid rain	4.3	Acidic
5	Clean rain	5.6	Acidic
6	Distilled water	7	Neutral
7	Blood	7.4	Alkaline
8	Sea water	8.1	Alkaline
9	Baking soda	9	Alkaline
10	Ammonia	11	Alkaline
11	Bleach	12.6	Alkaline
12	Liquid drain cleaner	14	Alkaline

acidic rain water pH is due to the presence of carbonic acid formed when CO_2 reacts with rain water. The pH of different substances is shown in Table 8.2.

8.3 Ozone

Ozone (O_3) is three oxygen atoms and a highly reactive gas. The concentration of O_3 in the earth's atmosphere (troposphere) changes with height and seasons, depending on atmospheric pollution by NO_2. NO_2 is a brown gas that contributes to urban haze when sufficiently present in the atmosphere. The NO_2 when absorbing sunlight break apart to produce oxygen atoms. The oxygen atoms are highly reactive and combine with O_2 molecules to produce O_3 (Das et al., 2016). O_3 is a toxic gas and powerful oxidizing agent (Godin-Beekmann, 2016). Its high concentration in the earth's atmosphere has a damaging effect on human health, vegetation, and ecosystem. The normal level of O_3 is 10–15 parts per billion (ppb) in the troposphere.

8.3.1 Ozone Layer

The atmospheric region extending up to 15–35 km above the earth's surface comprising a high concentration of ozone molecules is also called the ozonosphere. The layer above the earth's atmosphere (troposphere) extending from 18 km to 50 km is the stratosphere where 90% of ozone is present (Hegglin et al., 2014). The temperature in this layer is very high due to the absorption of solar radiation by ozone layer. The ozone layer blocks all the solar radiation of UV_B and UV_C (wavelength <290 nm) from reaching the earth's surface which could be damaging to living organisms on earth (Prölss, 2012; Bais et al., 2015). Therefore, the ozone layer is a protector of earth from harmful sun rays (Godin-Beekmann, 2016).

8.3.2 Ozone Cycle

The stratospheric ozone is called the good ozone and protects the earth from harmful solar radiation. The stratospheric ozone should not be confused with O_3 present in the earth's atmosphere which is considered a pollutant and causes many respiratory diseases (Hegglin et al., 2014).

O_3 is produced in the stratosphere in a cycle in which one million molecules of O_3 are formed and destroyed everyday naturally. The high energy solar photons of wavelength 242 nm dissociate the

O_2molecule into individual oxygen atoms. This process is called photo dissociation (Madronich and Flocke, 1999). The individual atom will either combine with O_2 molecule to form O_3 or recombine with O_3 molecule to form two O_2 molecules.

$$O_2 + \lambda \rightarrow O + O \tag{8.1}$$

$$O + O_2 \rightarrow O_3 \text{ (Formation)} \tag{8.2}$$

$$O + O_3 \rightarrow 2O_2 \tag{8.3}$$

$$O_3 + \lambda \rightarrow O + O_2 \text{ (destruction)} \tag{8.4}$$

where λ = High-frequency solar radiations.

The O_3 molecule in the presence of the solar radiation splits into O_2 molecules and reactive atomic oxygen (Madronich and Flocke, 1999). In this process, the equal amounts of O_3 are formed and destroyed daily and this process of formation and destruction of O_3 molecules initiated by solar radiation are referred to as 'chapman's reactions.' In this process, the UV radiations are converted into heat energy and increase the stratospheric temperature. It is believed by the scientists that the O_3 layer sustains life on earth because it screens the sun's rays with wavelength 315–280 nm.

8.3.3 Ozone Layer Depletion

The pollution released from anthropogenic activities, including industries and transportation, not only affects the living organisms on earth but also affects the ozone layer in the upper atmosphere (stratosphere) (Jenkin and Clemitshaw, 2000). The stable molecules of chlorofluorocarbon (CFC_4) used in the compressors are harmless on earth but when reaching the stratosphere, they interfere in the ozone cycle causing ozone layer depletion (Jenkin and Clemitshaw, 2000). The high UV radiation breaks the CFC molecules and releases free chlorine which then disturbs the normal ozone cycle (Wilmouth et al., 2018). The thinning of the ozone layer with the passage of time by the ozone-depleting chemicals, including chlorine and bromine, results in small holes over Antarctica. The depletion of ozone is a major environmental problem because the high frequency UV radiation directly pass through the holes and can damage the genetic and immune systems, and cause eye cataracts and skin cancer (Bais et al., 2015, Andrady et al., 2017). The Montreal Protocol, the comprehensive international agreement passed in 1987, was the first agreement enacted to stop the use and production of ozone-depleting substances (Wilmouth et al., 2018). The agreement was proved to be a successful global action signed by 197 countries including Pakistan. As a result of combined international efforts, the recovery of the ozone layer is expected over time (Chipperfield et al., 2015). The ozone-depleting substances used in Pakistan are chlorofluorocarbon, carbon tetra chloride, halons, oxides of nitrogen, and methyl bromide. Destruction of O_3 by CFC can be understood by the following reactions:

$$CCl_2F_2 + \lambda \rightarrow CClF_2 + Cl^* \tag{8.5}$$

$$Cl^* + O_3 \rightarrow ClO^* + O_2 \tag{8.6}$$

$$ClO^* + O \rightarrow Cl + O_2 \tag{8.7}$$

where λ = high frequency solar radiations.

The depletion of ozonecreated holes in the ozone layer and was first discovered in Antarctica in 1970. The depletion below 200 Dobson Units (DU) is considered to be an ozone hole (Hegglin et al., 2014). Ozone holes were also discovered in the arctic region and the depletion is increasing with the rate of 0.5% per year (Chipperfield et al., 2015, Wilmouth et al., 2018). Due to the ozone depletion, the harmful sun rays reach the earth's surface causing the formation of more ozone in the troposphere besides other health effects (Godin-Beekmann, 2016). A study observed that if the emission of ozone-depleting substances were aggressively reduced, then there would be possible recovery of the ozone layer (Andrady et al., 2017).

8.4 Smog

Smog that pollutes the air and greatly affects human health is of two types. In large cities, the burning of coal from factories emits fumes of sulphur dioxide (SO_2), CO_2, and other particulates combined with cold fog, forming industrial smog (also called London smog because it was first recorded in London in 1952, and which lasted for five days). London smog killed about 12,000 people due to respiratory illness (asthma), eye infections, pneumonia, and so forth. The major component of London smog is SO_2 and unburnt particulates released by the combustion of coal when mixed with air moisture forming acidic and caustic soup (Zhou et al., 2015). In November 2016, in the Lahore city of Pakistan, the particulate emissions from rice stubbles and factories resulted in the formation of smog which affected many people (Riaz and Hamid, 2018, Ali et al., 2019). Due to industrial promulgation and hot moist weather, this smog formation was also noted in Beijing and New Delhi (Chen et al., 2013).

The photochemical smog first noted in Los Angeles in 1950 was formed by the burning of fossil fuels and emission of vehicles on hot sunny days. The photochemical smog is also referred to as Los Angeles smog as it was first noted there (Rani et al., 2011). The pollutants responsible for the photochemical smog formation are hydrocarbons and nitric oxide (NO). When sufficient quantity of these pollutants present in the troposphere interact with sunlight during hot summer days, it forms a brown haze, which reduces visibility, and causes shortness of breath, lung infections, sickness, and premature death (Rani et al., 2011, Boningari and Smirniotis, 2016). Ozone is formed as a by-product of smog formation which is corrosive, causing damages to plants, tress, and even paints (Verstraeten et al., 2015). The photochemical smog affects thousands of people every year in Los Angeles, Denver, Vancouver, and cities in Mexico because it is formed in the low basin areas surrounded by mountains where the air gets trapped for days. The two types of smog rarely occur at once because both phenomena require different temperatures and weather conditions. London smog usually occurs in cold weather on cloudy nights and the main pollutant is SO_2 while photochemical smog requires hot, moist weather conditions in areas near mountains during the day time and the major pollutants are NO and hydrocarbons (Boningari and Smirniotis, 2016). The condition becomes worse when a layer of hot air covers a layer of cold air and the air becomes stagnant for days until the inversion layer break in the night by the circulation of air (Ma et al., 2012).

8.4.1 Photochemical Smog

The formation of photochemical smog is a complex chemical process that involves the formation of numerous pollutants. The fossil fuels (gasoline) burning in automobiles emitted NO, NO_2, and hydrocarbons as byproducts in the air. In the presence of sunlight, the NO converts into NO_2 which is a brown gas and causes urban haze (Ball, 2014). The problem becomes more serious when NO_2 in the presence of sunlight breaks apart and releases atomic oxygen (O^*). O^* combines with O_2 molecules to form O_3 which is another toxic pollutant and strong oxidizing agent. Thus, the automobile emissions convert the gaseous O_2 into O_3. The normal ozone level in the air is 10 ppb. O^* and O_3 can also react with water forming hydroxyl radicals (OH^*). OH^* oxidizes hydrocarbons to form aldehydes which then get oxidized to form peroxyacids which is a pollutant and causes numerous health effects. The peroxyacids can combine with NO_2 forming *peroxyacetyle nitrate* (PAN) which is the end product of photochemical reactions (Rani et al., 2011). The components of photochemical smog include different types of pollutants, that is, NO, O_3, NO_2, H_2SO_4, HNO_3, N_2O (nitrous acid), and PAN which are responsible for many adverse health and environmental effects (Ball, 2014). The chemical transformation involved in photochemical smog formation is illustrated in the reactions (5.1–5.8).

$$NO_2 + \lambda \rightarrow NO + \mathbf{O^*} \quad (8.8)$$

$$\mathbf{O^*} + \mathbf{O_2} + \mathbf{M} \rightarrow O_3 + \mathbf{M} \quad (8.9)$$

$$\mathbf{NO} + \mathbf{O_3} \rightarrow \mathbf{NO_2} + O_2 \quad (8.10)$$

$$\mathbf{O^*} + \mathbf{H_2O} \rightarrow 2OH^* \quad (8.11)$$

$$RH + OH^* \rightarrow H_2O + \mathbf{R^*}$$
$$\mathbf{R^*} + O_2 \rightarrow RO_2^* \text{ (fast reaction)} \quad (8.12)$$

$$RO_2^* + NO \rightarrow NO_2 + RO^*$$
$$RO^* + O_2 \rightarrow RCHO + HO_2^* \text{ (fast reaction)} \quad (8.13)$$

$$RCHO + OH^* \rightarrow RCO^* + H_2O$$
$$RCO^* + O_2 \rightarrow \mathbf{RC(O)O_2^*} \text{ (fast reaction)} \quad (8.14)$$

RCHO = aldehyde
RH = any Hydrocarbon i.e. (CH_3CH_3, $CH_3CH_2CH_3$)
RCO = acyl radicle
$RC(O)O_2$ = acylperoxy radicle
$RC(O)_2NO_2$ = acylperoxynitrate

$$\mathbf{RC(O)O_2^*} + \mathbf{NO_2} \rightarrow \mathbf{RC(O)O_2NO_2} \quad (8.15)$$

If R = Methyl (CH_3) group then the product is Peroxyacetyl nitrate ($CH_3COOONO_2$) (PAN)

8.5 Types of Atmosphere

The planet is surrounded by a gaseous area and is divided into several concentric spherical layers separated by a thin transition layer. The boundary where the gases disperse in the space is 1,000 km above sea level. The atmosphere is divided into five layers, the troposphere, stratosphere, mesosphere, thermosphere, and exosphere, and the division is called atmospheric stratification (Park and Park, 2006). The major parts (99%) of the atmospheric mass exist in the lower strata about 40 km from the earth's surface.

8.5.1 Troposphere

Troposphere is the region of 'turn' meaning mixing, and is the most important and first lower layer of atmosphere; the height extends up to 18 km from the earth's surface. The average thickness of this layer is about 12 km. All the gaseous mass, living organisms, and plants are present in these strata (Bergman et al., 2015). The normal average temperature is 15 °C regulated by the presence of gases and water vapours, suitable for the growth of living organisms and weather phenomena. The contents of water vapours and atmospheric temperature decrease with increasing altitude with 1 °C for 100 m rise in elevation resulting in the formation of clouds in the upper region of troposphere (Guha et al., 2017). Troposphere contains 99% of water vapours and some other gases like CO_2 which absorb solar radiations (UV_a) from the sun and emit back in the form of heat, which is called the greenhouse effect. If the greenhouse gases and water vapours were absent, life would not be possible because of freezing (Vargin et al., 2015). The concentration of water vapours changes with latitudinal position which is as high as 3% in the tropics while decreasing toward the polar regions. All weather phenomena, including storms, rainfall, lightening, clouds, and thundering, occur in the troposphere. The layer above the troposphere is the stratosphere and the transition thin belt of about two miles thickness between troposphere and stratosphere is the 'tropopause' where the temperature reaches to −51 °C (Guha et al., 2017).

8.5.2 Stratosphere

The stratosphere is the second layer above the troposphere and below the mesosphere and extends up to 50 km with thickness of about 35 km. There is a great temperature variation between the upper and lower layers of the stratosphere (Kodera et al., 2017). In contrast to the troposphere, the temperature increases with increasing altitude from −51 °C in the lower region of the stratosphere to −15 °C near the mesosphere

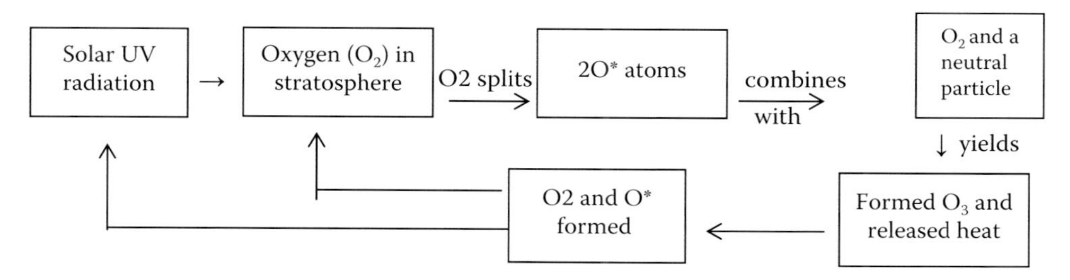

FIGURE 8.1 Ozone cycle.

(Kodera et al., 2017). The stratosphere is layered with cooler layers near the troposphere and warmer layers near the mesosphere because the important layer of ozone is present with a height of 20–34 km (Butler et al., 2015; Das et al., 2016). The presence of high concentrations of O_3 which absorbs solar radiations of 290–320 nm increases the temperature of the stratosphere (Figure 8.1) and protects life on earth (Godin-Beekmann, 2016). The solar radiations that are blocked by the ozone layer are very dangerous and if they reached the earth's surface would damage plant life and the nucleic acid of human cells would result in dramatic increase of cancer (Andrady et al., 2017). In comparison to the troposphere, most of the gases and water vapours are rarely present in this layer (Vargin et al., 2015). The upper layer of the stratosphere is the mesosphere and the transition layer between stratosphere and mesosphere is the stratopause where the temperature remains constant (Davis et al., 2017).

8.5.3 Mesosphere

The mesosphere is a Greek word derived from 'Mesos' meaning 'middle.' The mesosphere is the third layer of atmosphere which extends from the stratopause with an altitude of 50 km up to 85 km in the atmosphere and with characteristics of very low temperatures, lower than the troposphere and stratosphere (Lehmacher et al., 2018). This layer is the coldest part of the atmosphere and has a temperature of −85–120 °C (Wickwar et al., 2016). The mesosphere contains 0.1% of earth air mass and the water vapours are negligible. The photochemical species present in this layer are O_2, O_3, OH^*, H_2O_2, CH_4, O^*, HO_2, and some metals, including Na, K, Fe, and Ca, found by the ablation of meteors (Plane et al., 2016). A study conducted recently found that CO_2 is also present in this layer and increases in the troposphere. CO_2 causes global warming in the troposphere with a global cooling effect in space (Qian et al., 2017). The shortest UV radiations are absorbed by N_2 and O_2. The turbulence in the atmosphere thoroughly mixes the gases and different atoms together. The light gases escape to the universe due to the separation from earth's gravity. The weather balloons and other aircraft cannot reach this layer therefore scientists use sounding rockets to study this layer. The mesosphere has much less water vapours which make the highest clouds from the earth surface to the North or South Poles named as *noctilucent clouds* or polar mesospheric clouds (Nielsen et al., 2018). These clouds are made up of frozen water or crystals of ice (Baumgarten et al., 2018a, b). These clouds can be observed on earth in summer just after sunset and look like electric glows in blue and white colours (Baumgarten et al., 2018a, b). In recent years, these clouds can be seen very near to the poles due to global climate change and global warming due to human activities (Lübken et al., 2018).

8.5.4 Thermosphere

The layer above the mesosphere is the thermosphere and the layer between the mesosphere and thermosphere is the mesopause. The thermosphere is the fourth layer of atmosphere and extends from about 90 km to 1,000 km. The temperature of this layer is very high and sharply increases in the lower thermosphere below 200–300 km while increasing fairly steadily above this height (Ridley et al., 2006). The temperature is greatly influenced by solar radiation (Bates and Patterson, 1961). In the night time, the thermosphere has a temperature of about 200 °C while it reaches to 500 °C in the day time when the sun is active. The temperature of the thermosphere ranges from 200–2,000 °C or above.

The high energy X-ray photons and energetic UV radiation break apart the molecules, therefore atomic nitrogen, atomic oxygen, and helium are present in this layer (Bates and Patterson, 1961, Victor and Constantinides, 1979). Much of the sun radiation is absorbed in the thermosphere when the sun is active and the thermosphere gets hotter and expands. Many of the satellites orbit in this layer. The high energy solar photons emit electrons from the gas particles creating electrically charged ions and atoms. Therefore, this layer is also called ionosphere (Bates and Patterson, 1961; Ridley et al., 2006). In some parts of the thermosphere, the moving ions collide with the neutral gases and atoms produce powerful electrical currents. This layer is important for the radio engineer because it reflects radio waves. The aurora occurs in the thermosphere when the photons, electrons, and ions from space collide with the molecules and atoms of the thermosphere, excited to the higher state (Sinnhuber et al., 2012). The excited atoms release the excess energy in the form of photons of light therefore we see the colourful display of aurora (Seppälä et al., 2007).

8.5.5 Exosphere

The layer above the thermosphere is the exosphere and the boundary between the thermosphere and exosphere is thermopause. The exosphere is the uppermost layer of the atmosphere above the thermosphere. The altitude of the exosphere varies with sun radiation. When the sun radiation is active in the sunspot cycle, the thermosphere heats up and puffs up, which extends the thermopause to 1,000 km; when the sun is less active, the thermopause height reaches to 500 km. Some scientists considered the exosphere as part of outer space and the thermosphere the uppermost part of the atmosphere. In the exosphere, the atoms and ions rarely collide as in the thermosphere. They move in ballistic trajectory and curve back to the lower atmosphere due to the pull of gravity (Prölss, 2012). However, some of the particles do not return and move to space. It can be said that a small portion of the atmosphere leaks toward space each year.

8.6 Conclusion

Atmospheric chemistry is the branch of science which studies the chemistry of atmosphere and other planets from clean to polluted regions. The atmosphere is a mixture of gases in which nitrogen and oxygen are predominantly present as compared to other gases. These gases keep a normal temperature of earth's atmosphere of about 15 °C, which is important for the presence of life on earth. However the anthropogenic activities including fossil fuel combustion in automobiles and industrial emission of different gases (CO_2, SO_2, NO_2, NOM, etc.) change the composition of atmosphere, causing environmental pollution. The processes occurring in the atmosphere due to anthropogenic emission are acid rain, photochemical smog formation, and ozone layer depletion, which affect human health and the whole ecosystem. These gases not only affect the earth's atmosphere but when reaching the upper atmosphere cause ozone layer depletion. The atmosphere is divided into five layers on the basis of altitude from the earth surface, namely, troposphere, stratosphere, mesosphere, thermosphere, and exosphere. Each layer has specific features but the gaseous emission from anthropogenic activities is damaging the characteristics of these layers. Therefore, combined efforts are needed at the international level to overcome emerging environmental problems.

REFERENCES

Ali Y, Razi M, De Felice F, Sabir M, Petrillo A (2019) A VIKOR based approach for assessing the social, environmental and economic effects of 'smog' on human health. *Sci Total Environ* 650:2897–2905.

Andrady A, Aucamp PJ, Austin AT, Bais AF, Ballare CL, Barnes PW, Bernhard GH, Bjoern LO, Bornman JF, Congdon N (2017) Environmental effects of ozone depletion and its interactions with climate change: Progress report, 2016. *Photochem Photobiol Sci: Off J Eur Photochem Assoc Eur Soc Photobiol* 16(2):107.

Andreae MO, Crutzen PJ (1997) Atmospheric aerosols: Biogeochemical sources and role in atmospheric chemistry. *Science* 276(5315):1052–1058.

Bais A, McKenzie R, Bernhard G, Aucamp P, Ilyas M, Madronich S, Tourpali K (2015) Ozone depletion and climate change: impacts on UV radiation. *Photochem Photobiol Sci* 14(1):19–52.

Ball A (2014) *Air pollution, foetal mortality, and long-term health: Evidence from the Great London Smog.* MPRA Paper 63229, University Library of Munich, Germany.

Bates D, Patterson T (1961) Hydrogen atoms and ions in the thermosphere and exosphere. *Planet Space Sci* 5(4):257–273.

Baumgarten G, Fiedler J, Fritts D, Gerding M, Lübken F-J, Stober G (2018a) Noctilucent clouds and their link to atmospheric dynamics. EGU General Assembly Conference Abstracts. https://mpra.ub.uni-muenchen.de/63229/.

Baumgarten G, Fiedler J, Stober G, Gerding M, Luebken F-J, Fritts D (2018b) Small scale atmospheric dynamics revealed by noctilucent clouds. 42nd COSPAR Scientific Assembly.

Bergman J, Pfister L, Yang Q (2015) Identifying robust transport features of the upper tropical troposphere. *J Geophys Res: Atmos* 120(14):6758–6776.

Boningari T, Smirniotis PG (2016) Impact of nitrogen oxides on the environment and human health: Mn-based materials for the NOx abatement. *Curr Opin Chem Eng* 13:133–141.

Butler AH, Seidel DJ, Hardiman SC, Butchart N, Birner T, Match A (2015) Defining sudden stratospheric warmings. *Bull Am Meteorol Soc* 96(11):1913–1928.

Chen R, Zhao Z, Kan H (2013) Heavy smog and hospital visits in Beijing, China. *Am J Respir Crit Care Med* 188(9):1170–1171.

Chipperfield MP, Dhomse SS, Feng W, McKenzie R, Velders GJ, Pyle JA (2015) Quantifying the ozone and ultraviolet benefits already achieved by the Montreal Protocol. *Nat Commun* 6:7233.

Das SS, Ratnam MV, Uma KN, Subrahmanyam KV, Girach IA, Patra AK, Aneesh S, Suneeth KV, Kumar KK, Kesarkar AP (2016) Influence of tropical cyclones on tropospheric ozone: possible implications. *Atmos Chem Phys* 16(8):4837–4847.

Davis SM, Hegglin MI, Fujiwara M, Dragani R, Harada Y, Kobayashi C, Long C, Manney GL, Nash ER, Potter GL (2017) Assessment of upper tropospheric and stratospheric water vapor and ozone in re-analyses as part of S-RIP. *Atmos Chem Phys* 17(20):12743–12778.

Dove J (1996) Student teacher understanding of the greenhouse effect, ozone layer depletion and acid rain. *Environ Educ Res* 2(1):89–100.

Finlayson-Pitts BJ, Pitts JN Jr (1986) Atmospheric chemistry. Fundamentals and experimental techniques. https://www.osti.gov/biblio/6379212-atmospheric-chemistry-fundamentals-experimental-techniques.

Godin-Beekmann S (2016) Stratospheric and tropospheric ozone. *International Encyclopedia of Geography: People, the Earth, Environment and Technology: People, the Earth, Environment and Technology. John Wiley & Sons*, pp. 1–6.

Guha BK, Chakraborty R, Saha U, Maitra A (2017) Tropopause height characteristics associated with ozone and stratospheric moistening during intense convective activity over Indian sub-continent. *Glob Planet Chang* 158:1–12.

Haagen-Smit AJ (1952) Chemistry and physiology of Los Angeles smog. *Ind Eng Chem* 44(6):1342–1346.

Hegglin MI, Mack McFarland DWF, Montzka SA, Nash ER (2014) Scientific assessment of ozone depletion. *World Meteorological Organization, Geneva, Switzerland,* pp. 1–67.

Jacob, DJ (1999) *Introduction to atmospheric chemistry*, Princeton University Press.

Jenkin ME, Clemitshaw KC (2000) Ozone and other secondary photochemical pollutants: chemical processes governing their formation in the planetary boundary layer. *Atmos Environ* 34(16):2499–2527.

Kara F (2015) Knowledge level of prospective science teachers regarding formation and effects of acid rains on the environment and organisms. *Appl Sci Eng Prog* 5(4):128–131.

Kodera K, Eguchi N, Mukougawa H, Nasuno T, Hirooka T (2017) Stratospheric tropical warming event and its impact on the polar and tropical troposphere. *Atmos Chem Phys* 17: 615–625.

Lal N (2016) Effects of acid rain on plant growth and development. e-*J Sci Technol* 11:85–101.

Lehmacher GA, Larsen MF, Collins RL, Barjatya A, Strelnikov B (2018) On the short-term variability of turbulence and temperature in the winter mesosphere. *Ann Geophys* 36 (4):1099–1116.

Lübken FJ, Berger U, Baumgarten G (2018) On the anthropogenic impact on long-term evolution of noctilucent clouds. *Geophy Res Lett* 45(13):6681–6689.

Ma J, Xu X, Zhao C, Yan P (2012) A review of atmospheric chemistry research in China: Photochemical smog, haze pollution, and gas-aerosol interactions. *Adv Atmos Sci* 29(5):1006–1026.

Madronich S, Flocke S (1999) The role of solar radiation in atmospheric chemistry. *Environmental photochemistry, Springer,* 1–26.

McCormick J (2013) *Acid earth: The global threat of CCID pollution.* Routledge.

Nielsen K, Collins R, Negale M, Williams B, Otero S (2018) Atmospheric gravity wave propagation across the stratosphere, mesosphere, and thermosphere over Alaska, and the role of thermal gradients in the polar regions on vertical gravity wave propagation. *42nd COSPAR Scientific Assembly.*

Park M-S, Park S-U (2006) Effects of topographical slope angle and atmospheric stratification on surface-layer turbulence. *Boundary-Layer Meteor* 118(3):613–633.

Plane J, Marsh D, Höffner J, Janches D, Dawkins E, Gomez-Martin JC, Bones D, Feng W, Chipperfield M (2016) Metals in the mesosphere: chemistry and change. *41st COSPAR Scientific Assembly.*

Prölss G (2012) Physics of the earth's space environment: An introduction. Springer Science & Business Media.

Qian L, Burns AG, Solomon SC, Wang W (2017) Carbon dioxide trends in the mesosphere and lower thermosphere. *J Geophys Res: Space Phys* 122(4):4474–4488.

Rani B, Singh U, Chuhan A, Sharma D, Maheshwari R (2011) Photochemical smog pollution and its mitigation measures. *J Ad Sci Res* 2(4):28–33.

Riaz R, Hamid K (2018) Existing smog in Lahore, Pakistan: An alarming public health concern. *Cureus* 10(1):e2111.

Ridley A, Deng Y, Toth G (2006) The global ionosphere–thermosphere model. *J Atmos Solar-Terr Phys* 68(8):839–864.

Rubin MB (2001) The history of ozone: The Schönbein period, 1839–1868. *Bull Hist Chem* 26(1):40–56.

Seinfeld JH, Pandis SN (2016) *Atmospheric chemistry and physics: From air pollution to climate change.* John Wiley & Sons.

Seppälä A, Clilverd MA, Rodger CJ (2007) NOx enhancements in the middle atmosphere during 2003–2004 polar winter: Relative significance of solar proton events and the aurora as a source. *J Geophys Res: Atmos* 112(D23). https://doi.org/10.1029/2006JD008326.

Shukla J, Sundar S, Naresh R (2013) Modeling and analysis of the acid rain formation due to precipitation and its effect on plant species. *Nat Resour Model* 26(1):53–65.

Sinnhuber M, Nieder H, Wieters N (2012) Energetic particle precipitation and the chemistry of the mesosphere/lower thermosphere. *Surv Geophys* 33(6):1281–1334.

Vallero, DA (2014) *Fundamentals of Air Pollution.* Academic Press.

Vargin PN, Volodin EM, Karpechko AYE, Pogoreltsev AI (2015) Stratosphere-troposphere interactions. *Her Rus Acad Sci* 85(1):56–63.

Verstraeten WW, Neu JL, Williams JE, Bowman KW, Worden JR, Boersma KF (2015) Rapid increases in tropospheric ozone production and export from China. *Nat Geosci* 8(9):690.

Victor G, Constantinides E (1979) Double photoionization and doubly charged ions in the thermosphere. *Geophys Res Lett* 6:519–522.

Warneck P (2000) *Chemistry of the Natural Atmosphere.* 2nd edn. Academic Press, San Diego.

Wei H, Liu W, Zhang J, Qin Z (2017) Effects of simulated acid rain on soil fauna community composition and their ecological niches. *Environ Poll* 220:460–468.

Wickwar VB, Sox L, Emerick MT, Herron JP (2016) Seasonal temperatures from the upper mesosphere to the lower thermosphere obtained with the large, Alo-usu, Rayleigh LiDAR. 2016 Joint CEDAR-GEM Workshop, Santa Fe, NM.

Wilmouth DM, Salawitch RJ, Canty TP (2018) Stratospheric ozone depletion and recovery. *Green Chemistry, Elsevier,* 177–209.

Zhou M, He G, Fan M, Wang Z, Liu Y, Ma J, Ma Z, Liu J, Liu Y, Wang L (2015) Smog episodes, fine particulate pollution and mortality in China. *Environ Res* 136:396–404.

9

Ocean as the Driver of the Global Carbon Cycle

Nayab Gul
Climate Change Centre, The University of Agriculture Peshawar

9.1 Introduction

The ocean plays an important part in the carbon cycle which takes up carbon dioxide from the atmosphere through chemical and biological processes. Therefore, ocean is called a 'carbon sink'. Global warming is caused by the accumulation of CO_2 and other greenhouse gases in the earth atmosphere. Thus, ocean influences the climate by storing and absorbing CO_2 from the air. The natural sinks of carbon are oceans, plants, and soil (Bruhwiler et al., 2018). Carbon spends about 5 years in the air and 10 years in terrestrial vegetation, while in deep water and ocean, it spends about 380 years. Carbon after absorption in the ocean can be locked up for millions of years in the form of ocean sediments or fossil fuels. When the amount of CO_2 increases in the atmosphere through natural and anthropogenic activities, the ocean absorbs high concentration of atmospheric CO_2. CO_2 react with water-forming carbonic acid (H_2CO_3) which causes acidification of oceans.

CO_2 is highly different from other gasses due to its high solubility in water. Around 98.5% of the preindustrial CO_2 is embedded in the ocean while 1.5% in the atmosphere as compared to Oxygen (O_2) which is 99.4% in atmosphere and less than 0.6% in the ocean. The high amount of CO_2 is due to its high water solubility which is 30 times that of O_2 (Bruhwiler et al., 2018). The hydrolysis reaction of CO_2 forms H_2CO_3, carbonate, and bicarbonate in oceans. The solubility of dissolved inorganic carbon (DIC) is greatly influenced by biological pump (biological processes transport carbon in ocean) and solubility pump (physical and chemical processes that transport carbon to the interior ocean) (Figure 9.1) which reduce CO_2 in the atmosphere (Sarmiento, 2008). The solubility pump is driven by two processes namely solubility and thermohaline circulation. The solubility of CO_2 is inversely proportional with water temperature. At high latitude, the formation of deep water due to density gradient results in ocean circulation called thermohline circulation (Sigman et al., 2010).

Recent data suggest that 25–30% of CO_2 released in the atmosphere due to anthropogenic activities are removed by oceans (Landschuetzer et al., 2016). The largest inventory of carbon is in the ocean which exerts a major control over CO_2 in the atmosphere and helps reduce global warming. The anthropogenic activities, including fossil fuel burning, land use change (deforestation), and cement production increased the emission of CO_2 in the atmosphere in which 70% emission is due to fossil fuel burning. Before industrial revolution, the partial pressure ($p^{CO}{}_2$) of CO_2 increased from 280 ppm to 373 ppm in 2002 (Marinov and Sarmiento, 2004). Pressure ($p^{CO}{}_2$) and global climate are correlated with glacial cycle that $p^{CO}{}_2$ was lower in ice ages (Sigman et al., 2010). The evidence suggest that the southern ocean releases the deep sequestered CO_2 in the atmosphere but this leak was stopped in the ice age which led to CO_2 sequestration in the oceans (Sigman et al., 2010).

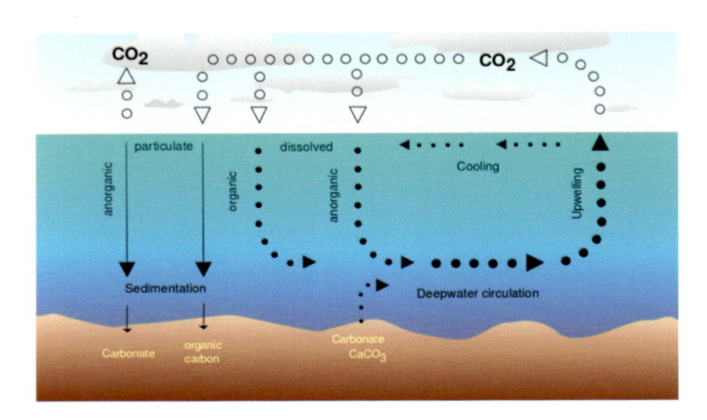

FIGURE 9.1 A diagram showing solubility of CO_2 (physical and chemical) and a biological pump of CO_2 (Grobe, 2006).

9.2 Carbon Cycle

Carbon is the backbone of life on earth. It provides food and fuel energy, and regulates the earth's temperature. The carbon cycle involves chemical, mechanical, and biological processes that transfer carbon between reservoirs (Figure 9.2). The carbon movement is a complex process through land, air, water, and living organisms. Carbon geologically transfers very slowly in comparison to the living organisms. In the earth system, the reservoir of carbon are referred as 'stocks' or 'pools' and the transfer between these stocks is called 'fluxes'. The fluxes of carbon are sensitive to climate. When the temperature rises, the water solubility of CO_2 decreases, resulting in more atmospheric CO_2 which leads to global warming. This response to climate change is referred as 'positive climate feedback'.

Carbon remains stored in the atmosphere, in water reservoirs, land, ocean sediments, and earth's interior for long periods of time. It is present in the atmosphere in the form of CO_2, essential for the process of photosynthesis. The exchanges of carbon occur between atmosphere and water reservoirs and affect each other reciprocally. CO_2 from the atmosphere is absorbed by the water through biological and solubility pump forming ionic and non-ionic species, collectively called DIC. The DIC present in the form of CO_2 (aqueous), H_2CO_3, carbonates, and bicarbonates which interact with water as follows (9.1–9.4):

$$\text{(atmospheric)}CO_2 \leftrightarrow CO_2\text{(aqueous)} \tag{9.1}$$

$$\text{(atmospheric)}CO_2 + H_2O \leftrightarrow H_2CO_3\text{(carbonic acid)} \tag{9.2}$$

$$H_2CO_3 \leftrightarrow H^+ + HCO_3^-\ \text{(bicarbonate ion)} \tag{9.3}$$

$$HCO^{-3} \leftrightarrow H^+ + CO_3^-\ \text{(carbonate ion)} \tag{9.4}$$

The evidence suggests that more than 90% of carbon is present in the form HCO_3^- in oceans. These bicarbonate ions form calcium carbonate ($CaCO_3$) when combine with calcium. $CaCO_3$ is the major component of organism's shells in the marine environment. These shells ultimately form ocean sediments and over geologic time, form the limestone which makes the largest reservoir of carbon on the earth.

The carbon is stored in the soil through decomposition of living organisms and weathering of rocks and minerals. Carbon gets leached to the water reservoir by the surface runoff and to the deeper underground. The fossil fuel is the *non-renewable* resource because it takes millions of years to form anaerobically by the decomposition of plants remains therefore their use exceeds their formation. Volcanic eruption is another way that brings carbon from the interior of the earth to the atmosphere in

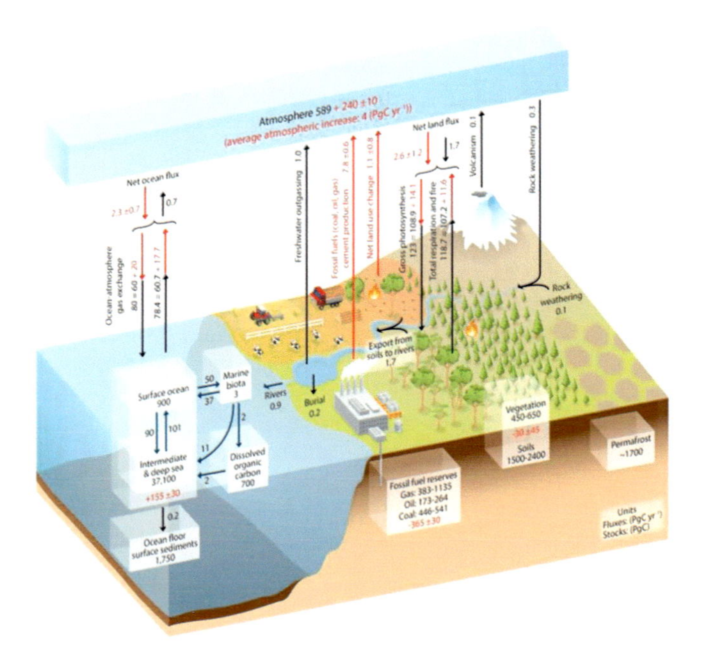

FIGURE 9.2 A diagram showing the carbon cycle on land and water. The number in the box represents Peta grams of carbon (Pg C). Arrows represent carbon exchange in Pg C/year. Black arrows represent preindustrial annual carbon exchange. Red arrows represent average annual anthropogenic fluxes from 2000–2009. Red numbers represent collective changes of anthropogenic carbon in an industrial period (Bruhwiler et al., 2018).

the form of CO_2. Carbon from the ocean sediments is taken within the earth by the process of *subduction.* Besides fossil fuel burning, other anthropogenic activities include adding CO_2 to the atmosphere through animal husbandry process, farming practices, respiration, and methane production.

9.2.1 Changes in the Carbon Cycle

Human activities make changes to the normal cycle of carbon. The burning of fossil fuels, deforestation, farming practices, and use of limestone transfer a significant amount of CO_2 to the atmosphere. Consequently the amount of CO_2 increases which is the main culprit of global warming. The major part of atmospheric CO_2 is absorbed by the ocean which lowers the pH of water causing ocean acidification. The ocean acidification greatly affects the marine organisms including corals, snails, crabs, fishes, and so forth. It is believed that about 1/3 (2 gigatons of CO_2/year) of atmospheric carbon is absorbed by the oceans (Takahashi et al., 2002). But the rate of this absorption in future is uncertain. The solubility of carbon decreases with increasing temperature or global warming which is slowing the ocean response to CO_2 absorption. Ocean warming also increases ocean stratification which isolates the upper layer of water from the deeper layers (Bryden et al., 2005). This stratification will eventually affect thrermohaline circulation (slowing) and will decrease the transfer of DIC into the deep water. However, oceans play an important role in decreasing climate change (Orr et al., 2005).

9.3 Biological Pump

The biological pump, also called soft-tissue pump, is the sequestration of carbon from the atmosphere in the ocean sediments. By this process, the organic matter moves from the surface to the interior of the ocean (Couldrey et al., 2019). The soft tissue pump can be divided into three phases (Figure 9.1).

In the primary phase, phytoplanktons use dissolved CO_2, Phosphorus (P), Nitrogen (N) and trace elements (Zinc, Barium, Iron) with a ratio of (106 C:16 N:1 P) in the surface water (where sunlight reaches) and make carbohydrates, proteins, and lipids through photosynthesis (Passow and De La Rocha, 2006).

$$CO_2 + H_2O + \text{light} \rightarrow CH_2O + O_2 \quad (9.5)$$

After fixation of carbon in hard and soft tissues, the organism may recycle in the surface water layer as part of the nutrient cycle or may enter to the second phase when dead. Some parts of the organic matter, about 30–35%, sink to the bottom layer as particulates. These particulate form aggregates in the ocean bottom; are decomposed by bacteria; and release CO_2, silicon, nitrogen, and phosphorus-bearing waste back into solution causing enrichment of DIC in the ocean floor (Couldrey et al., 2019). The marine phytoplanktons fixed about 50–60 Pg C/year despite the fact that these phytoplankton make less than 1% of the photosynthetic biomass of the earth. The majority (80%) of fixation occurs in the open ocean area which contributes to carbon fixation (Passow and De La Rocha, 2006).

The water that has recently been ventilated shows lower concentrations of DIC as compared to older water because the soft tissue pump has caused carbon to accumulate with water mass age. The third phase is the remineralization of sediments which convert the carbon back in the dissolved inorganic carbon (DIC) used in the primary production. Carbon remains there for millions of years and this is that sequestration which removes CO_2 from the atmosphere.

In the surface zone, calcifying planktons including coccolithophores and foraminifera produce calcium carbonates shells (Couldrey et al., 2019). When these shells sink in the bottom, some part gets dissolved back in the water solution, increasing the DIC pool, while some parts sink to the bottom and buried through sedimentation. The addition of carbon as $CaCO_3$ is referred as 'carbonate pump'. Like soft tissue pump carbonate pump also add carbon to water mass over time causing the enrichment of DIC in older water (Pilson, 2012).

$$CO_2 + H_2O \rightarrow H_2CO_3 \rightarrow H^+ + HCO_3^- \quad (9.6)$$

$$Ca^{2+} + 2HCO_3^- \rightarrow CaCO_3 + CO_2 + H_2O \quad (9.7)$$

9.4 Ocean Acidification

Ocean acidification is the decrease in ocean water pH by the uptake of CO_2 from the atmosphere. The ocean water is slightly basic having a pH more than 7 (neutral). The acidification of ocean is not the acidic condition of water but the transition from neutral pH level to acidic level (Caldeira and Wickett, 2003). Oceans dissolve about 30–40% of CO_2 from the atmosphere emitted from human activities and natural processes (Hall-Spencer et al., 2008). CO_2 reacts with water forming H_2CO_3 which then dissociates into HCO_3^- and H^+ which lowers the pH causing ocean acidity. The surface ocean pH has been decreased from 8.25 to 8.14 units between 1751 and 1996 which represents a 30% increase in the concentration of H^+ in the world's oceans and further increase was observed in 2008 (Millero, 1995). CO_2 dissolved in sea water releases H^+ which lowers the pH of sea water as follows:

$$CO_2 + H_2O \rightarrow H_2CO_3 \rightarrow HCO_3^- + H^+ \rightarrow CO_3^- + 2H^+ \quad (9.8)$$

It is estimated that since the industrial revolution, the absorption of about 1/3 of atmospheric CO_2 lowers the pH of surface ocean by more than 0.1 units (Caldeira and Wickett, 2003). It is expected that the surface ocean pH will further decrease by more than 0.3–0.5 units in 2100 because oceans absorb more CO_2. The lowering of ocean pH depends on the emission and mitigation pathway adopted by humans (Mora et al., 2013).

9.5 Impacts on Oceanic Organisms

The increasing acidity has greatly affected the marine ecosystem. The potential consequences of ocean acidity include weakness of the immune response of organisms, depressing metabolic rate and coral bleaching. Coral bleaching is a phenomenon in which the coral polyps release the algae, which lives inside their tissues in an endosymbiotic relationship which is crucial for the health of coral reef (Anthony et al., 2008). The increasing concentration of Hydrogen ions results in conversion of CO_3^- into HCO_3^-. The dissolution of CO_3^- for long period of time makes difficult for calcifying organisms such as planktons and corals to form biogenic $CaCO_3$ and such structures vulnerable to dissolution. The aragonite and calcite are at normal condition when the CO_3^- ions are at supersaturating condition. Ocean acidification also threatens the marine food chain (Dean, 2009). The science academies of Inter Academy Panel issued a statement that the level of ocean acidity needs to be reduced by 50% by 2050 in comparison to 1990. The origin of ocean acidification is largely anthropogenic that has occurred previously in earth's history. The notable examples is Paleocene-Eocene Thermal Maximum (PETM) when massive amount of CO_2 absorbed by the oceans from the atmosphere about 56 million years ago which led to carbonate dissolution in all ocean basins (IAP, 2009). The ocean acidity and anthropogenic climate change are called 'evil twins' of global warming because both arise due to CO_2 emission (Hull, 2016). The absorption of CO_2 by the oceans mitigates effects of climate change caused by anthropogenic emission of CO^2. The organisms, for example, foraminifera, shellfish, coralline algae, coccolithophore algae, and corals, experience reduced calcification when the ocean pH falls. Calcification is the process in which the marine organisms produce plates and shells out of $CaCO_3$. The aragonite is a specific form of $CaCO_3$ which is used by the marine organisms' for the production of shells. When CO_3^- is at supersaturated state, carbonate minerals (calcite, aragonite) are stable. However when the ocean pH lowers due to absorption of more CO_2, the concentration of CO_3^- increases, making the structures of $CaCO_3$ vulnerable to dissolution.

Some studies found that elevated CO_2 concentration in oceans has increased the photosynthetic and calcification rate of coccolithophorids (Mollica et al., 2018). A study in 2010 published by Stony Brook University showed that some areas are overharvested while some are being restored. Some species of shellfish have been completely extinct due to ocean acidification. It is evident from the study that the calcifying organisms are adversely affected. An experiment conducted on the larvae of sea star and brittle star, that at lower pH of 0.2 to 0.4 less than 0.1%, larvae survived for more than eight days. The fluid in the internal compartment of coral is important for the development of exoskeleton. When the level of aragonite in sea water is normal, the corals will grow aragonite crystals in their internal compartment but due to acidity the lower level of aragonite in sea water will be too low to make exoskeleton and the coral will work hard to maintain the right balance (Cohen and Holcomb, 2009). A study conducted in 2018 observed that the growth of corals to develop dense exoskeleton was greatly affected by ocean acidification. It is estimated that the density of corals could be reduced by 20% by the end of this century (Mollica et al., 2018). It was observed that CO_2 bubbles from the sea floor is changing the pH of water. The corals near CO_2 seep show different responses. The acidity affects the diversity and skeleton development of coral (Fabricius et al., 2011; Hull, 2016).

The ocean acidification affects the reproduction reef fish which normally spawn in late spring and fall. The acidic condition slows down the development of larvae hence increases the mortality rate. The evidence shows that larval and embryo stage fishes have not matured enough to show the normal level of acid base relation which is present in adults and hypoxia will result in which oxygen is driven off the hemoglobin. This will ultimately result in impaired growth and will increase mortality in acidic environment compared to normal level of acid in marine water. The fish larvae are sensitive to ocean water pH (Cripps et al., 2011). In acidic condition, the larval fish have lower survival rate and important species of fish will become extinct. The fish spend much of their energy to keep their osmoregulatory function in check. The cardiac failure in fishes and neurotransmitter interference between predator and prey fishes increases their mortality rate. Besides these, the increased level of CO_2 in marine water depresses the immune system of blue mussels and weakens the metabolic rate jumbo squid. Moreover, in acidic water, it becomes harder for juvenile clownfish to recognize the smell of non-predator and sound of predators.

Thus, acidity impacts the acoustic properties of water and the communication between animals. Another impact includes the accumulation of toxins in small organisms which leads to neurotoxic shellfish poisoning, amnesic shellfish poisoning, and paralytic shellfish poisoning (Orr et al., 2005).

The acidification impacts on human industry because the acidity affects the development of calcifying organisms by dissolving their shells which are the base of food chain, which ultimately affects commercial fisheries. Pteropods and brittle stars make the base of food chain serving as primary food for predators. The removal of these organisms will affect the whole ecosystem (Cripps et al., 2011).

9.6 Anthropogenic Emissions of CO_2

The normal concentration of CO_2 is 0.03% essential for photosynthetic activities on the earth. The solar radiation when reached to the earth surface reflects back into the atmosphere in the night time. CO_2 and some other gases that absorb the heat energy and reflected back in the atmosphere maintain a normal temperature on the earth required for the existence of living organisms (Kabir et al., 2018). The reserves of carbon are embedded in deep oceans and rocks in the form of fossil fuels and sediments. The carbon transfer between atmosphere and earth sediments in a natural way maintains a normal concentration of carbon in each element called carbon cycle (Figure 9.2) (Bruhwiler et al., 2018). The human activities including burning of fossil fuels (coal, gas, oil) disturb the natural carbon cycle, which accelerates the transfer of CO_2 from rocks and sediments to the atmosphere. The high concentration of CO_2 not only causes global warming but also can cause ocean acidification because major part of atmospheric CO_2 is absorbed by oceans (9.8). In the previous years from 1870 to 2017, the anthropogenic activities emitted about 430 Pg C (Pico gram of carbon) in the atmosphere (Figure 9.1). From 2000 to 2012, the global emission of CO_2 increased by a rate of 4%/year and then a decline in 2015 was observed. The emission of CO_2 from fossil fuel combustion and cement production industries is 9.9 Pg C/year which is hundred time faster than geological carbon fluxes. In 2017, the CO_2 emission increased by about 2% due to lower fuel prices and fast economic growth. Besides fossil fuel combustion, other human activities, including land use change, deforestation, and conversion of land into agriculture land, have increased CO_2 emissions in atmosphere. The reforestation with time causes the uptake of carbon. On average the CO_2 released since 1750 by the land use change was estimated as 225 Pg C (Le Quéré et al., 2018; Melieres and Marechal, 2015).

In 1992, at the United Nations Framework Convention on Climate Change (UNFCCC) an international environmental treaty was ratified by a sufficient number of countries. The objective of the treaty was to 'stabilize the anthropogenic emission of greenhouse gases to prevent the climate system from the dangerous anthropogenic interferences.' At the 'Conference of the Parties,' the parties at the convention meet yearly to assess the deal and progress on climate change. Later on, in 1997, the developed countries adapted the 'Kyoto Protocol' which is linked with the UNFCCC (Shishlov et al., 2016). It aims to set bindings on developed countries to limit emission reduction of greenhouse gases including CO_2 in the period of 2008–2012. Because developed countries are responsible for most of the greenhouse gases this protocol places a heavier burden on the developed nations under the 'common but not differentiated responsibilities.' The rules for the protocol were implemented in 2001 in Marrakesh, Morocco and referred to as the Marrakesh Accords, and its commitment starts in 2008. In 2015, within the UNFCCC, an agreement called the Paris Agreement was adopted to limit greenhouse gases with the goal of keeping the global temperature below 2 °C relative to preindustrial levels. In 2019, 195 members of the UNFCCC have signed the agreement (Hermwille et al., 2017).

9.6.1 Global Carbon Emission by Jurisdiction

Despite international agreements and contributions of member countries, the average global emission of CO_2 is increasing which is the main culprit for global warming because the sources of CO_2 emission are higher. Figure 9.3 shows an increasing concentration of CO_2 emission per year. If we observed the emission from 2000–2017, years 2016 and 2017 have the highest emissions as compared to other years because of industrial development. Developed countries, including the United States, Russia, China,

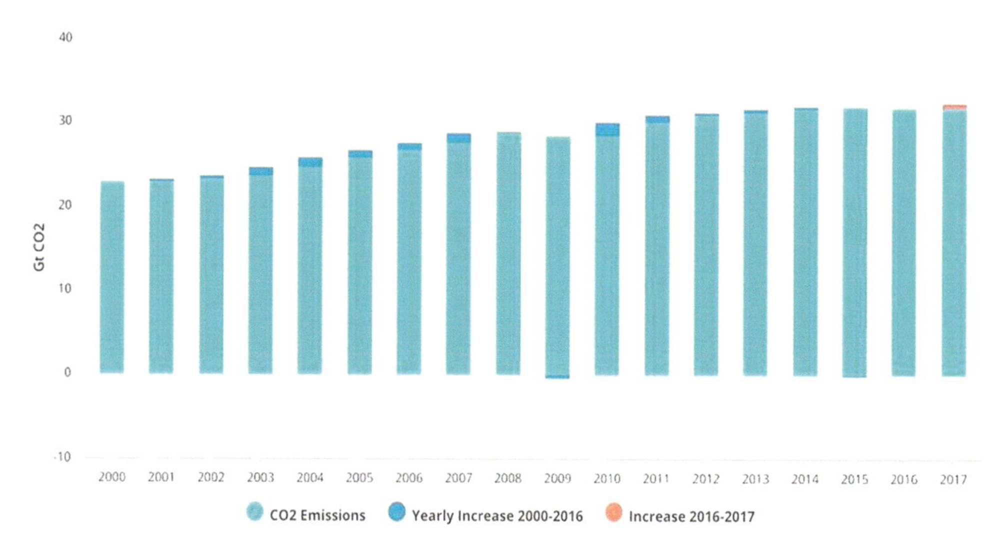

FIGURE 9.3 Showing yearly increasing concentrations of CO_2 emissions in gigatons.
Source: Bruhwiler et al., 2018.

Ukraine, Japan, and England, are the major contributors of CO_2 emissions (Han et al., 2018). CO_2 is released from fossil fuel combustion, land use change, cement manufacturing, emission from international shipping, and many other sources. CO_2 ends up in the three major carbon pools including atmosphere, ocean, and continental reservoir. Among the sources of CO_2, 92% (8.6 ± 0.4 gigatons of the CO_2/year) is released by the combustion of fossil fuel while 8% from other activities, including land use change and volcanic eruption. Further, 45% (4.3 ± 0.1 Gt C/year) of the total released CO_2 from different sources is absorbed by the atmosphere, where oceans absorb 27% (2.6 ± 0.5 Gt C/year), while continental reservoirs absorb 27% (2.6 ± 0.8 Gt C/year) (Bruhwiler et al., 2018).

9.7 Possible Responses

The member of the Inter Academy Panel suggested that the anthropogenic emission of CO_2 can be reduced to less than 50% of the level in 1990 through global action plans (IAP, 2009). This statement called the world leaders stated that the acidification of ocean is a real and direct consequence of CO_2 emission which changing the ocean PH and causing harm to the marine ecosystem because the CO_2 concentration reached to 450 ppm. The possible solution to prevent and reduce ocean acidification is to reduce the emission of CO_2 (Bruhwiler et al., 2018). Moreover management of the stressor, including overfishing, pollution, increased resilience of the marine biota, can mitigate the ocean acidification (Anthony et al., 2008).

The German Advisory Council on Global Change stated that to prevent the risks to change in fundamental food web and prevent the calcification of marine organisms, the surface ocean water pH should not drop more than 0.2 units below the pre-industrial average value. The parties of UNFCCC target to limit global warming to below 2 °C relative to pre-industrial level. Achieving the UNFCCC targets would be proved helpful in reduction of CO_2 emissions. Reducing global warming below 2 °C would reduce the ocean surface water pH of 0.16 relative to pre-industrial levels (UNFCCC, 2011). The USEPA denied the 30th June 2015 petition, which asked USEPA to regulate the emission of CO_2 for the mitigation of ocean acidification. The USEPA said that the risks of ocean acidification are effectively addressed through ‘President Climate Action Plan’ and multiple activities are working to reduce deforestation, clean energy actions, and reduce emission (IAP, 2009).

In June 2017, the United States withdrew from two important international efforts for reducing the emissions of CO_2: the Paris Accords and the G7 Climate Change Pledge (Viviani, 2015).

Climate engineering has been proposed as a policy response to mitigate ocean acidification. It involves addition of chemicals to treat acidification; however, the process is quite expensive and can be used at local scale (IAP, 2009). Moreover, there has been little research on the impacts and feasibility of these actions that how much they may pose unpredictable threats to marine biota. Therefore, extensive research is needed before applying these techniques because the Royal Society of UK and National Research Council of the United States warned about the risks to the marine biota by climate engineering (Hegerl and Solomon, 2009).

Iron fertilization is the mitigation approach for treating ocean acidification. It is the addition of iron in water to increase the photosynthetic activity of phytoplankton. In this activity, CO_2 in ocean water is absorbed by phytoplanktons in photosynthesis and gets converted into O_2 and carbohydrates. Many experiments conducted confirmed that addition of iron increase photosynthetic activity up to 30 times. The UK Royal Society 2009 reviewed the affordability and safety of the process for treating ocean acidification (Cao and Caldeira, 2010). The affordability of the process is a cost-effective process; however, the safety of the process is low because the addition can create anoxic condition in which the shortage of O_2 occurs because of excessive nutrient pollution (Denman, 2008).

9.8 Conclusion

Carbon is the main component of organic compounds essential for life on earth. The circulation of carbon between atmosphere and water is called the carbon cycle. The carbon cycle is disturbed by human activities. The anthropogenic activities interfere in the natural CO_2 cycle by releasing high concentrations of CO_2 in the atmosphere by fossil fuel combustion. Luckily oceans are helpful in keeping a normal balance of CO_2 cycles by absorbing about half of CO_2 from the atmosphere. Oceans are the important natural sink of CO_2; however, this absorption leads to acidification of ocean water which negatively affects the natural biota and normal balance of CO_3^- and HCO_3^-. Developed and developing countries through international policies are working to maintain the natural carbon cycle and protect oceans from acidification.

REFERENCES

Anthony KR, Kline DI, Diaz-Pulido G, Dove S, Hoegh-Guldberg O (2008) Ocean acidification causes bleaching and productivity loss in coral reef builders. *Proc Natl Acad Sci* 105:17442–17446.

Bruhwiler L, Michalak AM, Birdsey R, Fisher JB, Houghton RA, Huntzinger DN, Miller JB (2018) Overview of the global carbon cycle. In: *Second state of the carbon cycle report (SOCCR2): A sustained assessment report global change research program*, pp. 42–70. https://doi.org/10.7930/SOCCR2.2018.Ch1.

Bryden HL, Longworth HR, Cunningham SA (2005) Slowing of the Atlantic meridional overturning circulation at 25 N. *Nature* 438:655.

Caldeira, K, Wickett, ME (2003) Oceanography: Anthropogenic carbon and ocean pH. *Nature* 425:365.

Cao L, Caldeira K (2010) Can ocean iron fertilization mitigate ocean acidification? *Clim Chang* 99:303–311.

Cohen AL, Holcomb M (2009) Why corals care about ocean acidification: Uncovering the mechanism. *Oceanography* 22:118–127.

Couldrey MP, Oliver KI, Yool A, Halloran PR, Achterberg EP (2019) Drivers of 21st century carbon cycle variability in the North Atlantic Ocean. *Biogeosciences Discussions*: 1–33.

Cripps IL, Munday PL, McCormick MI (2011) Ocean acidification affects prey detection by a predatory reef fish. *PLoS One* 6:e22736.

Dean C (2009) Rising acidity is threatening food web of oceans, science panel says. *New York Times* 31.

Denman KL (2008) Climate change, ocean processes and ocean iron fertilization. *Mar Ecol Progr Ser* 364:219–225.

Fabricius KE, Langdon C, Uthicke S, Humphrey C, Noonan S, De'ath G, Okazaki R, Muehllehner N, Glas MS, Lough JM (2011) Losers and winners in coral reefs acclimatized to elevated carbon dioxide concentrations. *Nat Clim Chang* 1:165.

Grobe H (2006) *Air sea exchange of CO_2. Alfred Wegener Institute for Polar and Marine Research, Bremerhaven, Germany.*

Hall-Spencer JM, Rodolfo-Metalpa R, Martin S, Ransome E, Fine M, Turner SM, Rowley SJ, Tedesco D, Buia M-C (2008) Volcanic carbon dioxide vents show ecosystem effects of ocean acidification. *Nature* 454:96.

Han M, Yao Q, Liu W, Dunford M (2018) Tracking embodied carbon flows in the belt and road regions. *J Geogr Sci* 28:1263–1274.

Hegerl GC, Solomon S (2009) Risks of climate engineering. *Science* 325:955–956.

Hermwille L, Obergassel W, Ott HE, Beuermann C (2017) UNFCCC before and after Paris – What's necessary for an effective climate regime? *Clim Policy* 17:150–170.

Hull EV (2016) Ocean acidification: Legal and policy responses to address climate change's evil twin. *Wash J Environ Law Policy* 6, 349.

IAP (2009) *Interacademy Panel (IAP) Member Academies Statement on Ocean Acidification*, Secretariat: TWAS (The Academy of Sciences for the Developing World), Trieste, Italy.

Kabir E, Kumar P, Kumar S, Adelodun AA, Kim K-H (2018) Solar energy: Potential and future prospects. *Renew Sustain Energy Rev* 82:894–900.

Landschuetzer P, Gruber N, Bakker DC (2016) Decadal variations and trends of the global ocean carbon sink. *Glob Biogeochem Cycles* 30:1396–1417.

Le Quéré C, Andrew RM, Friedlingstein P, Sitch S, Hauck J, Pongratz J, Pickers PA, Korsbakken JI, Peters GP, Canadell JG (2018) Global carbon budget 2018. *Earth Syst Sci Data (Online)* 10. https://essd.copernicus.org/articles/10/2141/2018/.

Marinov I, Sarmiento JL (2004) The role of the oceans in the global carbon cycle: An overview. In: *The Ocean Carbon Cycle and Climate*, Springer, Kluwer Academic Publishers, pp. 251–295.

Melieres M, Marechal C (2015) Climate change: Past, present and future, 1st edn. Wiley, London, pp. 310–312.

Millero FJ (1995) Thermodynamics of the carbon dioxide system in the oceans. *Geochim Cosmochim Acta* 59:661–677.

Mollica NR, Guo W, Cohen AL, Huang K-F, Foster GL, Donald HK, Solow AR (2018) Ocean acidification affects coral growth by reducing skeletal density. *Proc Natl Acad Sci* 115:1754–1759.

Mora C, Wei C-L, Rollo A, Amaro T, Baco AR, Billett D, Bopp L, Chen Q, Collier M, Danovaro R (2013) Biotic and human vulnerability to projected changes in ocean biogeochemistry over the 21st century. *PLoS Biol* 11:e1001682.

Orr JC, Fabry VJ, Aumont O, Bopp L, Doney SC, Feely RA, Gnanadesikan A, Gruber N, Ishida A, Joos F (2005) Anthropogenic ocean acidification over the twenty-first century and its impact on calcifying organisms. *Nature* 437:681.

Passow U, De La Rocha CL (2006) Accumulation of mineral ballast on organic aggregates. *Glob Biogeochem Cycles* 20.

Pilson ME (2012) *An introduction to the chemistry of the sea.* Cambridge University Press, Cambridge.

Sarmiento RMAJL (2008) *Program in atmospheric and oceanic sciences. Princeton University,* New Jersey.

Shishlov I, Morel R, Bellassen V (2016) Compliance of the parties to the Kyoto Protocol in the first commitment period. *Clim Policy* 16:768–782.

Sigman DM, Hain MP, Haug GH (2010) The polar ocean and glacial cycles in atmospheric CO_2 concentration. *Nature* 466:47.

Takahashi T, Sutherland SC, Sweeney C, Poisson A, Metzl N, Tilbrook B, Bates N, Wanninkhof R, Feely RA, Sabine C, Olafsson J, NojiriYC (2002) Global sea-air CO_2 flux based on climatological surface ocean pCO_2, and seasonal biological and temperature effects. *Deep-Sea Res Part II* 49:1601–1622.

UNFCCC (2011) Addendum. Part two: Action taken by the Conference of the Parties at Its Sixteenth Session. Framework Convention on Climate Change, 29 November–10 December 2010. Geneva, Switzerland: United Nations. p. 3, paragraph 4.

Viviani DJ (2015) *Petition for rulemaking pursuant to section 21 of the toxic substances control act, 15 U.S.C. § 2620, concerning the regulation of carbon dioxide.* Center for Biological Diversity, Oakland.

10

Insect Pest Management in the Era of Climate Change

Mohd Abas Shah[1], Abdel Rahman Al-Tawaha[2], Kailash Chandra Naga[3], and Sanjeev Sharma[3]
[1]ICAR-Central Potato Research Station, Jalandhar, Punjab, India
[2]Department of Biological Sciences, Al-Hussein bin Talal University, Maan, Jordan
[3]ICAR-Central Potato Research Institute, Shimla, Himachal Pradesh, India

10.1 Introduction

Climate change is considered a major threat to ecosystems, water resources, agriculture, and food security (Adnan et al., 2018, 2020; Ahmad et al., 2019; Akram et al., 2018a,b; Depeng et al., 2018; Farhat et al., 2020; Gul et al., 2020; Habib ur Rahman et al., 2017; Hammad et al., 2016, 2018, 2019, 2020a,b; Huong et al., 2018; Hussain et al., 2019, 2020; Ilyas et al., 2020; Jan et al., 2019; Kamarn et al., 2017; Khan et al., 2017a,b; Mubeen et al., 2020; Muhammad et al., 2019; Myers et al., 2017; Naseem et al., 2017; Rehman et al., 2020; Saleem et al., 2020a,b,c; Saud et al., 2013, 2014, 2016, 2-17, 2020; Shafi et al., 2020; Shah et al., 2013; Subhan et al., 2020; Wahid et al., 2020; Wu et al., 2019, 2020; Yang et al., 2017; Zafar-ul-Hye et al., 2020a,b; Zahida et al., 2017; Zaman et al., 2017; Zamin et al., 2019). Climate change manifests itself as warming of atmosphere, rise in greenhouse gas (GHG) concentrations, decrease in snow and ice cover, and rise in sea levels (IPCC, 2014). It would aggravate the effect of biotic and abiotic stresses in agriculture and increase the resulting losses, thus threaten the food security and farmers' livelihoods (Alharby and Fahad, 2020; Fahad and Bano, 2012; Fahad et al., 2013, 2014a,b, 2015a,b, 2016a,b,c,d, 2017, 2018, 2019a,b). The most pronounced effect of climate change is expected to be on the developmental time and biogeographical distribution of insect pests. Many insect species are pests of crops, but they also play crucial roles as parasitoids and predators of key pest species. Changes in an insect population's physiology, biochemistry, biogeography, and population dynamics may occur among populations across their distribution, among the growing seasons, and crop types. An insect population's response to a rapidly changing climate may also be variable when insects interact with different competitors, predators, and parasitoids, and impose costs at different life stages. This also can influence the overall food production systems (IPCC, 2014). The major effects documented with or without evidence include changes in geographical distribution, asynchrony with host plants, increased voltinism, enhanced pest damage to crops, and changes in inter-species interactions; some effects more pronounced than others.

Studies indicate that climate change effects are mostly going to be species-specific (Lawton and Kinne, 2000). Therefore, the study of climate change-induced effects needs to process based on improving the understanding of such effects and accurately predicting the further consequences (Schurr et al., 2012).

This chapter provides a summary of the effects of climate change on insect pests and their response to it; mainly based on the accounts put forth by Hellmann et al. (2008), Cornelissen (2011), Yang and

Rudolf (2010), Robinet and Roques (2010), Torquebiau (2016), Thomson et al. (2010), Sharma (2014), Andrew and Hill (2017), Battisti and Larsson (2015), Kalinkat et al. (2015), Terblanche et al. (2015), Bjorkman and Niemela (2015), Dyderski et al. (2018), Macfadyen et al. (2018), Johnson and Züst (2018), Trębicki et al. (2017), and Cohen et al. (2018).

10.2 Climate Change and Insect Pest Responses

Insects are ectotherms hence more reliant on the external environment to maintain functions. As such, every stage of their life cycle and everyday functioning is generally subject to the abiotic environment that they inhabit. Insects function optimally within a certain thermal window, and this window can differ markedly among species, and even life stages. Key traits such as developmental time, longevity, and fecundity are influenced by temperature, with colder temperatures generally slowing an individual's growth rate and prolonging their life span and *vice versa* (Denlinger and Lee, 2010; Ryalls et al., 2017).

It's been realized that insect species can respond to climate change in a multitude of ways, mainly in anticipation of a generally warmer and drier environment rich in CO_2 predicted through climate change's general circulation models (IPCC, 2013), and insect species may shift their geographic distributions or phenology in an attempt to track changes in their optimal conditions. Within ectotherms and endotherms alike, there is substantial evidence of range shifts are already occurring, particularly towards the poles, where temperatures are increasing; with recent evidence also suggesting this is the case for pest species (Bebber et al., 2013). There are several reported cases of phenological change with the environment in insect pests. The broader impacts of such phenological change are, however, largely unclear. For example, a strong negative impact of climate change may occur when there is a mismatch between the phenology of insects and their host plants, that is, peak flowering and nutrient production (Visser and Holleman, 2001). Although these impacts are less well documented for pest damage on the crop or host, they may have subtle effects on other aspects of pest management, such as biocontrol efficacy, or predator-prey interactions. For agricultural pests, studies have detected a mismatch in phenology of the pest and their natural enemies (Evans et al., 2013).

Crop damage due to insect pests may increase due to higher numbers of generations in a season, that is, increased voltinism. Based on a day-degree model, an increased number of suitable days for development could allow faster generation time and, therefore, an additional generation (or possibly even two) to develop within a growing season (Barton and Terblanche, 2014). Model predictions for the non-diapausing moth, *Plutella xylostella*, indicate an additional two generations per year could be achieved under changing climate conditions (Morimoto et al., 1998). A potential increased growth season may, however, be limited by the timing of resources, although if resources persist for longer in agricultural environments, it can generally be expected that the pest may, too (Gregory et al., 2009). Increased carbon to nitrogen ratio in the plant tissues will make them poor in quality for the insects (Coviella and Trumble, 1999). This will require insects to escalate their food intake to maintain function, impacting their metabolic rate and resulting in fitness costs. Greater vegetative damage may also increase the potential for pathogen infection of crop plants, either through direct transmission of viral pathogens or through opportunistic infections, further impacting the sustainability of crop practices (Gregory et al., 2009). These results appear to paint a bleak future for agriculture; however, in warmer climates, an increase in temperature will also place species closer to their upper thermal limits, perhaps effectively limiting population persistence, despite the potential for increased voltinism. The interplay between faster generation times, reduced performance windows, and lower food quality indicates that predictions of agricultural pest damage may not be simple to make and are probably subject to change. With faster developmental rates and generation times comes the potential for an increase in the speed of the adaptive process. This could be beneficial for allowing species to undergo evolutionary changes to remain within their current native range, or even potentially expand their distributions.

Very few studies have measured the potential for adaptive and plastic change in physiological traits for ectotherms, particularly in light of a changing climate. What has been measured, however, indicates that there are latitudinal variation in the severity of the impact of climate change and the ability of species to evolve in response to climate variation. In general, mid-latitude populations or species experience greater heat stress conditions at present, and this is predicted to increase over time (Hoffmann et al., 2013). Current evidence suggests that these low latitude tropical species exist in habitats that experience conditions close to their upper temperature thresholds, indicating that they have a lower 'warming tolerance' (WT; the difference between current habitat temperature and thermal maxima for survival and performance). Comparisons of different insect species show that upper thermal limits are more constrained and tend to be more similar across closely related species. This indicates that there may be some limits to evolution that persist across all groups (Hoffmann et al., 2013; Mitchell and Hoffmann, 2010). In general, predictions for distributions and population dynamics of agricultural pest species indicate overall expansion or shifts of geographic limits towards higher latitudes or altitudes and a higher fecundity and shorter development time (and therefore increasing population sizes) under global climate change.

Other than elevated temperature scenarios due to climate change, studies have documented the drastic effects of CO_2 enrichment on the insect-plant system (reviewed in Stiling and Cornelissen, 2007). The concentration of CO_2 is expected to double or triple (540 to 970 ppm) by 2100, as compared to the pre-industrial era (280 ppm) (Stiling et al., 1999).

In general, the productivity of plants is known to increase in CO_2 enriched environments due to increased photosynthetic activity. Increased availability of carbon might alter the C/N ratio of the plant tissues and a potentially lower concentration of protein, often termed as 'nitrogen dilution effect.' Nitrogen (protein) poor diet has lower nutritional value for the feeding insects (Lincoln et al., 1986). Also, the secondary metabolite profile of the plants is expected to change (Bale, 2002).

It is generally believed that herbivore insects would respond to altered plant physiology in one or more of the following ways; increased food intake (Fajer, 1989; Marks and Lincoln, 1996), reduced growth rate and increased developmental time (Goverde and Erhardt, 2003), and reduced food conversion efficiency (Brooks and Whittaker, 1998). This could reduce the herbivore abundance, richness, and diversity, directly and indirectly. Stiling and Cornelissen (2007) reported several responses of herbivore insects to elevation in CO_2 in a meta-analytical study. They reported 22.0%, 9.0%, and 5% decrease in insect abundance, relative growth rate and pupal weight, respectively; and 17.0% and 4% increase in consumption rates and development time, respectively, at elevated CO_2 levels in comparison to ambient conditions.

Zvereva and Kozlov (2006) summarized the results of 42 studies in which the effect of increased carbon dioxide and the temperature was evaluated simultaneously. They reported that the leaf nitrogen reduced while the C/N ration increased under both conditions. The secondary metabolites did not show a significant change. The insect biology parameters, for example, relative growth rate, fecundity, survival, and pupal weight were affected positively and negatively by increased CO_2 and temperature alone, respectively, and not affected when the two varied together. However, further studies will clarify the phenomena (Zvereva and Kozlov, 2006).

10.3 Effect of Climate Change on Species Synchrony

Increased temperature will lead to the earliness of developmental events such as bud burst or the date of flowering in plants (Memmott et al., 2007). Synchronization to plant phenology is essential for many insect herbivores; therefore, development outside the optimal window can have fitness costs associated with it (Régnière and Nealis, 2018; Rosenblatt and Schmitz, 2016). In this context, the frequently reported effects are the advancement of bud burst and the date of first flowering in plants, which can have direct consequences for herbivores and flower visitors. At the same time, the interacting species may try to adjust to the altered phenology which could complicate the situation further; this could lead to spatial and temporal mismatches in between interacting species. Common examples include plants and insects feeding on them, plants and pollinators, and hosts and parasitoids. The effects are expected to be worst for the specialist insect herbivores and pollinators.

Among the clearest case studies of climate change-induced asynchrony is the winter moth, *Operophtera brumata*. Up to 90% of its eggs have been reported to hatch before the bud burst of its hosts, the oak trees (Visser and Holleman, 2001). Similarly, differential geographical range expansion between *Boloria titania* and its host, *Polygonum bistorta*, has led to a spatial mismatch between the two (Schweiger et al., 2008).

A multilevel phenological change was reported by Both et al. (2009) wherein differential phenological changes across four trophic levels were found. The first trophic level is the oak trees on moth caterpillars (second level) feed, which are prey for songbirds (third level), which are preyed upon by hawks (fourth level). An advancement over time has been shown across all the four trophic levels. The effects, however, were strongest for the caterpillars and weakest for the birds. This mismatch could lead to herbivore outbreaks and a decline in bird numbers which could be catastrophic for the whole food web. Similar climate change-induced phonological mismatches are known for the many other lepidopteran insects, including *Panolis flammea* (the pine beauty moth), *Tortrix viridana* (green oak tortrix), and *Choristoneura fumiferana* (eastern spruce budworm) (Dewar and Watt, 1992).

Memmott et al. (2007), based on simulation studies, reported that 1750% of all pollinators will suffer a reduction in food supply with two weeks' early peak flowering. Further, the specialist pollinators are assumed to be worst affected by this temporal mismatch and the consequent reduction in food resources.

10.4 Effects of Climate Change on Plant Volatile Compounds

Volatile organic compounds (VOCs) mediate various inter- and intra-specific interactions between insects and plants, including pollination and seed dispersal, defences against herbivory, allelopathy, and so on, by affecting the plant performance increase in atmospheric temperature, and concentration of CO_2 may affect the emission of VOCs (Peñuelas and Staudt, 2010).

In comparatively fewer studies, it is reported that the concentration and the duration of production of VOCs will increase under warmer conditions (Peñuelas and Llusià, 2003; Yuan et al., 2009), which could have mild to severe ecological consequences. The emission of higher concentration of VOCs, for example, ethyl jasmonate or methyl salicylate is likely to keep the neighbouring plants under alert which is an ecologically wasteful process and the damage due to herbivory may be higher. Also, a stronger cocktail of VOCs might interrupt inter-specific communication resulting in poor performance by pollinators and seed dispersers.

Like warmer temperature, the production and emission of VOCs is likely to increase under CO_2 enrichment due to the direct relation between carbon availability and production of VOCs (Yuan et al., 2009).

It has been suggested that the production of monoterpenes and sesquiterpenes will increase due to disturbed C/N ratio in plants due to CO_2 enrichment (Lerdau et al., 1994). Such effects are already documented for many cultivated plants, conifers, and oaks (Constable et al., 1999; Jasoni et al., 2004; Loreto et al., 2001; Tognetti et al., 1998). Although it is clear that the production of VOCs would increase under the climate change scenario, their effect on the interacting species pollinators, herbivores, and parasitoids has not been investigated fully as of yet.

10.5 Effect of Climate Change on Geographical Distribution of Insects

The thermal preference of insect species places important restrictions on the thermal conditions, and therefore habitats, in which the insect may function successfully. As such, they provide a useful tool for investigating thermal environmental thresholds that potentially could limit species distributions. Theoretically, the breadth of an individual's performance curve will match the average thermal fluctuations in the environment closely, as the costs to reproduction and survival of exceeding these critical thermal limits would not be sustainable (Gilchrist, 1995; Lynch and Gabriel, 1987). Using the same logic, species' optimum temperature requirement should also match the mean temperature of the habitat. It is logically believed that many insects respond to the effects of climate change by range expansion and many are likely to do so.

In a study by Battisti and Larsson (2015), the data on 50 species were summarized whose range is already affected or believed to change soon. They categorized the species into three groups; first group for which a plausible causal explanation exists (8 cases), a second group where the link with climate change is not clear yet but the range expansion is documented (17 cases), and third group where range expansion is expected based on modelling data and thermal optima of the insects (25 cases). The study revealed that a range expansion along latitudes was observed for all such species, and along with the altitude and longitudes for four and one species, respectively. The study also revealed that the most common reason for range expansion of insect species was the suitability of winter temperatures and thus reduced mortality in the novel areas (Battisti and Larsson, 2015). The third category includes many major crop pests, including *Chilo suppressalis*, *Cydia pomonella*, *Helicoverpa* spp., *Leptinotarsa decemlineata*, *Ostrinia nubilalis*, *Plutella xylostella*, *Rhopalosiphum padi*, *Bactrocera oleae*, and others.

Concerning climate change-induced range expansion, the case studies of some of the most studied insects are briefly described as follows. Among these, the pine processionary moth (*Thaumetopoea pityocampa*), for which range expansion has happened due to climate change, is the most studied (e.g. Battisti et al., 2005; Buffo et al., 2007; Robinet et al., 2007, 2013). Similarly, bark beetles (*Dendroctonus* spp.) in North America are the subjects of active research concerning climate change-induced range expansion (e.g. Seidl and Rammer, 2017; Weed et al., 2013). With range expansion, several *Dendroctonus* species are believed to be a threat to forests in new areas (Bentz et al., 2010). Increase in winter temperature enables the beetles to tolerate winter in previously unsuitable areas, for example, British Columbia and Alberta (Carroll et al., 2006). The mountain pine beetle can now be found at higher mountainous altitudes previously not inhabited (Raffa et al., 2013).

A similar phenomenon is operational with the hemlock woolly adelgid (*Adelges tsugae*). The insect is now found in the north eastern United States as well as away from its original locality in Virginia (Evans and Gregoire, 2007; Paradis et al., 2008). The aphid can now be found in the north and is a major pest in the already-colonized zones, resulting in a higher incidence of hemlock dieback (Fitzpatrick et al., 2012). The main reason reported for such changes is the increase in mean winter temperature. Similar phenomenon is reported for the green stink bug (*Nezara viridula*) in Japan (Musolin and Saulich, 2012; Tougou et al., 2009). For gypsy moth, *Lymantria dispar*, range expansion along the altitude is reported with the outbreak area shifted to a higher elevation (Liebhold et al., 1992).

Thus, it has become clear that expansion along the latitudes is the most likely, along with elevational and longitudinal expansion for some.

10.6 Potential Consequences of Climate Change for Invasive Species

Climate change is believed to affect the spread and invasiveness of invasive species (Colautti et al., 2017; Hulme, 2017; Qian et al., 2018). Also, the invasive species might need redefinition in the era of climate change because many non-invasive species might become invasive; range expansion might enable some species to move into new areas where the impact of already invasive species may decrease; the last one being the least probable as the already invasive species have expanded their range of distribution considerably (Qian and Ricklefs, 2006).

In a theoretical account put forth by Hellmann et al. (2008), many specific points have been identified in the invasion biology of species that may be impacted by climate change. For a better understanding of the potential impact of climate change on invasive species, the process of invasion must be known clearly. For a species to become invasive, it must pass through a number of ecological barriers (Theoharides and Dukes, 2007; Vermeij, 1996; Williamson, 2006; Williamson and Fitter, 1996), and each such step may be affected by climate change, some more than others (Rahel and Olden, 2008).

Common steps include the crossing of geographical barriers and arrival at the new site, tolerance of new environmental conditions, and survival in them; interactions include trophic (food and natural enemies) and mutualistic; establishment and multiplication at the new site; and finally, dispersal and expansion in the new landscape. Climate change enables a higher multiplication rate of species and range expansion which is likely to be helpful for the invasive species. Climatic suitability at newer locations and its mode of dispersal is also likely to be affected by climate change.

Based on the invasion pathway, Hellmann et al. (2008) specified five possible effects of climate change on invasion biology. These are (1) changes in transportation and introduction; these could include range expansion, a higher rate of multiplication, and so forth; (2) changed climatic factors at each location, in terms of more suitable climate for the invasive species; (3) changes in spread of existing invasive species, range expansion, and a higher rate of multiplication expected to enable the already-invasive species to become more aggressive; (4) changes in the impacts of invasion, the faster displacement of native species which lead to faster local extinction of species; and (5) revision of management tactics, most probably the management strategies that are expected be less effective while some might become more effective.

10.7 Effect of Climate Change on Natural Enemies and Biological Control

Climate change would have both direct and indirect (plant mediated via herbivores or even directly) effects on natural enemies. Among the documented effects are the changes in synchronization of insect pests and their natural enemies, reduction in the quality of the hosts/prey, effect on the predation and parasitization efficiency, and effect on the host searching behaviour. Effect of climate change on predators and parasitoids is double-pronged as there are host-mediated effects as well as direct effects to be expected; which makes the evaluation of effects complicated.

The elevated CO_2 level is known to reduce the nutritional quality of plant tissues by lowering the nitrogen concentration to which herbivores respond by increased consumption rate (Bezemer et al., 1998). An important outcome of this is the reduced nutritional value of the herbivores for the third trophic level as well as the difference in size and weight (Zvereva and Kozlov, 2006). The fitness and performance of predators and parasitoids declines as the nutritional quality of their herbivore hosts decreases (Bilde and Toft, 1999; Butler and O'Neil, 2007; Furlong and Zalucki, 2017; Hvam and Toft, 2005; Tougeron et al., 2017; Wang et al., 2008). Similarly, the growth and development of parasitoids is dependent upon the host size, stage of development, and quality of diet (Godfray and Godfray, 1994; Haeckermann et al., 2007; Harvey et al., 1999, 2004).

Prey consumption rates are reported to increase under CO_2 enrichment conditions. For example, pentatomid bug, *Oechalia schellenbergii*, while preying upon cotton bollworm, *Helicoverpa armigera* caterpillars (Coll and Hughes, 2008); coccinellid predator, *Leis axyridis*, preying upon *Aphis gossypii* (Chen et al., 2005); higher parasitization of aphid pest, *Sitobion avenae* F., by the braconid parasitoid, *Aphidius picipes* (Chen Jun et al., 2007).

In addition to increased food consumption by herbivores in response to reduced food quality, they may reduce the developmental rate (Lindroth et al., 1993). All these factors have a direct and variable bearing on the third trophic level. Longer developmental time increases the chances of predation. The abundance of herbivores may also vary (Watt et al., 1995).

Other than these effects, some less studied aspects include the changes in gene regulation in response to elevated CO_2 or elevated temperature, changes in the expression profile of VOCs and their effect on the predators and parasitoids, and difference in the synchronization of the host and natural enemies.

10.8 The Omics Solutions to New Challenges

The omics (genomics, transcriptomics, proteomics, metabolomics, etc.) fields of scientific investigation bring forth new opportunities and insights to decipher the evolution and dynamics of crop-pest interactions. Breeding of climate resilient varieties, genome-wide search of trait loci, and mechanisms of gene expression and character regulation are improving our understanding of the crop-pest interaction. It is expected that such technological advancements will be useful to overcome the novel issues due to climate change scenarios.

10.9 Conclusion

The response of organisms to climate change is believed to be species-specific, with variable rates. Agriculture and food security are expected to be the worst affected. It's likely that the crop damage due to herbovory will increase, pest species may expand in distribution, and pollination and other ecosystem functions would get affected, with more frequent pest outbreaks (Coley, 1998). It is expected that the omics studies could be useful in providing some insight into helping reduce climate change-induced stresses.

REFERENCES

Adnan M, Shah Z, Fahad S, Arif M, Alam M, Khan IA, Mian IA, Basir A, Ullah H, Arshad M, Rahman I-U, Saud S, Ihsan MZ, Jamal Y, Amanullah, Hammad HM, Nasim W (2018) Phosphate-solubilizing bacteria nullify the antagonistic effect of soil calcification on bioavailability of phosphorus in alkaline soils. *Sci Rep* 8:4339. https://doi.org/10.1038/s41598-018-22653-7.

Adnan M, Fahad S, Zamin M, Shah S, Mian IA, Danish S, Zafar-ul-Hye M, Battaglia ML, Naz RMM, Saeed B, Saud S, Ahmad I, Yue Z, Brtnicky M, Holatko J, Datta R (2020) Coupling phosphate-solubilizing bacteria with phosphorus supplements improve maize phosphorus acquisition and growth under lime induced salinity stress. *Plants* 9(900). doi: 10.3390/plants9070900.

Ahmad S, Kamran M, Ding R, Meng X, Wang H, Ahmad I, Fahad S, Han Q (2019) Exogenous melatonin confers drought stress by promoting plant growth, photosynthetic capacity and antioxidant defense system of maize seedlings. *Peer J* 7:e7793 http://doi.org/10.7717/peerj.7793.

Akram R, Turan V, Hammad HM, Ahmad S, Hussain S, Hasnain A, Maqbool MM, Rehmani MIA, Rasool A, Masood N, Mahmood F, Mubeen M, Sultana SR, Fahad S, Amanet K, Saleem M, Abbas Y, Akhtar HM, Waseem F, Murtaza R, Amin A, Zahoor SA, ul Din MS, Nasim W (2018a) Fate of organic and inorganic pollutants in paddy soils. In: Hashmi MZ, Varma A (eds.) *Environmental pollution of paddy soils*. Springer International Publishing, Cham, Switzerland, pp. 197–214.

Akram R, Turan V, Wahid A, Ijaz M, Shahid MA, Kaleem S, Hafeez A, Maqbool MM, Chaudhary HJ, Munis, MFH, Mubeen M, Sadiq N, Murtaza R, Kazmi DH, Ali S, Khan N, Sultana SR, Fahad S, Amin A, Nasim W (2018b) Paddy land pollutants and their role in climate change. In: Hashmi MZ, Varma A (eds.) *Environmental pollution of paddy soils*. Springer International Publishing, Cham, Switzerland, pp. 113–124.

Alharby AF, Fahad S (2020) Melatonin application enhances biochar efficiency for drought tolerance in maize varieties: Modifications in physio-biochemical machinery. *Agron J*: 1–22.

Andrew NR, Hill SJ (2017) Effect of climate change on insect pest management. In: Coll M, Wajnberg E (eds.) *Environmental pest management: Challenges for agronomists, ecologists, economists and policymakers*. John Wiley & Sons, pp. 195–224.

Bale JS (2002) Insects and low temperatures: From molecular biology to distributions and abundance. *Philos Trans R Socf London. Ser B: Biol Sci* 357(1423):849–862.

Barton MG, Terblanche JS (2014) Predicting performance and survival across topographically heterogeneous landscapes: The global pest insect Helicoverpa armigera (Hübner, 1808) (Lepidoptera: Noctuidae). *Austral Entomol* 53(3):249–258.

Battisti A, Stastny M, Netherer S et al. (2005) Expansion of geographic range in the pine processionary moth caused by increased winter temperatures. *Ecol Appl* 15(6):2084–2096.

Battisti A, Larsson S (2015) Climate change and insect pest distribution range. *Climate Change and Insect Pests*. CABI, Wallingford, pp. 1–15.

Bebber DP, Ramotowski MA, Gurr SJ (2013) Crop pests and pathogens move pole wards in a warming world. *Nat Clim Chang* 3(11):985.

Bentz BJ, Régnière J, Fettig CJ et al. (2010) Climate change and bark beetles of the western United States and Canada: Direct and indirect effects. *BioScience* 60(8):602–613.

Bilde T, Toft S (1999) Prey consumption and fecundity of the carabid beetle, Calathus melanocephalus on diets of three cereal aphids: High consumption. *Pedobiologia* 43(5):429.

Bjorkman C, Niemela P (eds.) (2015) *Climate change and insect pests* (Vol. 8). CABI.

Both C, Van Asch M, Bijlsma RG et al. (2009) Climate change and unequal phenological changes across four trophic levels: Constraints or adaptations? *J Anim Ecol* 78(1):73–83.

Brooks GL, Whittaker JB (1998) Responses of multiple generations of Gastrophysa viridula, feeding on Rumex obtusifolius, to elevated CO_2. *Glob Chang Biol* 4(1):63–75.

Buffo E, Battisti A, Stastny M, Larsson S (2007) Temperature as a predictor of survival of the pine processionary moth in the Italian Alps. *Agric Forest Entomol* 9(1):65–72.

Butler CD, O'Neil RJ (2007) Life history characteristics of Orius insidiosus (Say) fed diets of soybean aphid, Aphis glycines Matsumura and soybean thrips, Neohydatothrips variabilis (Beach). *Biol Control* 40(3):339–346.

Carroll AL, Régnière J, Logan JA et al. (2006) *Impacts of climate change on range expansion by the mountain pine beetle*. Pacific Forestry Centre, Canada.

Chen F, Ge F, Parajulee MN (2005) Impact of elevated CO_2 on tri-trophic interaction of Gossypium hirsutum, Aphis gossypii, and Leis axyridis. *Environ Entomol* 34(1):37–46.

Chen Jun F, Wu G, Parajulee MN et al. (2007) Impact of elevated CO_2 on the third trophic level: A predator Harmonia axyridis and a parasitoid Aphidius picipes. *Biocontrol Sci Technol* 17(3):313–324.

Cohen JM, Lajeunesse MJ, Rohr JR (2018) A global synthesis of animal phenological responses to climate change. *Nat Clim Chang* 8(3):224.

Colautti RI, Ågren J, Anderson JT (2017) Phenological shifts of native and invasive species under climate change: Insights from the Boechera–Lythrum model. *Philos Trans R Soc B: Biol Sci* 372(1712):20160032. http://dx.doi.org/10.1098/rstb.2016.0032.

Coley PD (1998) Possible effects of climate change on plant/herbivore interactions in moist tropical forests. *Clim Chang* 39(2–3):455–472.

Coll M, Hughes L (2008) Effects of elevated CO_2 on an insect omnivore: A test for nutritional effects mediated by host plants and prey. *Agric Ecosyst Environ* 123(4):271–279.

Constable JV, Litvak ME, Greenberg JP, Monson RK (1999) Monoterpene emission from coniferous trees in response to elevated CO_2 concentration and climate warming. *Glob Chang Biol* 5(3):252–267.

Cornelissen T (2011) Climate change and its effects on terrestrial insects and herbivory patterns. *Neotropical Entomol* 40(2):155–163.

Coviella CE, Trumble JT (1999) Effects of elevated atmospheric carbon dioxide on insect-plant interactions. *Conserv Biol* 13(4):700–712.

Denlinger DL, Richard E, Lee Jr (2010) *Low temperature biology of insects*. Cambridge University Press, Cambridge.

Dewar RC, Watt AD (1992) Predicted changes in the synchrony of larval emergence and budburst under climatic warming. *Oecologia* 89(4):557–559.

Dyderski MK, Paź S, Frelich LE et al. (2018) How much does climate change threaten European forest tree species distributions? *Glob Chang Biol* 24(3):1150–1163.

Evans AM, Gregoire TG (2007) A geographically variable model of hemlock woolly adelgid spread. *Biol Invasions* 9(4):369–382.

Evans EW, Carlile NR, Innes MB et al. (2013) Warm springs reduce parasitism of the cereal leaf beetle through phenological mismatch. *J Appl Entomol* 137(5):383–391.

Fahad S, Bano A (2012) Effect of salicylic acid on physiological and biochemical characterization of maize grown in saline area. *Pak J Bot* 44:1433–1438.

Fahad S, Chen Y, Saud S, Wang K, Xiong D, Chen C, Wu C, Shah F, Nie L, Huang J (2013) Ultraviolet radiation effect on photosynthetic pigments, biochemical attributes, antioxidant enzyme activity and hormonal contents of wheat. *J Food Agric Environ* 11(3&4):1635–1641.

Fahad S, Hussain S, Bano A, Saud S, Hassan S, Shan D, Khan FA, Khan F, Chen Y, Wu C, Tabassum MA, Chun MX, Afzal M, Jan A, Jan MT, Huang J (2014a) Potential role of phytohormones and plant growth-promoting rhizobacteria in abiotic stresses: Consequences for changing environment. *Environ Sci Pollut Res* 22(7):4907–4921. https://doi.org/10.1007/s11356-014-3754-2.

Fahad S, Hussain S, Matloob A, Khan FA, Khaliq A, Saud S, Hassan S, Shan D, Khan F, Ullah N, Faiq M, Khan MR, Tareen AK, Khan A, Ullah A, Ullah N, Huang J (2014b) Phytohormones and plant responses to salinity stress: A review. *Plant Growth Regul* 75(2):391–404. https://doi.org/10.1007/s10725-014-0013-y.

Fahad S, Hussain S, Saud S, Tanveer M, Bajwa AA, Hassan S, Shah AN, Ullah A,Wu C, Khan FA, Shah F, Ullah S, Chen Y, Huang J (2015a) A biochar application protects rice pollen from high-temperature stress. *Plant Physiol Biochem* 96:281–287.

Fahad S, Nie L, Chen Y, Wu C, Xiong D, Saud S, Hongyan L, Cui K, Huang J (2015b) Crop plant hormones and environmental stress. *Sustain Agric Rev* 15:371–400.

Fahad S, Hussain S, Saud S, Hassan S, Chauhan BS, Khan F et al. (2016a) Responses of rapid viscoanalyzer profile and other rice grain qualities to exogenously applied plant growth regulators under high day and high night temperatures. *PLoS One* 11(7):e0159590. https://doi.org/10.1371/journal.pone.0159590.

Fahad S, Hussain S, Saud S, Hassan S, Ihsan Z, Shah AN, Wu C, Yousaf M, Nasim W, Alharby H, Alghabari F, Huang J (2016b) Exogenously applied plant growth regulators enhance the morphophysiological growth and yield of rice under high temperature. *Front Plant Sci* 7:1250. https://doi.org/10.3389/fpls.2016.01250.

Fahad S, Hussain S, Saud S, Hassan S, Tanveer M, Ihsan MZ, Shah AN, Ullah A, Nasrullah KF, Ullah S, AlharbyH NW, Wu C, Huang J (2016c) A combined application of biochar and phosphorus alleviates heat-induced adversities on physiological, agronomical and quality attributes of rice. *Plant Physiol Biochem* 103:191–198.

Fahad S, Hussain S, Saud S, Khan F, Hassan S, Jr A, Nasim W, Arif M, Wang F, Huang J (2016d) Exogenously applied plant growth regulators affect heat-stressed rice pollens. *J Agron Crop Sci* 202:139–150.

Fahad S, Bajwa AA, Nazir U, Anjum SA, Farooq A, Zohaib A, Sadia S, Nasim W, Adkins S, Saud S, Ihsan MZ, Alharby H, Wu C, Wang D, Huang J (2017) Crop production under drought and heat stress: Plant responses and management options. *Front Plant Sci* 8:1147. https://doi.org/10.3389/fpls.2017.01147.

Fahad S, Ihsan MZ, Khaliq A, Daur I, Saud S, Alzamanan S, Nasim W, Abdullah M, Khan IA, Wu C, Wang D, Huang J (2018) Consequences of high temperature under changing climate optima for rice pollen characteristics-concepts and perspectives. *Archives Agron Soil Sci* DOI: 10.1080/03650340.2018.1443213.

Fahad S, Adnan M, Hassan S, Saud S, Hussain S, Wu C, Wang D, Hakeem KR, Alharby HF, Turan V, Khan MA, Huang J (2019a) Rice responses and tolerance to high temperature. In: Hasanuzzaman M, Fujita M, Nahar K, Biswas JK (eds.) *Advances in rice research for abiotic stress tolerance*. Woodhead, Cambridge, pp. 201–224.

Fahad S, Rehman A, Shahzad B, Tanveer M, Saud S, Kamran M, Ihtisham M, Khan SU, Turan V, Rahman MHU (2019b) Rice responses and tolerance to metal/metalloid toxicity. In: Hasanuzzaman M, Fujita M, Nahar K, Biswas JK (eds.) *Advances in rice research for abiotic stress tolerance*. Woodhead, Cambridge, pp. 299–312.

Fajer ED (1989) The effects of enriched CO_2 atmospheres on plant-insect herbivore interactions: Growth responses of larvae of the specialist butterfly, Junonia coenia (Lepidoptera: Nymphalidae). *Oecologia* 81(4):514–520.

Farhat A, Hafiz MH, Wajid I, Aitazaz AF, Hafiz FB, Zahida Z, Fahad S, Wajid F, Artemi C (2020) A review of soil carbon dynamics resulting from agricultural practices. *J Environ Manag* 268(2020):110319.

Fitzpatrick MC, Preisser EL, Porter A et al. (2012) Modeling range dynamics in heterogeneous landscapes: Invasion of the hemlock woolly adelgid in eastern North America. *Ecol Appl* 22(2):472–486.

Furlong MJ, Zalucki MP (2017) Climate change and biological control: The consequences of increasing temperatures on host–parasitoid interactions. *Curr Opin Insect Sci* 20:39–44.

Gilchrist GW (1995) Specialists and generalists in changing environments. I. Fitness landscapes of thermal sensitivity. *Am Nat* 146(2):252–270.

Godfray HCJ, Godfray HCJ (1994) *Parasitoids: Behavioral and Evolutionary Ecology*. Princeton University Press.

Goverde M, Erhardt A (2003) Effects of elevated CO_2 on development and larval food-plant preference in the butterfly Coenonympha pamphilus (Lepidoptera, Satyridae). *Glob Chang Biol* 9(1):74–83.

Gregory PJ, Johnson SN, Newton AC et al. (2009) Integrating pests and pathogens into the climate change/food security debate. *J Exp Bot* 60(10):2827–2838.

Gul F, Ahmad I, Ashfaq M, Jan D, Shah F, Li X, Wang D, Fahad M, Fayyaz M, Shah SA (2020) Use of crop growth model to simulate the impact of climate change on yield of various wheat cultivars under different agro-environmental conditions in Khyber Pakhtunkhwa, Pakistan. *Arabian J Geosci* 13:112 https://doi.org/10.1007/s12517-020-5118-1.

Habib ur Rahman M, Ahmad A, Wajid A, Hussain M, Rasul F, Ishaque W, Islam A, Shelia V, Awaish M, Ullah A, Wahid A, Sultana SR, Saud S, Khan S, Fahad S, Hussain M, Hussain S, Nasim W (2017) Application of CSM-CROPGRO-Cotton model for cultivars and optimum planting dates: Evaluation in changing semi-arid climate. *Field Crops Res* http://dx.doi.org/10.1016/j.fcr.2017.07.007.

Haeckermann J, Rott AS, Dorn S (2007) How two different host species influence the performance of a gregarious parasitoid: Host size is not equal to host quality. *J Anim Ecol* 76(2):376–383.

Hammad HM, Farhad W, Abbas F, Fahad S, Saeed S, Nasim W, Bakhat HF (2016) Maize plant nitrogen uptake dynamics at limited irrigation water and nitrogen. *Environ Sci Pollut Res* 24(3):2549–2557. https://doi.org/10.1007/s11356-016-8031-0.

Hammad HM, Abbas F, Saeed S, Shah F, Cerda A, Farhad W, Bernado CC, Wajid N, Mubeen M, Bakhat HF (2018) Offsetting land degradation through nitrogen and water management during maize cultivation under arid conditions. *Land Degrad Dev* 1–10. DOI: 10.1002/ldr.2933.

Hammad HM, Ashraf M, Abbas F, Bakhat HF, Qaisrani SA, Mubeen M, Shah F, Awais M (2019) Environmental factors affecting the frequency of road traffic accidents: A case study of sub-urban area of Pakistan. *Environ Sci Pollut Res.* https://doi.org/10.1007/s11356-019-04752-8.

Hammad HM, Abbas F, Ahmad A, Bakhat HF, Farhad W, Wilkerson CJ W, Shah F, Hoogenboom G (2020a) Predicting kernel growth of maize under controlled water and nitrogen applications. *Int J Plant Prod.* https://doi.org/10.1007/s42106-020-00110-8.

Hammad HM, Khaliq A, Abbas F, Farhad W, Shah F, Aslam M, Shah GM, Nasim W, Mubeen M, Bakhat HF (2020b) Comparative effects of organic and inorganic fertilizers on soil organic carbon and wheat productivity under arid region. *Commun Soil Sci Plant Anal.* DOI: 10.1080/00103624.2020.1763385.

Harvey JA, Bezemer TM, Elzinga JA et al. (2004) Development of the solitary endoparasitoid *Microplitis demolitor*: Host quality does not increase with host age and size. *Ecol Entomol* 29(1):35–43.

Harvey JA, Jervis MA, Gols R et al. (1999) Development of the parasitoid, Cotesia rubecula (Hymenoptera: Braconidae) in Pieris rapae and Pieris brassicae (Lepidoptera: Pieridae): Evidence for host regulation. *J Insect Physiol* 45(2):173–182.

Hellmann JJ, Byers JE, Bierwagen BG et al. (2008) Five potential consequences of climate change for invasive species. *Conserv Biol* 22(3):534–543.

Hoffmann AA, Chown, SL, Clusella-Trullas S (2013) Upper thermal limits in terrestrial ectotherms: How constrained are they? *Funct Ecol* 27(4):934–949.

Hulme PE (2017) Climate change and biological invasions: Evidence, expectations, and response options. *Biol Rev* 92(3):1297–1313.

Huong NTL, Bo YS, Fahad S (2018) Economic impact of climate change on agriculture using Ricardian approach: A case of northwest Vietnam. *J Saudi Soc Agric Sci* 18(4):449–457.

Hussain S, Mubeen M, Ahmad A, Akram W, Hammad HM, Ali M, Masood N, Amin A, Farid HU, Sultana SR, Shah F, Wang D, Nasim W (2019) Using GIS tools to detect the land use/land cover changes during forty years in Lodhran district of Pakistan. *Environ Sci Pollut Res.* https://doi.org/10.1007/s11356-019-06072-3.

Hussain MA, Fahad S, Sharif R, Jan MF, Mujtaba M, Ali Q, Ahmad A, Ahmad H, Amin N, Ajayo BS, Sun C, Gu L, Ahmad I, Jiang Z, Hou J (2020) Multifunctional role of brassinosteroid and its analogues in plants. *Plant Growth Regul.* https://doi.org/10.1007/s10725-020-00647-8.

Hvam A, Toft S (2005) Effects of prey quality on the life history of a harvestman. *J Arachnol* 33(2):582–590.

Ilyas M, Mohammad N, Nadeem K, Ali H, Aamir HK, Kashif H, Fahad S, Aziz K, Abid U, (2020) Drought tolerance strategies in plants: A mechanistic approach. *J Plant Growth Regulation.* https://doi.org/10.1007/s00344-020-10174-5.

Intergovernmental Panel on Climate Change (2013). *Climate change 2013: Mitigation of climate change.* Vol. 3. Cambridge University Press, Cambridge.

Intergovernmental Panel on Climate Change (2014) Climate change 2014: Mitigation of climate change. *Contribution of working group III to the fifth assessment report of the intergovernmental panel on climate change.* Cambridge University Press, Cambridge.

Jan M, Anwar-ul-Haq M, Shah AN, Yousaf M, Iqbal J, Li X, Wang D, Shah F (2019) Modulation in growth, gas exchange, and antioxidant activities of salt-stressed rice (Oryza sativa L.) genotypes by zinc fertilization. *Arabian J Geosci* 12:775 https://doi.org/10.1007/s12517-019-4939-2.

Jasoni R, Kane C, Green C et al. (2004). Altered leaf and root emissions from onion (Allium cepa L.) grown under elevated CO_2 conditions. *Environ Exp Bot* 51(3):273–280.

Johnson SN, Züst T (2018) Climate change and insect pests: Resistance is not futile? *Trends Plant Sci* 23(5):367–369.

Kalinkat G, Rall BC, Björkman C et al. (2015) Effects of climate change on the interactions between insect pests and their natural enemies. *Climate change and insect pests.* Wallingford: CABI, pp. 74–91.

Kamarn M, Wenwen C, Irshad A, Xiangping M, Xudong Z, Wennan S, Junzhi C, Shakeel A, Fahad S, Qingfang H, Tiening L (2017) Effect of paclobutrazol, a potential growth regulator on stalk mechanical strength, lignin accumulation and its relation with lodging resistance of maize. *Plant Growth Regul* 84:317–332. https://doi.org/10.1007/ s10725-017-0342-8.

Khan A, Tan DKY, Munsif F, Afridi MZ, Shah F, Wei F, Fahad S, Zhou R (2017a) Nitrogen nutrition in cotton and control strategies for greenhouse gas emissions: A review. *Environ Sci Pollut Res* 24:23471–23487. https://doi.org/10.1007/s11356-017-0131-y.

Khan A, Kean DKY, Afridi MZ, Luo H, Tung SA, Ajab M, Fahad S (2017b) Nitrogen fertility and abiotic stresses management in cotton crop: A review. *Environ Sci Pollut Res* 24:14551–14566. https://doi.org/10.1007/s11356-017-8920-x.

Lawton JH, Kinne O (2000) *Community Ecology in a Changing World* (Vol. 11). Ecology Institute, Oldendorf, Germany.

Lerdau M, Litva M, Monson R (1994) Plant chemical defense: Monoterpenes and the growth-differentiation balance hypothesis. *Trends Ecol Evolution* 9(2):58–61.

Liebhold AM, Halverson JA, Elmes GA (1992) Gypsy moth invasion in North America: A quantitative analysis. *J Biogeogr* 19(5):513–520.

Lincoln DE, Couvet D, Sionit N (1986) Response of an insect herbivore to host plants grown in carbon dioxide enriched atmospheres. *Oecologia* 69(4):556–560.

Lindroth RL, Kinney KK, Platz CL (1993) Responses of diciduous trees to elevated atmospheric CO_2: productivity, phytochemistry, and insect performance. *Ecology* 74(3):763–777.

Loreto F, Fischbach RJ, Schnitzler J et al. (2001) Monoterpene emission and monoterpene synthase activities in the Mediterranean evergreen oak Quercus ilex L. grown at elevated CO_2 concentrations. *Glob Change Biol* 7(6):709–717.

Lynch M, Gabriel W (1987) Environmental tolerance. *Am Nat* 129:283–303.

Macfadyen S, Mc Donald G, Hill MP (2018) From species distributions to climate change adaptation: Knowledge gaps in managing invertebrate pests in broad-acre grain crops. *Agric Ecosyst Environ* 253:208–219.

Marks S, Lincoln DE (1996) Antiherbivore defense mutualism under elevated carbon dioxide levels: A fungal endophyte and grass. *Environ Entomol* 25(3):618–623.

Memmott J, Craze PG, Waser NM et al. (2007) Global warming and the disruption of plant–pollinator interactions. *Ecol Lett* 10(8):710–717.

Mitchell KA, Hoffmann AA (2010) Thermal ramping rate influences evolutionary potential and species differences for upper thermal limits in Drosophila. *Funct Ecol* 24(3):694–700.

Morimoto N, Imura O, Kiura T (1998) Potential effects of global warming on the occurrence of Japanese pest insects. *Appl Entomol Zool* 33(1):147–155.

Mubeen M, Ahmad A, Hammad HM, Awais M, Farid H, Saleem M, Sami ul Din M, Amin A, Ali A, Shah F, Nasim W (2020) Evaluating the climate change impact on water use efficiency of cotton-wheat in semi-arid conditions using DSSAT model. *J Water Clim Chang.* doi/10.2166/wcc.2019.179/622035/jwc2019179.pdf.

Muhammad B, Adnan M, Munsif F, Fahad S, Saeed M, Wahid F, Arif M, Jr. Amanullah, Wang D, Saud S, Noor M, Zamin M, Subhan F, Saeed B, Raza MA, Mian IA (2019) Substituting urea by organic wastes for improving maize yield in alkaline soil. *J Plant Nutr.* https://doi.org/10.1080/01904167.2019.1659344.

Musolin DL, Saulich AK (2012) Responses of insects to the current climate changes: From physiology and behavior to range shifts. *Entomol Rev* 92(7):715–740.

Myers SS, Smith MR, Guth S et al. (2017) Climate change and global food systems: Potential impacts on food security and under nutrition. *Annu Rev Public Health* 38:259–277.

Nasim W, Ahmad A, Amin A, Tariq M, Awais M, Saqib M, Jabran K, Shah GM, Sultana SR, Hammad HM, Rehmani MIA, Hashmi MZ, Habib Ur Rahman M, Turan V, Fahad S, Suad S, Khan A, Ali S (2017) Radiation efficiency and nitrogen fertilizer impacts on sunflower crop in contrasting environments of Punjab. *Pak Environ Sci Pollut Res* 25:1822–1836. https://doi.org/10.1007/s11356-017-0592-z.

Paradis A, Elkinton J, Hayhoe K et al. (2008) Role of winter temperature and climate change on the survival and future range expansion of the hemlock woolly adelgid (Adelges tsugae) in eastern North America. *Mitig Adapt Strateg Glob Chang* 13(5–6):541–554.

Peñuelas J, Llusià J (2003) BVOCs: Plant defense against climate warming? *Trends Plant Sci* 8(3):105–109.

Peñuelas J, Staudt M (2010) BVOCs and global change. *Trends Plant Sci* 15(3):133–144.

Qian H, Ricklefs RE (2006) The role of exotic species in homogenizing the North American flora. *Ecol Lett* 9(12):1293–1298.

Qian L, He S, Liu X et al. (2018) Effect of elevated CO_2 on the interaction between invasive thrips, Frankliniella occidentalis, and its host kidney bean, Phaseolus vulgaris. *Pest Manag Sci* 74(12):2773–2782.

Raffa KF, Powell EN, Townsend PA (2013). Temperature-driven range expansion of an irruptive insect heightened by weakly coevolved plant defenses. *Proc Natl Acad Sci* 110(6):2193–2198.

Rahel FJ, Olden JD (2008) Assessing the effects of climate change on aquatic invasive species. *Conserv Biol* 22(3):521–533.

Régnière J, Nealis VG (2018) Two sides of a coin: Host-plant synchrony fitness trade-offs in the population dynamics of the western spruce budworm. *Insect Sci* 25(1):117–126.

Rehman M, Fahad S, Saleem MH, Hafeez M, Habib ur Rahman M, Liu F, Deng G (2020) Red light optimized physiological traits and enhanced the growth of ramie (Boehmeria nivea L.). *Photosynthetica* 58(4):922–931.

Robinet C, Baier P, Pennerstorfer J et al. (2007) Modelling the effects of climate change on the potential feeding activity of Thaumetopoea pityocampa (Den. & Schiff.) (Lep., Notodontidae) in France. *Glob Ecol Biogeogr* 16(4):460–471.

Robinet C, Roques A (2010) Direct impacts of recent climate warming on insect populations. *Integr Zool* 5(2):132–142.

Robinet C, Rousselet J, Pineau P et al. (2013) Are heat waves susceptible to mitigate the expansion of a species progressing with global warming? *Ecol Evol* 3(9):2947–2957.

Rosenblatt AE, Schmitz OJ (2016) Climate change, nutrition, and bottom-up and top-down food web processes. *Trends Ecol Evol* 31(12):965–975.

Ryalls JM, Moore BD, Riegler M et al. (2017) Climate and atmospheric change impacts on sap-feeding herbivores: A mechanistic explanation based on functional groups of primary metabolites. *Funct Ecol* 31(1):161–171.

Saleem MH, Fahad S, Adnan M, Mohsin A, Muhammad SR, Muhammad K, Qurban A, Inas AH, Parashuram B, Mubassir A, Reem MH (2020a) Foliar application of gibberellic acid endorsed phytoextraction of copper and alleviates oxidative stress in jute (Corchorus capsularis L.) plant grown in highly copper-contaminated soil of China. *Environ Sci Pollut Res* https://doi.org/10.1007/s11356-020-09764-3.

Saleem MH, Fahad S, Shahid UK, Mairaj D, Abid U, Ayman ELS, Akbar H, Analía L, Lijun L (2020b) Copper-induced oxidative stress, initiation of antioxidants and phytoremediation potential of flax (Linum usitatissimum L.) seedlings grown under the mixing of two different soils of China. *Environ Sci Poll Res* https://doi.org/10.1007/s11356-019-07264-7.

Saleem MH, Rehman M, Fahad S, Tung SA, Iqbal N, Hassan A, Ayub A, Wahid MA, Shaukat S, Liu L, Deng G (2020c) Leaf gas exchange, oxidative stress, and physiological attributes of rapeseed (Brassica napus L.) grown under different light-emitting diodes. *Photosynthetica* 58(3):836–845.

Saud S, Chen Y, Long B, Fahad S, Sadiq A (2013) The different impact on the growth of cool season turf grass under the various conditions on salinity and drought stress. *Int J Agric Sci Res* 3:77–84.

Saud S, Li X, Chen Y, Zhang L, Fahad S, Hussain S, Sadiq A, Chen Y (2014) Silicon application increases drought tolerance of Kentucky bluegrass by improving plant water relations and morph physiological functions. *SciWorld J* 2014:1–10. https://doi.org/10.1155/2014/ 368694.

Saud S, Chen Y, Fahad S, Hussain S, Na L, Xin L, Alhussien SA (2016) Silicate application increases the photosynthesis and its associated metabolic activities in Kentucky bluegrass under drought stress and post-drought recovery. *Environ Sci Pollut Res* 23(17):17647–17655. https://doi.org/10.1007/s11356-016-6957-x.

Saud S, Fahad S, Yajun C, Ihsan MZ, Hammad HM, Nasim W, Amanullah Jr, Arif M and Alharby H (2017) Effects of nitrogen supply on water stress and recovery mechanisms in Kentucky bluegrass plants. *Front Plant Sci* 8:983. doi: 10.3389/fpls.2017.00983.

Saud S, Fahad S, Cui G, Chen Y, Anwar S (2020) Determining nitrogen isotopes discrimination under drought stress on enzymatic activities, nitrogen isotope abundance and water contents of Kentucky bluegrass. *Sci Rep* 10:6415. https://doi.org/10.1038/s41598-020-63548-w.

Schurr FM, Pagel J, Cabral JS et al. (2012) How to understand species' niches and range dynamics: A demographic research agenda for biogeography. *J Biogeogr* 39(12):2146–2162.

Schweiger O, Settele J, Kudrna O et al. (2008) Climate change can cause spatial mismatch of trophically interacting species. *Ecology* 89(12):3472–3479.

Seidl R, Rammer W (2017) Climate change amplifies the interactions between wind and bark beetle disturbances in forest landscapes. *Landsc Ecol* 32(7):1485–1498.

Shafi MI, Adnan M, Fahad S, Fazli W, Ahsan K, Zhen Y, Subhan D, Zafar-ul-Hye M, Brtnicky M, Datta R (2020) Application of single superphosphate with humic acid improves the growth, yield and phosphorus uptake of wheat (Triticum aestivum L.) in calcareous soil. *Agronomy* (10):1224. doi:10.3390/agronomy10091224.

Shah F, Lixiao N, Kehui C, Tariq S, Wei W, Chang C, Liyang Z, Farhan A, Fahad S, Huang J (2013) Rice grain yield and component responses to near 2°C of warming. *Field Crop Res* 157:98–110.

Sharma HC (2014) Climate change effects on insects: Implications for crop protection and food security. *J Crop Improv* 28(2):229–259.

Stiling P, Rossi AM, Hungate B et al. (1999) Decreased leaf-miner abundance in elevated CO_2: Reduced leaf quality and increased parasitoid attack. *Ecol Appl* 9(1):240–244.

Stiling P, Cornelissen T (2007) How does elevated carbon dioxide (CO_2) affect plant–herbivore interactions? A field experiment and meta-analysis of CO_2-mediated changes on plant chemistry and herbivore performance. *Glob Chang Biol* 13(9):1823–1842.

Subhan D, Zafar-ul-Hye M, Fahad S, Saud S, Brtnicky M, Hammerschmiedt T, Datta R (2020) Drought stress alleviation by ACC Deaminase Producing Achromobacter xylosoxidans and Enterobacter cloacae, with and without timber waste biochar in maize. *Sustainability* 12(6286). doi:10.3390/su12156286.

Terblanche JS, Karsten M, Mitchell KA et al. (2015) Physiological variation of insects in agricultural landscapes: Potential impacts of climate change. *Clim Chang Insect Pests* 8:92.

Theoharides KA, Dukes JS (2007) Plant invasion pattern and process: Factors affecting plant invasion at four spatio-temporal stages. *New Phytol* 176:256–273.

Thomson LJ, Macfadyen S, Hoffmann AA (2010) Predicting the effects of climate change on natural enemies of agricultural pests. *Biol Control* 52(3):296–306.

Tognetti R, Johnson JD, Michelozzi M, Raschi A (1998) Response of foliar metabolism in mature trees of Quercus pubescens and Quercus ilex to long-term elevated CO_2. *Environ Exp Bot* 39(3):233–245.

Torquebiau E (Ed.) (2016) *Climate Change and Agriculture Worldwide*. Springer.

Tougeron K, Le Lann C, Brodeur J et al. (2017) Are aphid parasitoids from mild winter climates losing their winter diapause? *Oecologia* 183(3):619–629.

Tougou D, Musolin DL, Fujisaki K (2009) Some like it hot! Rapid climate change promotes changes in distribution ranges of Nezara viridula and Nezara antennata in Japan. *Entomol Exp Appl* 130(3):249–258.

Trębicki P, Dáder B, Vassiliadis S, Fereres A (2017) Insect–plant–pathogen interactions as shaped by future climate: Effects on biology, distribution, and implications for agriculture. *Insect Sci* 24(6):975–989.

Vermeij GJ (1996) An agenda for invasion biology. *Biol Conserv* 78(1–2):3–9.

Visser ME, Holleman LJ (2001) Warmer springs disrupt the synchrony of oak and winter moth phenology. *Proc R Soc London, Ser B: Biol Sci* 268(1464):289–294.

Wahid F, Fahad S, Subhan D, Adnan M, Zhen Y, Saud S, Manzer HS, Brtnicky M, Hammerschmiedt T, Datta R (2020) Sustainable management with mycorrhizae and phosphate solubilizing bacteria for enhanced phosphorus uptake in calcareous soils. *Agriculture* 10(334). doi:10.3390/agriculture10080334.

Wang XY, Yang ZQ, Wu H, Gould JR (2008) Effects of host size on the sex ratio, clutch size, and size of adult Spathius agrili, an ectoparasitoid of emerald ash borer. *Biol Control* 44(1):7–12.

Wang D, Shah F, Shah S, Kamran M, Khan A, Khan MN, Hammad HM, Nasim W (2018) Morphological acclimation to agronomic manipulation in leaf dispersion and orientation to promote 'Ideotype' breeding: Evidence from 3D visual modeling of 'super' rice (Oryza sativa L.). *Plant Physiol Biochem.* https://doi.org/10.1016/j.plaphy.2018.11.010.

Watt AD, Whittaker JB, Docherty M et al. (1995) The impact of elevated atmospheric CO_2 on insect herbivores. *Insects in a Changing Environment.* London: Academic Press, pp. 197–217.

Weed AS, Ayres MP, Hicke JA (2013) Consequences of climate change for biotic disturbances in North American forests. *Ecol Monogr* 83(4):441–470.

Williamson M, Fitter A (1996) The varying success of invaders. *Ecology* 77(6):1661–1666.

Williamson M (2006) Explaining and predicting the success of invading species at different stages of invasion. *Biol Invasions* 8(7):1561–1568.

Wu C, Tang S, Li G, Wang S, Fahad S, Ding Y (2019) Roles of phytohormone changes in the grain yield of rice plants exposed to heat: A review. *PeerJ* 7:e7792. DOI 10.7717/peerj.7792.

Wu C, Kehui C, She T, Ganghua L, Shaohua W, Fahad S, Lixiao N, Jianliang H, Shaobing P, Yanfeng D (2020) Intensified pollination and fertilization ameliorate heat injury in rice (Oryza sativa L.) during the flowering stage. *Field Crops Res* 252:107795.

Yang LH, Rudolf VHW (2010) Phenology, ontogeny and the effects of climate change on the timing of species interactions. *Ecol Lett* 13(1):1–10.

Yang Z, Zhang Z, Zhang T, Fahad S, Cui K, Nie L, Peng S, Huang J (2017) The effect of season-long temperature increases on rice cultivars grown in the central and southern regions of China. *Front Plant Sci* 8:1908. https://doi.org/10.3389/fpls.2017.01908.

Yuan JS, Himanen SJ, Holopainen JK et al. (2009) Smelling global climate change: Mitigation of function for plant volatile organic compounds. *Trends Ecol Evol* 24(6):323–331.

Zafar-ul-Hye M, Naeem M, Danish S, Fahad S, Datta R, Abbas M, Rahi AA, Brtnicky M, Holatko, Jiri, Tarar ZH, Nasir M (2020a) Alleviation of cadmium adverse effects by improving nutrients uptake in bitter gourd through cadmium tolerant rhizobacteria. *Environments* 7(54). doi:10.3390/environments7080054

Zafar-ul-Hye M, Tahzeeb-ul-Hassan M, Abid M, Fahad S, Brtnicky M, Dokulilova T, Datta RD, Danish S (2020b) Potential role of compost mixed biochar with rhizobacteria in mitigating lead toxicity in spinach. *Sci Rep* 10:12159. https://doi.org/10.1038/s41598-020-69183-9.

Zahida Z, Hafiz FB, Zulfiqar AS, Ghulam MS, Fahad S, Muhammad RA, Hafiz MH,Wajid N, Muhammad S (2017) Effect of water management and silicon on germination, growth, phosphorus and arsenic uptake in rice. *Ecotoxicol Environ Saf* 144:11–18.

Zaman QU, Aslam Z, Yaseen M, Ishan MZ, Khan A, Shah F, Basir S, Ramzani PMA, Naeem M (2017) Zinc biofortification in rice: Leveraging agriculture to moderate hidden hunger in developing countries. *Arch Agron Soil Sci* 64:147–161. https://doi.org/10.1080/03650340.2017.1338343.

Zamin M, Khattak AM, Salim AM, Marcum KB, Shakur M, Shah S, Jan I, Fahad S (2019) Performance of Aeluropus lagopoides (mangrove grass) ecotypes, a potential turfgrass, under high saline conditions. *Environ Sci Pollut Res*. https://doi.org/10.1007/s11356-019-04838-3.

Zvereva EL, Kozlov MV (2006) Consequences of simultaneous elevation of carbon dioxide and temperature for plant–herbivore interactions: A meta-analysis. *Glob Change Biol* 12(1):27–41.

11

Insect–Plant Interactions

Abdel Rahman Al-Tawaha[1], Anamika Sharma[2], Dhriti Banerjee[3], and Jayita Sengupta[3]
[1]Department of Biological Sciences, Al Hussein Bin Talal University, Ma'an, Jordan
[2]Western Triangle Agriculture Research Center, Montana State University Bozeman, USA
[3]Zoological Survey of India, Ministry of Environment, Forests and Climate Change (Govt. of India) 'M' Block, New Alipore, Kolkata

11.1 Introduction

Insects share antagonism, commensalism, and mutualism relationships with plants and all these interactions play a major role in food production in agriculture, horticulture, and forestry (Raman, 1997; Schoonhoven et al., 2005). This interaction happens on the whole-plant and community levels and on the morphological and cellular levels of insects and plants (Kessler and Baldwin, 2002; Sharma et al., 2014). Primarily, only 9 out of 30 orders are phytophagous (Gullan and Cranston, 2000). Nonetheless, members of Orthoptera, Hemiptera, Hymenoptera, Diptera, Lepidoptera, Coleoptera, Thysanoptera, Phasmida, Trichoptera, Plecoptera, and Ephemeroptera, Acarina (Arachnida) are plant feeding. Phytophagous insects seem to show a certain degree of coevolution with plants and exhibit two modes of co-evolution: specific and diffuse (Gullan and Cranston, 2000). Phytophagous insects have chewing and biting, sap-sucking, and siphoning types of mouth parts. Insects feed on plants to consume primary metabolites (e.g. carbohydrates, lipids, and proteins) for their growth and development. Plants also produce a high diversity of secondary metabolites (e.g. alkaloids, terpenoids, and phenolics) and insects become continuously challenged with counter mechanisms to detoxify plant defense mechanisms (Nishida, 2014). Insects are polyphagous, oligophagous, and monophagous due to variation in their requirement for primary metabolites for food, reproduction, habitat, and microclimate (Gullan and Cranston, 2000; Sharma et al., 2015).

Insect and plant interactions become complicated and necessary since plants serve as food for insects but at the same time have an array of secondary metabolites which can function as phago-stimulant and deterrent, and they also get sequestered by insects and used for their own defense mechanisms. A plant's secondary metabolites are also used in communication with insects. Several secondary metabolites also enable the management of insect pests through the incorporation of these compounds in attract-and-kill strategies and push-and-pull strategies (Nishida, 2014; Khan et al., 2016). Insect attack also induces volatile production in plants, which, in turn, attracts the predators and parasites of insects. In short, from finding the plant to feed and lay eggs to coping with the plant's secondary metabolites, both partners in insect–plant interaction show an array of strategies that benefit from this interaction.

Insects' sensory structures enable them to find the host plant which mainly includes optical and odor clues. However, optical clues for insects remain fairly constant since light intensity is the only factor which influences the optical clues. In contrast, chemical clues or odorous clues vary immensely due to variation in wind speed, temperature, and plant's physiological state especially when the plants are under attack (Blaakmeer et al., 1994; Takabayashi et al., 1994). Insects' endocrine system also plays an integral part in host plant selection. Choice of annual and perennial plants or a particular developmental stage depends on the insect's physiological requirement. Hormones and pheromones required for molting, morphism, diapause, maturation, and mating in insects depend on the quantity and quality of host plant (Goehring and Osberhauser, 2002; Ochieng et al., 2002; Reddy and Guerrero, 2010).

Symbionts play a major role in allowing phytophagous insects to feed on plants with several secondary metabolites which can be harmful to insects. Endosymbionts live in insects extracellularly and intracellularly and improve the capacity of insects to digest food, supply essential amino acids which otherwise cannot be obtained by insects, and detoxify plant's allelochemicals which can be harmful to insects (Douglas, 1998; Moran, 2002). Along with endosymbionts, for detoxification of secondary metabolites of plants, insects excrete them physiologically and also detoxify the metabolites with the help of enzymes. The enzymatic detoxification happens before food ingestion, that is, through salivary proteins and also during digestion (Sharma et al., 2014).

On one hand, insects possess various features to successfully consume plants, and on the other hand, plants comprise several characteristics to avoid getting attacked by insects. Plants imply 'phenological escape' to avoid getting attacked by specific insects (Visser and Holleman, 2001); plants imply secondary metabolites which can be toxic and digestibility reducer to insects (Baldwin, 1998); emit insects-induced volatiles which attract natural enemies of insects (Landolt, 1993); can increase nectar secretion to provide food to natural enemies (Heil et al., 2001); and has morphological traits such as surface structure and trichomes (Panda and Kush, 1995).

11.2 Stratum in Insect–Plant Interaction: Tri-Trophic Interactions or Three-Way Interactions

Three-way interactions occur when another component adds up in the insect-plant interaction. Tritrophic interaction occurs between insects–plants–biotic factors (natural enemies, pathogens, endophytic fungus, endosymbionts, genetic variability) and insect–plants–abiotic factors (soil, drought, light intensity, wind speed, air pollution, and other temporal variants and genetic variability). These participants profoundly influence the dynamics of insect–plant interaction (Mooney et al., 2010; Shikano et al., 2017). Increase in nitrogen due to nitrogen fertilizers in soil decreases carbon-based secondary metabolites in plant's foliage (Koricheva, 1999) and hence can impact the feeding behaviour of insects.

Microbes induce changes in plant defenses and nutritive quality of plants and hence as the consequences also change the behaviour and fitness of phytophagous insects. Insect (herbivore and parasitoid)-associated microbes can favourably impact the fitness of insects by suppressing and detoxifying plant defenses and phytochemicals. Whereas phytopathogens (biotrophs and necrotrophs) can alter plant quality and hence can influence insect behaviour and fitness. Similarly beneficial microbes for plants promote the plant nutritional quality and phytochemical composition. Plant-beneficial microbes can promote plant growth and influence plant nutritional and phytochemical composition that can positively or negatively influence insect fitness (Shikano et al., 2017; Raman and Suryanarayanan, 2017). The entomopathogenic fungus can act as an endophyte and can influence and improve plant defenses and hence influence insect physiology and also uses plants to aid their dispersal to insects (McKinnon et al., 2018). The fungus interaction with gall-inducing (especially gall-inducing Cecidomyiidae) insects shows a mutualistic relationship with insects where insects (Raman and Suryanarayanan, 2017).

11.3 Resistance

Plant-derived insecticides have been extensively used to manage insects (e.g. nicotine, rotenone, pyrethrins, and *Azadirachta indica*). Bracken fern (*Pteridium aquilinum*: Pteridophyta) strikingly has fewer arthropods which are due to the presence of toxic components such as indanones, cyanogenic glycosides, and tannins, and these can be used as indirect defenses against insects (Radhika et al., 2012). Plants involve such chemical compounds to avoid insect attack which is known as 'induced direct resistance', whereas herbivory-induced production of carnivore-attracting volatiles creates 'induced indirect resistance'. Other than induced resistance, plants also have constitutive, genetically inherent qualities which make plants susceptible and resistant. Plants show three categories of resistance antibiosis (adverse effects on insect survival), antixenosis (adverse effects on insect behaviour), and tolerance (ability to withstand, repair, or recover from insect damage) (categorized by Painter, 1951) (Smith, 2005). Two classes of plant genes contribute to abiotic and biotic resistance in plants. Resistance (R) and defence response (DR) genes are responsible for defense responses in plants and resistant gene analogs (RGAs) get used to identifying the arthropod resistant genes (Smith, 2005).

In the next section, we compile recent major advances, challenges, and future perspectives of this interaction.

11.4 Advances, Future Perspectives, and Challenges

In the past three decades, various aspects of insect–plant interaction have been explored extensively, particularly in plant biochemistry and evolution. Molecular ecology of insect–plant interaction is being extensively studied. However, to date, well-characterized mutants and transgenic lines are only available for model species such as *Arabidopsis thaliana* (Brassicaceae) in plants and *Drosophila melanogaster* (Diptera: Drosophilidae) in insects. We are at present gaining knowledge on techniques allowing us to study the presence or absence of a functional expression of the gene of interest, nevertheless, a greater understanding of quantitative variation is required by exploiting natural variation in finding genotypes that differ quantitatively in the functional expression of genes of interest (Schoonhoven et al., 2005; Zheng and Dicke, 2008).

Molecular aspects of the mechanism of insect and plant interaction in terms of insect effectors and plant responses are studied extensively and as of now we know about damage-associated molecular patterns (DAMPs) and herbivore-associated molecular patterns (HAMPs) called elicitors and effectors (Giron et al., 2018). Recently several effectors have been identified from chewing and sap-sucking insects (Acevedo et al., 2017; Kaloshian and Walling, 2016; McKenna et al., 2016; Sharma et al., 2014). However, a functional characterization and molecular targets of insect elicitors and effectors and also a thorough understanding of links between the induced signalling cascades and the diversity in plant responses to insects are required (Giron et al., 2018).

Lately, the molecular aspects of identification and role of symbionts in plant-feeding insects have been explored and the difference in symbiont titer in sex and different orders of insects (especially in holometabolous and hemimetabolous insects) have been explored further. However, regulation of these symbionts and evolution of putative regulatory genes remains a mystery this date (Skidmore and Hansen, 2017). The recent advancement in network analysis tools enabled us to study the integration of mechanism of interaction into ecological networks. These tools are efficient in quantifying relevant parameters and incorporate various types of traits to study community ecology. Two approaches are used to study the interaction: autecological approach (the study of individual interaction in the laboratory) and synecological approach (the study of entire multitrophic webs) (Richards et al., 2015; Scherrer et al., 2016). Nevertheless, appropriate temporal and spatial scales and natural history of the ecosystem is mandatory to yield an applicable design to understand the network. Also, greater work on the role of microorganisms in multitrophic interaction, the connection between plant defense and reproduction, and use of this tool in practical aspects such as managing pest pressure, and biological conservation is direly needed (Giron et al., 2018). Another aspect where we advanced in past few years is the relationship between community ecology and phylogenetics, including abiotic factors and both mutualistic and antagonistic insect–plant relationships (Kergoat et al., 2017; Trivellone et al., 2017). However future studies are focusing on understanding the beneficial and detrimental interactions, combining the effect of multiple factors, and integrating history into present insect-plant interaction (Giron et al., 2018). Evolutionary genomics (the study of genomic changes over the course of evolution) is establishing a better link between genotypes and phenotypes (Nosil and Feder, 2011). In past few years, molecular mechanism of plant choice has been studied and key genes are revealed responsible for these interactions (Kaloshian and Walling, 2016); however, the establishment of a stronger link between genotype and phenotype, and an intensive comparative genomics study is further required (Giron et al., 2018).

The omics techniques (genomics, transcriptomics, and proteomics) have provided answers to several puzzling queries in ecological interactions, especially in terms of the revealing regulatory genes and processes involved in the adaptation of plants to abiotic and biotic stresses (Dyer et al., 2018). Today, several advanced techniques are available to study the omics and chemical ecology, but the metabolomics step is still mainly unclear due to the vast variety of plant secondary compounds and several biosynthetic pathways. Characterization of individual compounds, structure determination, and examination of the bioactivity of pure compounds, will reveal the long-standing questions in chemistry, ecology, and evolutionary biology (Dyer et al., 2018). Impact of climate change on this interaction is another facet which recently has emerged and gained a lot of attention. Study of microclimatic conditions is required to be integrated into species distribution models (SDMs) and environmental niche model (ENMs). Study of trait measurement and study of modulation of chemical communication between plants-herbivores-predators, and parasitoids due to microclimate will improve our understanding (Giron et al., 2018).

At present, a major challenge we face is to integrate the research approaches addressing different levels of biological organization and their functions in ecological communities (Kessler and Baldwin, 2002; Bezemer and van Dam, 2005; Kessler and Halitschke, 2007; Schoonhoven et al., 2005; Snoeren et al., 2007); connecting these research fields is required at present. The ever-changing insect–plant system shows immense variation in time and space. The changes also occur in the plant and insect behaviours due to learning and mutation (insect) induced, epigenetic, and gene mutations (plant). These variations provide greater chances of survival in a constantly changing environment. The interaction between insects and plants in the dynamic environment also depends on the natural and cultivated planting. The unpredictable genetic and temporal variability in both ecosystems also depends on the history and habitat diversity of the system. Greater habitat diversity and natural resistant traits in natural ecosystem get lost in the selected cultivating system and makes the use of pesticide necessary in the cultivated ecosystem and that adds a new factor in insect-plant interaction (Bruce, 2015; Schoonhoven et al., 2005; Sharma et al., 2018a,b).

Reduction of insect pest pressure in a natural and cultivated ecosystem is the need of the era. Although several facets such as development of resistance (Smith, 2005), role of symbionts in pest host range and providing a better survival to the insect pest (Hansen and Moran, 2014), and development of reduced risk insecticide have been explored, various other aspects are still understudied and require immediate attention, such as greater molecular understanding of benefits of plant diversification, safe incorporation of new techniques (e.g. gene editing), involvement of plant breeders, and most importantly incorporating 'omics' in each and every step of study is required (Giron et al., 2018).

Current investigations of plant–insect interactions hold promise for using the variation in polyphagous, oligophagous, and monophagous insects and their evolution, host specificity of specific insects, inter- and intra-specific relationship between insects and plants, and most importantly the integration of various streams (functional and applied), and incorporation of these techniques in the present agricultural system is direly needed.

11.5 Insects: Omnipresent and Versatile

When Pandora, out of ceaseless curiosity, opened her 'Box,' out came a multitude of plagues for the ill-fated men. What came out of the box were similar totiny buzzing moths that stung Pandora all over. When she quickly slammed the lid shut, all that was left behind, fluttering inside the box, was the only good thing–hope–which could heal all ills. Possibly Pandora had released all the disease-causing insects and kept behind the small chocolate midges which could pollinate the cacao plant, thus producing cocoa, which brought hope and happiness to the hapless mankind. So goes the legend.

With approximately 1 million described species, and an estimated more than 4 million undescribed species, the insect fauna has always been both the good and the evil in the Pandora box of animal diversity (Stork et al., 2015). With an extensive range of ecosystem services (Dangles and Casas, 2019), insects have always emerged as superheroes, rendering services varying from nutrient cycling, soil turnover, and decomposition of biological byproducts (Allsopp et al., 2008) to blossoming of chocolate plants (Trivellone et al., 2017). Because of their significant economic and aesthetic benefits as well as cultural values to human society (Elhassan et al., 2019), insects have always appeared as a globally important faunal group (Gill et al., 2016). Out of the manifold responsibilities (Garibaldi et al., 2011), insects carry out in their surrounding ecosystems, one of the utmost important service is pollination of flowering plants (Prado et al., 2019). It is precisely found that three-fourth of the world's total flowering plants, 35% of world's food crops and 85% of total described flowering plant species (Powney et al., 2019) of world are reliant on animal pollination (Kevan, 1972; Ollerton et al., 2011). Out of this, a significant percentage is dependent on mobile pollinating activity of insects (Kevan and Baker, 1983; Kremen et al., 2007). The increasing anthropogenic pressure is already raising the level of calorific intake of humans with every passing year. This in turn will definitely hoist a question regarding food security of upcoming generations (Godfray et al., 2010). Though the increasing pressure of the billions of mouths to feed is exhausting both our natural and manmade agro-ecosystems (Powney et al., 2019), unfortunately, the role of insects in pollination and subsequent mankind's development has still remained a neglected issue of underestimated economics.

11.6 Insect Pollinators: The Unsung Superheroes!

It is the general consensus that the honeybee is the most versatile, ubiquitous, and commonly used managed pollinator (Kremen et al., 2007; Requier et al., 2019). However, entomophily or insect pollination is not restricted only to the bees (Venkatraman, 2013). Apart from honeybees, there are also other insects (Kremen et al., 2007), both free living and wild, providing allied ecosystem services to crops (Garibaldi et al., 2013). Melittophily or the pollination by honeybees and psychrophilic (pollination by butterflies) were the only popular insect pollination mechanisms known over a long period of time (Venkatraman, 2013). But there are several other lesser known insects groups like wasps, moths, beetles, flies, midges, gnats, ants, and bugs which aid in pollination (Hall et al., 2017; Kevan et al., 1990). Apart from honeybees, butterflies and beetles have also acquired the position of first line pollinators (Perrot et al., 2019). Hover flies, fruit flies, and bumble bees are also getting the fame as alternative pollinators in recent era (Sengupta et al., 2019). Wild pollinators other than the few managed hymenopteran taxa recently have been recognized for their role in increasing and stabilizing crop pollination services (Garibaldi et al., 2011). Among the diverse group of insects, since only few are well established as pollinators (Majewska and Altizer, 2019), while few are still struggling for their identity as pollinators (Jauker et al., 2019; Orford et al., 2015), pollination services rendered by all of them should be properly assessed and addressed (Winder, 1978).

While discussing about pollinators, the foremost group that appears in the research history of pollination worldwide is the bee (Aizen and Harder, 2009; Requier et al., 2019). Presently 25,000 species of bees are known (Kroupa et al., 2019). The bee pollination as well as the bee pollinators are always prominent in research headlines across the globe. Honeybees are the most commonly exploited bees (Watanabe1994; Alaux et al., 2019) for commercial pollination purposes followed by bumblebees (Crowther et al., 2019; Galen and Stanton, 1989; Vaudo et al., 2018;). On the other hand, wild bees also share an important proportion in the pollination network (Greenleaf and Kremen, 2006; Klatt et al., 2014). As a result of this, managed bee hives are increasing worldwide (Joseph et al., 2019). No doubt honeybee pollination services are rising globally (Hill et al., 2019) with an approximate 2.4 million colonies in the United States presently (Goodrich, 2019; Rucker et al., 2012). Two-thirds of these are used for pollination of agricultural crops only (Levin, 1983, Grab et al., 2019). Surprisingly, out of these, 1 million colonies are being used every year for pollination of almond crops alone (Lazaro and Alomar, 2019). Globally, the main food and agricultural crops which are being pollinated by honeybees are tomatoes, cabbages, peppers, melons, watermelons, apple, avocado, blueberry, cranberry, strawberry, kiwi, pumpkin, zucchini, squash, currants, raspberries, blackberries, and many others (Blair and Sampson, 2000; Macinnis and Forrest, 2019). *Apis mellifera* Linnaeus,1758, *Apis (Apis) dorsata* Fabricius,1793, *Apis (Apis) cerana* Fabricius, 1793, *Bombus atripes* Smith, 1852, *Bombus impatiens*

Cresson, 1863, *Bombus ruderatus* Fabricius, 1775, *Bombus dahlbomii* Guerin Meneville, 1835 (Paini, 2004) are the first benchers in the class of honeybee and bumble bee pollination (Ye et al., 2018). Economic benefits derived from pollination services by honeybees can be summarized in one sentence: honeybees pollinate one-third of the food we eat (Marshman et al., 2019). Only in Nearctic realm, it is estimated that pollination services of honeybee are estimated more than $1.46 billion (Saez et al., 2019). Whereas, worldwide the economic value is as high as $90 billion approximately (Goodrich, 2019). Unfortunately, according to a latest survey, 41% of global insect species have declined (Potts et al., 2010, Jacobson et al., 2018) over the past decade (Gangwani and Landin, 2018), and these make the bees one of the most affected taxa (Cameron et al., 2011; Watanabe, 1994) among the range of declining pollinators (Lebuhn et al., 2013).

The insect world always reserves a special place for the charismatic butterflies (Ray et al., 2019). Butterflies and moths together share a bulk credit for pollination (Guenat et al., 2019). Presently, 174,240 species of moth and butterflies are known worldwide, which include 17,698 species of butterflies across the globe (Lahondère et al., 2019). Being nocturnal in nature, the phalaenophilic moths can also mediate the night time pollination services (Trunschke et al., 2019) unlike the day flying bees. The main family of flowers which are being aided with pollination services by butterflies (Balducci et al., 2019) are Lamiaceae, Aslepiadaceae, Asteraceae, and so on. Apart from this, different groups of wild flowers, including evergreen violet, western verbena, and bush money flowers, along with food crops, including peas, beans, cilantro, parsnip, celery, dill, lavender, basil, rosemary, thyme, oregano, broccoli, and cauliflower, are also being served by pollination services from butterflies (Bartomeus and Dicks, 2019; Berkley et al., 2018). Some of the notable lepidopteran pollinators are *Melantis lada* (Linnaeus 1758), *Ypthima huebneri* (Kirby 1871), *Hebomoia glaucipe* (Linnaeus 1758), *Ixias pyrene* (Linnaeus 1758) (Duara and Kalita 2014). Among moths, *Macroglossum corythuus* (Walker 1856), *Macroglossum alluaudi* (De Joannis 1893), *Macroglossum hirundo* (Boisduval 1832) are some important plant pollinators (Dotterl et al., 2005).

One of the most ancient pollinators of pre-historic time were the cantharophilic beetles (Johnson et al., 2007; Sayers et al., 2019). They are reported to have pollinated flowers about 150 million years ago (Carmichael, 2019). This happened to be 50 years earlier than bees (Gill et al., 2016). Fossil evidences have also suggested that they were more active as mess and soil pollinators (Jiménez et al., 2019; Straarup et al., 2018). Approximately four lac species of beetles are known worldwide (Roskov et al., 2018) of which a good percentile of species are involved in pollination, of which the most notable are *Cyclocephala borealis* Arrow, 1911, *Cyclocephala melanocephala* Fabricius, 1775, *Erioscelis columbica* Endrodi, 1966 (Young, 1988). Along with regular food crops, the additional crops for which beetles mediate pollination are magnolia, spicebush, macadamia, nuts, paw paws, and so forth (Straarup et al., 2018).

Another major group of pollinators includes wasps (Souto et al., 2018). Pollination research have always prioritized them as a group of hard-working pollinators. The most common pollinators among wasps are fig wasps, common wasps, and European wasps (Hu et al., 2013). Among wasps some of the most notable pollinators are *Vespa gormanica* Fabricius, 1793, *Vespa vulgaris* Linnaeus, 1758, *Pseudomasaris vespoides* Cresson, 1863, *Pseudomasaris zonalis* Cresson, 1864, and *Pseudomasaris marginalis* Cresson, 1864 (Tepedino et al., 2006).

Ants are a group of hymenopteran pollinators whose contribution in pollination history has been rarely admired (Majetic et al., 2019; Petanidou and Ellis, 1993; Rostás et al., 2018). Although because of supreme mobility, they can promote genetic diversity via pollination (Longino and Branstetter, 2019). In Australia, a large proportion of orchid and lilies are effectively pollinated by ants (Thurman et al., 2019). One of the effective pollinating ants of Brazilian Savana is *Paepalanthus lundii* (De Claro et al., 2019).

A group of pollinating insects without whom the story of pollination will always be incomplete are the dipteran pollinators (Sengupta et al., 2016; Sengupta et al., 2018a). Presently it has been reported that out of 154 dipteran families known worldwide (Evenhuis and Pape, 2019), 71 families have flower-visiting species, aiding regular pollination to at least 555 plant species which includes more than 100 cultivated plant species all over the world (Sengupta et al., 2019). These plant species includes a wide range of economically important food crops like mango, cocoa, onion, cacao, tea, cashew, cauliflower, mustards, carrots, apples, oil seed, and rapeseeds (Manobanda et al., 2018; Mitra et al., 2005; Orford et al., 2015). The giant chocolate industry across the world, worth nearly $ 100 billion (Claus et al., 2018) in a year depends on the pollination network of tiny biting midges (Diptera: Ceratopogonidae) which are responsible for pollination of the cacao flower (*Theobroma cacao*, Linnaeus). Undeniably flies are thus recognized as second most important flower visitor and pollinator (Inouye et al., 2015). Different families specifically Syrphidae, Bombyliidae, Tephritidae, Dorsophilidae, Calliphoridae, and Tachinidae have been established as good pollinators (Banerjee, 2005) because they are assorted, common, and ubiquitous in both natural and managed habitations (Vanbergen and Initiative, 2013). Research has shown that the dipteran pollinators transmit a good percentile of total pollen carried out by farmland pollinators (Biesmeijer et al., 2006). Across all habitat scales, dipteran flies are available as pollinators, but their efficiency as a pollination leader is mainly accredited in higher elevational and latitudinal areas (Sengupta et al., 2018b) like arctic, sub-arctic, and alpine types of ecosystems (Sengupta et al., 2019) where profusion of bees and other hymenoptera is limited due to environmental adversity. *Episyrphus (Episyrphus) balteatus* (De Geer, 1776), *Ischiodon scutellaris* (Fabricius, 1805), *Sphaerophoria (Sphaerophoria) indiana* (Bigot, 1884), *Melanostoma orientale* (Wiedemann 1824), *Bactrocera (Zeugodacus) scutellaris* (Bezzi, 1909), *Bombylius (Bombylius) major* (Linnaeus, 1758), and many others are some pollinators among dipteran flies (Sengupta et al., 2018a).

Surprisingly, the blood sucker mosquitoes also have a share in contribution toward pollination (Gill et al., 2016). The male mosquitoes who are entirely dependent on pollen and nectar for food resources (Dangles and Casas, 2019) help in mediating pollen networks, mainly in certain groups of rare orchids (Lahondère et al., 2019). As loss of diversity of pollinating flies will lead to shortage of pollination services particularly in cooler and wetter habitats where flies replace bees (Devoto et al., 2005). Shortage of this pollination will enhance the demand for agricultural land, a trend that will be more detrimental in the developing world (Guenat et al., 2019) potentially leading to decreased food security for future.

11.7 The Silent Alarm: The Unacknowledged Extinction Threat!

The United Nations in October 2010 released a report stating that insect pollination was valued at $134 billion. Although research has shown that the role of pollination is priceless (Decourtye et al., 2019), unfortunately it is still underrated (Chain et al., 2019). Earth's entomo-fauna is in an ongoing state of collapse, including the pollinators (Rhodes, 2019). An increased rate of threats upon insect pollinators (Campbell, 2013) will appear soon as a serious global issue (Potts et al., 2010) if not properly addressed in the very near future. In Europe, 38% of bee and hoverfly species have been in decline (Biesmeijer et al., 2006) and 24% of the European bumble bee population is currently facing serious risk of extinction (Goulson et al., 2008). More precisely in the UK, two-thirds of the moth species are in a long-term declining process (Goulson et al., 2015), where 75% of the total butterfly species are in decline (Powney et al., 2019). Three species of bumble bee have already become extinct, two species are critically endangered, and ten species are facing serious threats of extinction. This includes five species that were even common in the 1980s (Gill et al., 2016). Further application of genetically modified crops carrying traits of herbicide tolerance or insect resistance is diminishing the food resources for the insect pollinators (IPBES et al., 2016).

Things are equally alarming in other parts of the world. In the Nearctic, USA, 57 species of bees are already red listed (Gill et al., 2016) for being seriously endangered or for being nearly extinct (Pettis and Delaplane, 2010). Fifty-eight species of butterfly and moths are also the US red list (Wallisdevries and Van Swaay, 2006). In Canada, the geographic range of pollinating insects have contracted by 23–87% within the last 20 years (Cameron et al., 2011). Even managed honeybees have shown rapid rate of declination both in the United States and Europe (Potts et al., 2010). In China and Argentina, unfortunately, the rate of employment in human hand pollination has been increased noticeably in the past 15 years. The rest of the zoogeographical realms are extremely data deficient, but still stray reports support the pollinator's declination hypothesis (Goulson et al., 2015). Scattered studies have shown that in Asian countries like India, Thailand, Indonesia, and Sumatra, due to extensive habitat destruction, loss of natural vegetation, aggressive use of pesticides as well as altered land usage pattern, the abundance of insect pollinators has nosedived in recent years (Mariau and Genty, 1988; Sengupta et al., 2019; Siregar et al., 2016). Over the past few years, dramatic decline in the ratio of insects to insect pollinators has occurred across the globe (Lebuhn et al., 2013). The absence of appropriate habitat and landscapes are also acting as a driving force in this declination (Kevan and Viana, 2003). The overall decreased pollination of food crops will no doubt lead to a serious threat to nutritional values (Steffan et al., 2005). As the crisis is going to worsen, the problem of hidden hunger can erode ecosystem resilience and can destabilize ecosystem services (Tscharntke et al., 2012). The projected world population is going to be 9–10 billion by 2050 (Butchart et al., 2010), leading to more negative climatic impact, decreased food security, and a challenging global hunger index (Jansson and Polasky, 2010). This will probably decimate the insect pollinator populations globally.

11.8 To Conserve, To Survive, To Sustain

No doubt, it's high time for policy makers to reverse the clock of destruction (Tylianakis, 2013). For a more precise management approach (Lewis et al., 2019), a more proper understanding of insect pollinators is urgently needed (Moron et al., 2019; Vanbergen and Initiative, 2013). In the framework of pollination protection, the decisions of policymakers have always been criticized since they are mainly intended for the building of human benefits from biodiversity as a rationale for conservation (Senapathi et al., 2015). Recent reports on the declining of insect pollinators globally (Powney et al., 2019) are pointing toward a necessary urgency (Hall and Steiner, 2019) of review of conservation management policies (Basset and Lamarre, 2019), regarding insect pollinators (Requier et al., 2019). A better understanding of the driving factors should be taken care of by policymakers while planning conservation strategies (Byrne and Fitzpatrick, 2009; Hall and Steiner, 2019) to address the adversity.

While discussing conservation approaches, the status of most of the current insect pollinators, and the future shortage of insect pollinators, pollination-friendly agricultural practices, sustainable use of landscape patterns, and changed approaches in farming methods should be taken into consideration as well (Kearns and Inouye, 1997). Further to addressing the issue of the rising global hunger index, technological knowhow like development of pathogen- and parasite-resistant stock of honeybees (Ramadani et al., 2019), modern approaches by commercial queen producers (Requier et al., 2019) and adoption of practices of managed bee hives, increased attention for butterflies and hoverflies in managed and unmanaged ecosystems are to be implemented. For conservation of wild and unmanaged pollinators, the main approach to be taken is the maintenance of natural habitats. Along with this regulation on the level of anthropogenic interference, artificial pollution should also be brought under consideration and controlled. Also, economic and policy incentives will inspire the extensive variety of urban and rural areas to accept the pollination friendly practices (Bartomeus and Dicks, 2019). Earlier studies on pollinator communities had focused mainly on primary pollinators and management strategies were primarily designed to conserve these taxa (Lowenstein et al., 2019; Ramírez and Kallarackal, 2018; Sengupta et al., 2019). Consequentially agro-ecosystems tended to rely on them entirely. Today, classical pollinator decline and degrading landscapes are making it imperative to search for non-traditional pollinators to further augment the global pollination network.

11.9 Conclusion

So, in conclusion, it can be said that insect pollination declination is a global phenomenon, which is an unfortunate outcome of multiple factors like overusage of chemical pesticides, habitat alteration, degradation of landscapes, climate change, and failed conservation policies. To address these issues at the ground level, only scientific calls are not enough. Multi-state and multi-nation agreements should be legalized and revised. Apart from this, governments are required to deploy active policies for conservation management, global monitoring as well as conservation agreements urgently. It's time for not only the policy makers but also for people at the grassroot level to spin the wheels. Because conservation of pollinating insects is no more confined as an elegant topic of discussion in International conferences, it's no more a luxury, but a necessity for survival of mankind. The takehome message for all people out there is that both wild and managed insect pollinators have a globally acknowledged role in generating a yearly estimate of $235–$577 billion through crop production (IPBES et al., 2016). Apart from this, pollinators serve as important spiritual symbols in many cultures. Mention of sacred passages about bees in the world's major religions actually highlight their significance in human societies over the millennia. Implications of conservation of insect pollinators for safeguarding of intangible cultural habitats of our world were understood and accepted even in ancient times. So both insect pollinators and pollinating services, which have evolved through our civilizations, need to be conserved critically.

REFERENCES

Acevedo FE, Stanley BA, Stanley A et al. (2017) Quantitative proteomic analysis of the fall armyworm saliva. *Insect Biochem Mol Biol* 86:81–92.

Aizen MA, Harder LD (2009) The global stock of domesticated honey bees is growing slower than agricultural demand for pollination. *Curr Biol* 19(11):915–918.

Alaux C, Le Conte Y, Decourtye A (2019) Pitting wild bees against managed honey bees in their native range a losing strategy for the conservation of honey bee biodiversity. *Front Ecol Evol* 7:60.

Allsopp MH, De Lange WJ, Veldtman R (2008) Valuing insect pollination services with cost of replacement. *PLoS One* 3(9):e3128.

Balducci MG, Van der Niet T, Johnson SD (2019) Butterfly pollination of *Bonatea cassidea* (Orchidaceae): Solving a puzzle from the Darwin era. *South African J Bot* 123:308–316.

Baldwin IT (1998) Jasmonate-induced responses are costly but benefit plants under attack in native populations. Proceedings of the National Academy of Sciences: 95:8113–8118.

Banerjee D (2005) A report on flies (Diptera: Insecta) as flower visitors and pollinators of Kolkata and its adjoining areas. *Rec Zool Surv India* 105(3-4):1–20.

Bartomeus I, Dicks LV (2019) The need for coordinated transdisciplinary research infrastructures for pollinator conservation and crop pollination resilience. *Environ Res Lett* 14(4):045017.

Basset Y, Lamarre GP (2019) Toward a world that values insects. *Science* 364(6447):1230–1231.

Beccaloni G, Scoble M, Kitching I et al. Lep Index: The Global Lepidoptera Names Index (version 12.3, Jan 2012). In: Bailly N, Kirk PM, Bourgoin T, DeWalt RE, Decock W, Nieukerken E van, Zarucchi J, Penev L (eds.) Species 2000 and IT IS. Leiden, Netherlands: Naturalis. ISSN 2405-884X. www.catalogueoflife.org/annual-checklist/2019

Berkley NA, Hanley ME, Boden R et al. (2018) Influence of bioenergy crops on pollinator activity varies with crop type and distance. *GCB Bioenergy* 10(12):960–971.

Bezemer TM, van Dam NM (2005) Linking aboveground and belowground interactions via induced plant defenses. *Trends Ecol Evol* 20:617–624.

Biesmeijer JC, Roberts SP, Reemer M et al. (2006) Parallel declines in pollinators and insect-pollinated plants in Britain and the Netherlands. *Science* 313(5785):351–354.

Blaakmeer A, Geervliet JBF, van Loon JJA et al. (1994) Comparative headspace analysis of cabbage plants damaged by two species of Pieris caterpillars: Consequences for in-flight host location by Cotesia parasitoids. *Entomol Exp Appl* 73:175–182.

Blair J, Sampson JH (2000) Cane Pollination Efficiencies of Three Bee (Hymenoptera: Apoidea) Species Visiting Rabbiteye Blueberry. *J Econ Entomol* 93(6):1726–1731.

Bruce TJA (2015) Interplay between insects and plants: Dynamic and complex interactions that have coevolved over millions of years but act in milliseconds. *J Exp Bot* 66:455–465.

Butchart SH, Walpole M, Collen B et al. (2010) Global biodiversity: indicators of recent declines. *Science* 328(5982):1164–1168.

Byrne A, Fitzpatrick U (2009) Bee conservation policy at the global regional and national levels. *Apidologie* 40(3):194–210.

Cameron SA, Lozier JD, Strange JP et al. (2011) Patterns of widespread decline in North American bumble bees. Proceedings of the National Academy of Sciences 108(2):662–667.

Campbell PJ (2013) Declining European bee health: Banning the neonicotinoids is not the answer. *Outlooks Pest Manag* 24(2):52–57.

Carmichael SW (2019) Did beetles pollinate ancient plants? *Microsc Today* 27(1):8–11.
Catalogue of Life (2019) Annual checklist. www.catalogueoflife.org/annual-checklist/2019. ISSN 2405-884X.
Chain GA, Martínez SA, Aristizábal N et al. (2019) Ecosystem services by birds and bees to coffee in a changing climate: A review of coffee berry borer control and pollination. *Agric Ecosyst Environ* 280:53–67.
Claus G, Vanhove W, Van Damme P et al. (2018) Challenges in Cocoa Pollination: The Case of Côte d'Ivoire. In: *Pollination in Plants*. pp. 39–40.
Crowther LP, Wright DJ, Richardson DS et al. (2019) Spatial ecology of a range-expanding bumble bee pollinator. *Ecol Evol* 9(3):986–997.
Dangles O, Casas J (2019) Ecosystem services provided by insects for achieving sustainable development goals. *Ecosyst Serv* 35:109–115.
Decourtye A, Alaux C, Le Conte Y et al. (2019) Toward the protection of bees and pollination under global change: Present and future perspectives in a challenging applied science. *Current Opin Insect Sci* 35:123–131.
Del-Claro K, Rodriguez-Morales D, Calixto ES et al. (2019) Torezan-Silingardi Ant pollination of Paepalanthus lundii (Eriocaulaceae) in Brazilian savanna Annals of Botany 123(7):1159–1165.
Delphia CM, Griswold T, Reese EG et al. (2019) Checklist of bees (Hymenoptera: Apoidea) from small diversified vegetable farms in south-western Montana. *Biodivers Data J* DOI: 10.3897/BDJ.7.e30062.
Devoto M, Medan D, Montaldo NH (2005) Patterns of interaction between plants and pollinators along an environmental gradient. *Oikos* 109(3):461–472.
Dotterl S, Wolfe LM, Jurgens A (2005) Qualitative and quantitative analyses of flower scent in Silene latifolia. *Phytochemistry* 66(2):203–213.
Douglas AE (1998) Nutritional interactions in insect-microbial symbioses: aphids and their symbiotic bacteria Buchnera. *Annu Rev Entomol* 43:17–37.
Dyer LA, Philbin CS, Ochsenrider KM et al. (2018) Modern approaches to study plant–insect interactions in chemical ecology. *Nat Rev Chem* 2:50–64.
Elhassan M, Wendin K, Olsson V et al. (2019) Quality aspects of insects as food – Nutritional sensory and related concepts. *Foods* 8(3):95.
Evenhuis NL, Pape T (eds.) (2019) Systema dipterorum, version (2.3). http://www.diptera.dk/, accessed July29, 2019.
Galen C, Stanton ML (1989) Bumble bee pollination and floral morphology: factors influencing pollen dispersal in the alpine sky pilot Polemonium viscosum (Polemoniaceae). *Am J Bot* 76(3):419–426.
Gangwani K, Landin J (2018) The decline of insect representation in biology textbooks over time. *Am Entomol* 64(4):252–257.
Garibaldi LA, Kitzberger T, Chaneton EJ (2011) Environmental and genetic control of insect abundance and herbivory along a forest elevational gradient. *Oecologia* 167(1):117–129.
Garibaldi LA, Steffan DI, Winfree R et al. (2013) Wild pollinators enhance fruit set of crops regardless of honey bee abundance. *Science* 339(6127):1608–1611.
Gill R, Baldock K, Brown Mark et al. (2016) Protecting an ecosystem service: Approaches to understanding and mitigating threats to wild insect pollinators. *Adv Ecol Res* 54. DOI: 10.1016/bs.aecr.2015.10.007.
Giron D, Dubreuil G, Bennett A et al. (2018) Promises and challenges in insect-plant interactions. *Entomol Exp Appl* 166:319–343.
Godfray HCJ, Beddington JR, Crute IR et al. (2010) Food security: the challenge of feeding 9 billion people. *Science* 327:812–818.
Goehring L, Osberhauser KS (2002) Effects of photoperiod, temperature and host-plant age on induction of reproductive diapause and development time in Danus plexippus. *Ecol Entomol* 27:674–685.
Goodrich BK (2019). Do more bees imply higher fees? Honey bee colony strength as a determinant of almond pollination fees. *Food Policy* 83:150–160.
Goulson D, Lye GC, Darvill B (2008) Decline and conservation of bumble bees. *Annu Rev Entomol* 53:191–208.
Goulson D, Nicholls E, Botías C et al. (2015) Bee declines driven by combined stress from parasites pesticides and lack of flowers. *Science* 347(6229):1255957.
Grab H, Branstetter MG, Amon N et al. (2019) Agriculturally dominated landscapes reduce bee phylogenetic diversity and pollination services. *Science* 363(6424):282–284.

Greenleaf SS, Kremen C (2006) Wild bees enhance honey bees' pollination of hybrid sunflower. Proceedings of the National Academy of Sciences 103(37):13890–13895.

Guenat S, Kunin WE, Dougill AJ et al. (2019) Effects of urbanisation and management practices on pollinators in tropical Africa. *J Appl Ecol* 56(1):214–224.

Gullan PJ, Cranston PS (2000) *The Insects: An Outline of Entomology*. 5th edn. Blackwell Science, Oxford, p. 624.

Hall DM, Camilo GR, Tonietto RK et al. (2017) The city as a refuge for insect pollinators. *Conserv Biol* 31(1):24–29.

Hall DM, Steiner R (2019) Insect pollinator conservation policy innovations: Lessons for lawmakers. *Environ Sci Policy* 93:118–128.

Hansen AK, Moran NA (2014) The impact of microbial symbionts on host plant utilization by herbivorous insects. *Mol Ecol* 23:1473–1496.

Heil M, Koch T, Hilpert A et al. (2001) Extra floral nectar production of the ant associative plants, Macaranga tanarius, is an induced, indirect, defensive response elicited by jasmonic acid. Proceedings of the National Academy of Sciences 98:1083–1088.

Hernandez T, Wanger TC, Tscharntke T (2017) Neglected pollinators: Can enhanced pollination services improve cocoa yields? A review. *Agric Ecosyst Environ* 247:137–148.

Hill R, Nates PG, Quezada EJ et al. (2019) Biocultural approaches to pollinator conservation. *Nat Sustain* 2(3):214.

Hu HY, Chen ZZ, Jiang ZF et al. (2013) Pollinating fig wasp Ceratosolen solmsi adjusts the offspring sex ratio to other foundresses. *Insect Sci* 20(2):228–234.

Inouye DW, Larson BM, Symank A et al. (2015) Flies and flowers III: Ecology of foraging and pollination. *J Pollinat Ecol* 16(16):115–133.

Jacobson MM, Tucker EM, Mathiasson ME et al. (2018) Decline of bumble bees in northeastern North America with special focus on Bombus terricola. *Biol Conserv* 217:437–445.

Jansson A, Polasky S (2010) Quantifying biodiversity for building resilience for food security in urban landscapes: getting down to business. *Ecol Soc* 15(3) DOI: 10.5751/ES-03520-150320.

Jauker F, Jauker B, Grass I et al. (2019) Partitioning wild bee and hoverfly contributions to plant–pollinator network structure in fragmented habitats. *Ecology* 100(2):e02569.

Jiménez PD, Hentrich H, Aguilar RP et al. (2019) A review on the pollination of aroids with bisexual flowers. *Ann Mo Bot Gard* 104(1):83–104.

Johnson SD, Ellis A, Dotterl S (2007) Specialization for pollination by beetles and wasps: the role of lollipop hairs and fragrance in Satyrium microrrhynchum (Orchidaceae). *Am J Bot* 94(1):47–55.

Joseph J, Santibanez F, Laguna MF et al. (2019) *The role of landscape structure on the pollination service of Apis mellifera*. 1–9.

Kaloshian I, Walling LL (2016) Hemipteran and dipteran pests: Effectors and plant host immune regulators. *J Integr Plant Biol* 58:350–361.

Kearns CA, Inouye DW (1997) Pollinators flowering plants and conservation biology. *Bioscience* 47(5):297–307.

Kergoat GJ, Meseguer AS, Jousselin E (2017) Evolution of plant–insect interaction. *Adv Bot Res* 81:25–53.

Kessler A, Baldwin IT (2002) Plant responses to insect herbivory: the emerging molecular analysis. *Annu Rev Plant Biol* 53:299–328.

Kessler A, Halitschke R (2007) Specificity and complexity: the impact of herbivore-induced plant responses on arthropod community structure. *Curr Opin Plant Biol* 10:409–414.

Kevan PG (1972) Floral colors in the high arctic with reference to insect–flower relations and pollination. *Can J Bot* 50(11):2289–2316.

Kevan PG, Baker HG (1983) Insects as flower visitors and pollinators. *Annu Rev Entomol* 28(1):407–453.

Kevan PG, Viana BF (2003) The global decline of pollination services. *Biodiversity* 4(4):3–8.

Kevan PG, Clark EA, Thomas VG (1990) Insect pollinators and sustainable agriculture. *Am J Altern Agric* 5(1):13–22.

Khan Z, Midega CA, Hooper A, Pickett J (2016) Push-pull: chemical ecology-based integrated pest management technology. *J Chem Ecol* 42:689–697.

Klatt BK, Holzschuh A, Westphal C et al. (2014) Bee pollination improves crop quality shelf life and commercial value. Proceedings of the Royal Society B: Biological Sciences 281(1775):2013–2440.

Koricheva J (1999) Interplanting phenotypic variation in plant allelochemistry: problems with the use of concentrations. *Oecologia* 119:467–473.

Kovacs HA, Espíndola A, Vanbergen A et al. (2017) Ecological intensification to mitigate impacts of conventional intensive land use on pollinators and pollination. *Ecol Lett* 20(5):673–689.

Kremen C, Williams NM, Aizen MA et al. (2007) Pollination and other ecosystem services produced by mobile organisms: a conceptual framework for the effects of land-use change. *Ecol Lett* 10(4):299–314.

Kroupa AS, Lohrmann V, Pulawski WJ et al. (2019) HymIS: Hymenoptera Information System (version Jul 2017). www.catalogueoflife.org/annual-checklist/2019

Lahondère C, Vinauger C, Okubo RP et al. (2019) The olfactory basis of orchid pollination by mosquitoes. bioRxiv 643510.

Landolt PJ (1993) Effects of host plant leaf damage on cabbage looper moth attraction and oviposition. *Entomol Exp Appl* 67:79–85.

Lazaro A, Alomar D (2019) Landscape heterogeneity increases the spatial stability of pollination services to almond trees through the stability of pollinator visits. *Agric Ecosyst Environ* 279:149–155.

Lebuhn G, Droege S, Connor EF et al. (2013) Detecting insect pollinator declines on regional and global scales. *Conserv Biol* 27(1):113–120.

Levin MD (1983) Value of bee pollination to US agriculture. *Am Entomol* 29(4):50–51.

Lewis AD, Bouman MJ, Winter A et al. (2019) Does nature need cities? Pollinators reveal a role for cities in wildlife conservation. *Front Ecol Evol* 7:220.

Longino JT, Branstetter MG (2019) The truncated bell: an enigmatic but pervasive elevational diversity pattern in Middle American ants. *Ecography* 42(2):272–283.

Lowenstein DM, Matteson KC, Minor ES (2019) Evaluating the dependence of urban pollinators on ornamental, non-native, and 'weedy'floral resources. *Urban Ecosyst* 22(2):293–302.

Macinnis G, Forrest JR (2019) Pollination by wild bees yields larger strawberries than pollination by honey bees. *J Appl Ecol* 56(4):824–832.

Majetic CJ, Castilla AR, Levin DA (2019) Losing a scent of one's self: Is there a reduction in floral scent emission in self-pollinating Phlox cuspidata versus outcrossing Phlox drummondii. *Int J Plant Sci* 180(1):86–92.

Majewska AA, Altizer S (2019) Planting gardens to support insect pollinators. *Conservation Biology*. DOI: 10.1111/cobi.13271.

Manobanda M, Vasquez CL, Pérez SM (2018) Entomo fauna diversity in sunflower crop in association with attractant plants. *Trop Subtrop Agroecosystems* 21(1):30–38.

Mariau D, Genty P (1988) IRHO contribution to the study of oil palm insect pollinators in Africa South America and Indonesia. *IRHO contribution to the study of oil palm insect pollinators in Africa South America and Indonesia* 43(6):233–240.

Marshman J Blay PA, Landman K (2019) Anthropocene crisis: Climate change pollinators and food security. *Environments* 6(2):22.

McKenna DD, Scully ED, Pauchet Y et al. (2016) Genome of the Asian longhorned beetle (*Anoplophora glabripennis*), a globally significant invasive species, reveals key functional and evolutionary innovations at the beetle-plant interface. *Genome Biology* 17:227.

McKinnon AC, Glare TR, Ridgway HJ et al. (2018) Detection of the entomopathogenic fungus Beauveria bassiana in the rhizosphere of wound-stressed Zea mays plants. *Front Microbiol* 9:1161.

Mitra B, Parui P, Banerjee D et al. (2005) A report on flies (Diptera: Insecta) as flower visitors and Pollinators of Kolkata and its adjoining areas. *Rec Zool Surv India* 105(3-4):1–20.

Mooney KA, Halitschke R, Kessler A, Agrawal AA (2010) Evolutionary trade-offs in plants mediate the strength of trophic cascades. *Science* 327:1642–1644.

Moran N (2002) The ubiquitous and varied role of infection in the lives of animals and plants. *Am Nat* 160:1–8.

Moron D, Skorka P, Lenda M (2019) Disappearing edge: The flowering period changes the distribution of insect pollinators in invasive goldenrod patches. *Insect Conserv Divers* 12(2):98–108.

Nishida R (2014) Chemical ecology of insect–plant interactions: ecological significance of plant secondary metabolites. *Biosc Biotechnol Biochem* 78:1–13.

Nosil P, Feder JL (2011) Genomic divergence during speciation: Causes and consequences. *Philos Trans R Soc B* 367:332–342.

Ochieng SA, Park KC, Baker TC (2002) Host plant volatiles synergise responses of sex pheromone-specific olfactory receptor neurons in male Helicoverpa Zea. *J Comp Physiol* 188:325–333.

Ollerton J, Winfree R, Tarrant S (2011) How many flowering plants are pollinated by animals? *Oikos* 120(3):321–326.

Orford KA, Vaughan IP, Memmott J (2015) The forgotten flies: The importance of non-syrphid Diptera as pollinators. Proceedings of the Royal Society B: Biological Sciences 282(1805) 2014–2934.

Paini DR (2004) Impact of the introduced honey bee (Apis mellifera) (Hymenoptera: Apidae) on native bees: A review. *Austral Ecol* 29:399–407.

Painter RH (1951) *Insect resistance in crop plants*. University of Kansas Press, Lawrence, p. 520.

Panda N, Kush GS (1995) *Host plant resistance to insects*. CAB International, Oxon.

Perrot T, Gaba S, Roncoroni M et al. (2019). Experimental quantification of insect pollination on sunflower yield reconciling plant and field scale estimates. *Basic Appl Ecol* 34:75–84.

Petanidou T, Ellis WN (1993) Pollinating fauna of a phryganic ecosystem: composition and diversity. *Biodivers Lett* 1(1):9–22.

Pettis JS, Delaplane KS (2010) Coordinated responses to honey bee decline in the USA. *Apidologie* 41(3):256–263.

Potts SG, Biesmeijer JC, Kremen C et al. (2010) Global pollinator declines: trends impacts and drivers. *Trends Ecol Evol* 25(6):345–353.

IPBES (2016) Summary for policymakers of the assessment report of the Intergovernmental Science-Policy Platform on Biodiversity and Ecosystem Services on pollinators, pollination and food production. In: Potts SG, Imperatriz-Fonseca VL, Ngo HT, Biesmeijer JC, Breeze TD, Dicks LV, Garibaldi LA, Hill R, Settele J, Vanbergen AJ, Aizen MA, Cunningham SA, Eardley C, Freitas BM, Gallai N, Kevan PG, Kovács-Hostyánszki A, Kwapong PK, Li J, Li X, Martins DJ, Nates-Parra G, Pettis JS, Rader R, BF Viana (eds.) *Secretariat of the intergovernmental science-policy platform on biodiversity and ecosystem services*. Bonn, Germany. 36 pages.

Powney GD, Carvell C, Edwards M et al. (2019) Widespread losses of pollinating insects in Britain. *Nat Commun* 10(1):1018.

Prado A, Marolleau B, Vaissiere BE et al. (2019) Insect pollination is an ecological process involved in the assembly of the seed microbiota. DOI: 10.1101/626895.

Radhika V, Kost C, Bonaventure G et al. (2012) Volatile emission in bracken fern is induced by Jasmonates but not by Spodoptera littoralis or Strongylogaster multifasciata herbivory. *PLoS One* 7:e48050.

Ramadani V, Hisrich RD, Dana LP et al. (2019) Beekeeping as a family artisan entrepreneurship business. *Int J Entrep Behav Res* 25(4):717–730.

Raman A (Ed.) (1997) *Ecology and evolution of plant-feeding insects in natural and man-made environments*. Backhuys, Leiden, p. 245.

Raman A, Suryanarayanan TS (2017) Fungus-plant interaction influences plant-feeding insects. *Fungal Ecol* 29:123–132.

Ramírez F, Kallarackal J (2018) Climate change tree pollination and conservation in the tropics: A research agenda beyond IPBES. *Wiley Interdiscip Rev Clim Chang* 9(1):e502.

Ray HA, Stuhl CJ, Kane ME et al. (2019) Aspects of the pollination biology of Encyclia tampensis the commercially exploited butterfly orchid and Prosthechea cochleata the endangered clamshell orchid in south Florida. *Florida Entomol* 102(1):154–160.

Reddy GVP, Guerrero A (2010) New pheromones and insect control strategies. *Vitam Horm* 83, 493–519.

Requier F, Garnery L, Kohl PL et al. (2019) The conservation of native honey bees is crucial. Trends in Ecology and Evolution. DOI 10.1016/j.tree.2019.04.008.

Rhodes CJ (2019) Are insect species imperilled? Critical factors and prevailing evidence for a potential global loss of the entomofauna: A current commentary. *Sci Progr* 102(2):181–196.

Richards LA, Dyer LA, Forister ML et al. (2015) Phytochemical diversity drives tropical plant insect community diversity. Proceedings of the National Academy of Sciences 112:10973–10978.

Roskov Y, Zarucchi J, Novoselova M, Bisby F (eds). (2018) *Species 2000 & ITIS Catalogue of Life, 2018 Annual Checklist*. Naturalis: Species 2000.

Rostás M, Bollmann F, Saville D et al. (2018) Ants contribute to pollination but not to reproduction in a rare calcareous grassland forb. *Peer J* 6:e4369.

Rucker RR, Thurman WN, Burgett M (2012) Honey bee pollination markets and the internalization of reciprocal benefits. *Am J Agric Econ* 94(4):956–977.

Saez A, Negri P, Viel M et al. (2019) Pollination efficiency of artificial and bee pollination practices in kiwifruit. *Sci Hortic* 246:1017–1021.

Sayers TD, Steinbauer MJ, Miller RE (2019) Visitor or vector? The extent of rove beetle (Coleoptera: Staphylinidae) pollination and floral interactions. *Arthropod-Plant Interactions* DOI: 10.1016/j.tree. 2019.04.008.

Scherrer S, Lepesqueur C, Vieira MC et al. (2016) Seasonal variation in diet breadth of folivorous Lepidoptera in the Brazilian cerrado. *Biotropica* 48:491–498.

Schoonhoven LM, van Loon JJA, Dicke M (2005) *Insect–plant biology*. New York: Oxford University Press, pp. 1–421.

Senapathi D, Carvalheiro LG, Biesmeijer JC et al. (2015) The impact of over 80 years of land cover changes on bee and wasp pollinator communities in England. Proceedings of the Royal Society B: Biological Sciences, 282(1806):20150294.

Sengupta J, Naskar A, Maity A et al. (2016) New distributional records and annotated keys of hover flies (Insecta: Diptera: Syrphidae) from Himachal Pradesh India. *Jf Adv Zool* 37(1):31–54.

Sengupta J, Naskar A, Maity A et al. (2018a) Distributional scenario of hover flies (Diptera: Syrphidae) from the state of West Bengal. *Munis Entomol Zool* 13(2):447–457.

Sengupta J, Naskar A, Maity A et al. (2018b) Effects of selected environmental variable upon the distribution of Hover Fly (Insecta: Diptera: Syrphidae) along with an altitudinal gradient. *Int J Adv Life Sci Res* 1(3):1–6.

Sengupta J, Naskar A, Banerjee D (2019) *Pollinating Diptera: The forgotten Superhero ENVIS Newsletter* 1(3):1–6.

Sharma A, Khan AN, Subrahmanyam S et al. (2014) Salivary proteins of plant-feeding hemipteroids – Implication in phytophagy. *Bull Entomol Res* 104:117–136.

Sharma A, Allen J, Madhavan S et al. (2015) How do free-living, lerp-forming, and gall-inducing Aphalaridae (Hemiptera: Psylloidea) affect the nutritional quality of Eucalyptus leaves? *Ann Entomol Soc Am* 109:127–135.

Sharma A, Shrestha G, Reddy GVP (2018a) Trap Crops: How far we are from using them in cereal crops. *Ann Entomol Soc Am*. https://doi.org/10.1093/aesa/say047.

Sharma A, Jha P, Reddy GVP (2018b) Multidimensional relationships of herbicides with insect-crop food webs. *Sci Total Environ* 643:1522–1532.

Shikano I, Rosa C, Tan C et al. (2017) Tritrophic Interactions: Microbe-mediated plant effects on insect herbivores. *Annu Rev Phytopathol* 55:313–331.

Siregar EH, Atmowidi T, Kahono S (2016) Diversity and abundance of insect pollinators in different agricultural lands in Jambi Sumatera. HAYATI *J Biosci* 23(1):13–17.

Skidmore IH, Hansen AK (2017) The evolutionary development of plant-feeding insects and their nutritional endosymbionts. *Insect Sci* 24:910–928.

Smith CM (2005) *Plant Resistance to Arthropods*. Netherlands: Springer, p. 421.

Snoeren TAL, De Jong PW, Dicke M (2007) Ecogenomic approach to the role of herbivore-induced plant volatiles in community ecology. *J Ecol* 95:17–26.

Souto VD, Proffit M, Buatois B et al. (2018) Pollination along an elevational gradient mediated both by floral scent and pollinator compatibility in the fig and fig-wasp mutualism. *J Ecol* 106(6):2256–2273.

Steffan DI, Potts SG, Packer L (2005) Pollinator diversity and crop pollination services are at risk. *Trends Ecol Evol* 20(12):651–652.

Stork NE, McBroom J, Gely C et al. (2015) New approaches narrow global species estimates for beetles, insects, and terrestrial arthropods. Proceedings of the National Academy of Sciences 112(24): 7519–7523.

Straarup M, Hoppe LE, Pooma R et al. (2018) The role of beetles in the pollination of the mangrove palm Nypa fruticans. *Nordic J Bot* 36(9):e01967.

Takabayashi J, Dicke M, Posthumus MA (1994) Volatile herbivore-induced terpenoids in plant-mite interactions: variation caused by biotic and abiotic factors. *J Chem Ecol* 20:1329–1354.

Tepedino VJ, Bowlin WR, Griswold TL (2006) Pollination biology of the endangered Blowout Penstemon (Penstemon haydenii S. Wats.: Scrophulariaceae) in Nebraska1. *J Torrey Bot Soc* 133(4):548–560.

Thurman JM, Northfield TD, Snyder WE (2019) Weaver ants provide ecosystem services to tropical tree crops. *Front Ecol Evol* 7:120.

Trivellone V, Bougeard S, Giavi S et al. (2017) Factors shaping community assemblages and species co-occurrence of different trophic levels. *Ecol Evol* 7:4745–4754.

Trunschke J, Sletvold N, Agren J (2019) The independent and combined effects of floral traits distinguishing two pollination ecotypes of a moth-pollinated orchid. *Ecol Evol* 9(3):1191–1201.

Tscharntke T, Clough Y, Wanger TC et al. (2012) Global food security biodiversity conservation and the future of agricultural intensification. *Biol Conserv* 151(1):53–59.

Tylianakis JM (2013) The global plight of pollinators. *Science* 339(6127):1532–1533.

Vanbergen AJ, Initiative TIP (2013) Threats to an ecosystem service: pressures on pollinators. *Front Ecol Environ* 11(5):251–259.

Vaudo AD, Farrell LM, Patch HM et al. (2018) Consistent pollen nutritional intake drives bumble bee (Bombus impatiens) colony growth and reproduction across different habitats. *Ecol Evol* 8(11):5765–5776.

Veldtman R (2018) Are managed pollinators ultimately linked to the pollination ecosystem service paradigm? *South African J Sci* 114(11-12):1–4.

Venkatraman K (2013) Role of insect pollinators on the conservation of the major mangrove species in Sundarban Islands West Bengal. DOI: 10.13140/RG.2.1.5170.1364.

Visser ME, Holleman LJM (2001) Warmer springs disrupt the synchrony of Oak and winter moth phenology. Proceedings of the Royal Society B: Biological Sciences 268:289–294.

Wallisdevries MF, Van Swaay CA (2006) Global warming and excess nitrogen may induce butterfly decline by microclimatic cooling. *Glob Change Biol* 12(9):1620–1626.

Watanabe ME (1994) Pollination worries rise as honey bees decline. *Science* 265(5176):1170–1171.

Winder JA (1978) The role of non-dipterous insects in the pollination of cocoa in Brazil. *Bull Entomol Res* 68(4):559–574.

Ye ZM, Jin XF, Inouye DW et al. (2018) Variation in composition of two bumble bee species across communities affects nectar robbing but maintains pollinator visitation rate to an alpine plant Salvia przewalskii. *Ecolo Entomol* 43(3):363–370.

Young RM (1988) A monograph of the genus Polyphylla Harris in America north of Mexico (Coleoptera: Scarabaeidae: Melolonthinae).

Zheng SJ, Dicke M (2008) Ecological genomics of plant-insect interactions: from gene to community. *Plant Physiol* 146:812–817.

12

Human-Induced Climate Change

Qurat ul Ain Farooq[1], Noor ul Haq[2], and Zeeshan Shaukat[3]
[1]*College of Life Science and Bioengineering, Beijing University of Technology, Beijing, China*
[2]*Department of Computer Science and BioInformatics, Khushal Khan Khattak University, Karak, Khyber-Pakhtunkhwa, Pakistan*
[3]*Faculty of Information Technology, Beijing University of Technology, Beijing, China*

12.1 Introduction

Social and physical systems on earth have been seriously affected by change in the global environment, especially climatic change caused by anthropogenic activities. Human society and its sustainable development are at serious risk because of the negative impact these environmental changes are causing. In our academic and political community, the issue of how to deal with and adapt to the climate changes has become the topic of great interest (Wang et al., 2014) (Table 12.1). Recent climate change has been caused due to global warming through certain human activities, including the burning of fossil fuels and pollution due to increase in population and industrialization in the past few decades (Adnan et al., 2018, 2020; Ahmad et al., 2019; Akram et al., 2018a,b; Wang et al., 2018; Farhat et al., 2020; Gul et al., 2020; Habib ur Rahman et al., 2017; Hammad et al., 2016, 2018, 2019, 2020a,b; Hussain et al., 2020; Ilyas et al., 2020; Jan et al., 2019; Kamarn et al., 2017; Khan et al., 2017a,b; Mubeen et al., 2020; Muhammad et al., 2019; Nasim et al., 2017; Rehman, 2020; Sajjad et al., 2019; Saleem et al., 2020a,b,c; Saud et al., 2013, 2014, 2016, 2017, 2020; Shafi et al., 2020; Shah et al., 2013; Subhan et al., 2020; Wahid et al., 2020; Wu et al., 2019, 2020; Yang et al., 2017; Zafar-ul-Hye et al., 2020a,b; Zahida et al., 2017; Zaman et al., 2017; Zamin et al., 2019). United Nations Environmental Programme (UNEP) and the World Meteorological Organization have been established in 1998. The Intergovernmental Panel on Climate Change (IPCC) is an organization whose purpose is to assess scientific, technical, and socioeconomic knowledge that is relevant in perceiving human-induced climate change, its potential effects, and options for reduction in climatic change and adaptation. Current global warming is mainly caused by humans as per the reports of Intergovernmental Panel on Climate Change (IPCC) (Alharby and Fahad, 2020; Anderegg et al., 2010; Fahad and Bano, 2012; Fahad et al., 2013, 2014a,b, 2015a,b, 2016a,b,c,d, 2017, 2018, 2019a,b; Pachauri et al., 2014). Anthropogenic emissions of greenhouse gases are at highest level in the history as per the report which has been summarized for policy makers. IPCC in its fourth assessment report states that anthropogenic greenhouse gases may be considered as the major reason to increase the global temperature in the last half of the 20th century (Solomon et al., 2007).

12.2 Role of Humans in Worldwide Climate Change

Greenhouse gases like CO_2 are mainly released by energy production processes by humans and more precisely by combustion of fossil fuel. Consequently, human-induced global warming and anthropogenic greenhouse gases are substantially associated with future energy production. Some statistics predict a tenfold growth in production of world gas, while others illustrate future oil production by 2100

TABLE 12.1

Top 20 high-yielding institutes based on the total number of articles

Rank	Institute	Articles
1	National Center of Atmospheric Research (USA)	28
2	Stanford University (USA)	28
3	Chinese Academy of Sciences (China)	28
4	University of Southampton (UK)	28
5	Arizona State University (USA)	29
6	University of Cape Town (South Africa)	29
7	Pennsylvania State University (USA)	30
8	University of Washington (USA)	31
9	National Autonomous University of Mexico (Mexico)	31
10	VU University Amsterdam (Netherlands)	32
11	United States Geological Survey (USA)	35
12	University of Melbourne (Australia)	36
13	Australian National University (Australia)	40
14	University of Leeds (UK)	42
15	University of Oxford (UK)	46
16	University of Guelph (Canada)	52
17	McGill University (Canada)	53
18	James Cook University (Australia)	56
19	Potsdam Institute for Climate Impact Centre (Germany)	57
20	University of East Anglia (UK)	83

Source: Wang et al., 2014.

to 300 million barrels per day (Dudley, 2012). According to 2010 statistics, oil production in the world remains around 3,900 million tons annually or 85 million barrels per day of oil equivalents. Currently, around 80% of primary energy is obtained from fossil fuels with natural gas (20.9%), oil (32.8%), and coal (27.2%) (Khatib, 2012). Higher production of CO_2 shows substantial growth in production of oil, coal, and natural gas. Almost 70% of anthropogenic greenhouse gases emanations are originated from the energy sector where fossil fuel burning contributes CO_2 significantly. Fossil fuels become a driving force behind industrialization of the world and its economic evolution. Based on present dominance regarding energy production, fossil fuel will remain on the top of the world's energy system. Moreover, global dependence on energy from fossil fuel can cause an associated problem, that is, emissions. Energy creation is a leading source of CO_2 and other greenhouse gases.

Oceans cover approximately 72% of the earth's surface and hence are a substantial component of the global climate system. They maintain and improve the climate variability by their thermal inertia and heat capacity. Many researchers have used the ocean's surface temperature in their detection and attribution studies, but evidently, temperature at depth of the ocean has not been considered. During the past 45 years, temperature of the higher ocean has been rising in all worlds' oceans as per the recent observational research (Levitus et al., 2000). However, the warming rate differs significantly among different ocean basins. According to the National Academy of Sciences report, human activities are the main contributors in the accumulation of greenhouse gases in earth's atmosphere and the successive rise in temperature of surface air and oceans (Council, 2001).

During the Industrial Revolution, the CO_2 level has increased in the atmosphere by more than 40% because of burning coal, gas, and oil and also because of the cutting of forests. So as a result, it has been acknowledged that in the past two centuries, this CO_2 entraps heat. Another addition to the burden caused by heat-trapping gases is the emission of nitrous oxide and methane from human activities and agriculture. Human-induced climate change does not only refer to hotter weather; rather it means much more than that. The rise in the temperature of oceans, freshwater, frost-free days, and heavy rainstorms has been reported. There has been a huge depletion in sea ice, glaciers, and snow-covered areas and a high rise in global sea level. Human health and many sectors, including energy, agriculture, livestock, coastal areas, transportation, and water supply, have been affected by these climatic changes. These adverse impacts will continue affecting the quality of life day by day.

The rise in temperature caused by methane is more than carbon dioxide, and the proportion of methane in the earth's atmosphere, like that of CO_2, is also increased at a level that has not been seen in 800 millennia. Human activities such as deforestation; increase in farm animals; industrialization; decomposition of garbage; and production of gas, oil, and sewage contribute to the two-thirds of present-day emission of methane.

Humans are so tiny in comparison with the planet. Hence, it is difficult to believe that such a tiny creature's activities can affect a system as massive as earth's climate. However, if one notices that tiny human beings do many activities on daily basis that releases different greenhouse gases into the earth's atmosphere. We cool or heat up our offices or houses, turn on lights, drive cars, watch television; eat food transported to us by different means, including trucks, planes, and ships; and do deforestation as well as act in innumerable other ways through which indirectly or directly we contribute greenhouse gases into the environment. All these cause 29.3 billion metric tons of CO_2 which is 3.66 times more than carbon emitted into the environment. It is evident that different activities of human beings are responsible for the global release of greenhouse gases including CO_2 and methane while increase of these gases is directly associated with global warming (Thompson, 2010).

12.3 Effect of Environmental Changes on Human Health

Global environmental changes may have adverse effects on human health including some positive effects additional to the mostly negative effects. The health of the population gets affected directly due to change in the frequencies of intense cold and heat, the prevalence of droughts and floods, due to local air pollution and aeroallergens. The majority of areas that get affected by climate change belong to the developing countries. In the result, developing countries may face adverse effects on their economy, investment, and life losses. The prevalence of heat-related and infectious diseases would increase in the tropical and temperate Asia due to warmer and wetter conditions. Human health gets adversely affected by increase in temperature on surface air and change in precipitation in Asia. Even though warming of the atmosphere or environment of the planet would decrease the rate of deaths caused during wintertime in temperate countries, but consequently, the occurrence and period of heat stress would be greater in megalopolises during summers. In some parts of Asia, including arid, semi-arid, temperate and tropical regions, the frequency of respiratory and cardiovascular diseases may also rise due to global warming (McCarthy et al., 2001). More than two billion people have been affected over recent years mainly by weather-related disasters, including floods, cyclones, drought, and fires while this number is expected to increase with continuous change in climate (Cannon, 2014).

Sea level increases gradually which is among the impacts of global warming and half of this rise is caused by thermal expansion, that is, with a rise in ocean temperature, the water gets warmer and expands. Moreover, the temperature of oceans is also rising because of ice. Additionally, ice on mountains and glaciers are melting and water gets its way towards the sea so its level is increased (Meier et al., 2007). If 80% of the total ice present on earth melts, then it will show drastic effects on the some coastal regions. For example, much of Louisiana, including New Orleans and some parts of the Florida peninsula, would be under water while low-lying cities, for example, Shanghai, New York, and London, would be in danger of extinction (Thompson, 2010).

The temperature of earth has increased for almost 0.6°C during the past three decades and it has been predicted that by the end of the 21st century, it will rise further to 1°C–6.4°C (Gregory et al., 2007). The rise in temperature is very harmful to human health as it may cause many climate-related problems, including heatstroke caused by heat waves, malaria due to an increased number of mosquitoes, dengue fever, and several other infectious diseases caused by climate. As per different models, there may be 300 million more people affected due to malaria by 2080 (Lindsay and Martens, 1998). In 2000, the World Health Organization (WHO) estimated that 5.5 million Disability-Adjusted Life Years (DALYs) were lost as a result of increased cardiovascular disease, diarrhoea, malaria, malnutrition, and injuries from flooding due to climatic changes (McMichael et al., 2003).

12.4 What Can Possibly Be Done?

For proper sustenance of life on earth, different factors are involved, including the accessibility of clean water, air, food, and sanitation, while the health of populations relies on different environmental hazards, exposure to pathogens and toxins, as well as various other genetic and social factors. It is now obvious that the planet's surface has been affected and its temperature is increased through emission of greenhouse gases from the combustion of fossil fuel. These all processes are causing adverse changes in atmospheric and ocean systems and also disturb the weather and hydrological patterns of earth. Human health is in constant risk because of the impact of climate change on food and water security, floods, drought and violent storms, the prevalence of infectious diseases, and rise in sea levels (Barrett et al., 2015).

12.5 Conclusion

Climate, sea level, and the earth's temperature have been adversely affected through destruction of forests and burning of fossil fuels. Human-induced environment transformation is expected to continue, and if worldwide emissions of heat-trap gases remain surging, it would hasten expressively. Over the next few decades, heat-trap gases in the atmosphere have committed us to a hotter environment with more climate-linked impacts. Now and in the future, the magnitude of climate changes depends primarily on the amount of heat-trap gases that human events emanate worldwide. Different amounts of heat-trap gases emitted into the atmosphere by human actions produce different anticipated surges in the terrain's temperature. A lot of research is done on this topic and additional study is required to make these technologies reduce leakage risks and costs, minimize any adverse impacts on the environment, and enhance safety.

REFERENCES

Adnan M, Shah Z, Fahad S, Arif M, Alam M, Khan IA, Mian IA, Basir A, Ullah H, Arshad M, Rahman I-U, Saud S, Ishan MZ, Jamal Y, Amanullah, Hammad HM, Nasim W (2018) Phosphate-solubilizing bacteria nullify the antagonistic effect of soil calcification on bioavailability of phosphorus in alkaline soils. *Sci Rep* 8:4339. https://doi.org/10.1038/s41598-018-22653-7.

Adnan M, Fahad S, Zamin M, Shah S, Mian IA, Danish S, Zafar-ul-Hye M, Battaglia ML, Naz RMM, Saeed B, Saud S, Ahmad I, Yue Z, Brtnicky M, Holatko J, Datta R (2020) Coupling phosphate-solubilizing bacteria with phosphorus supplements improve maize phosphorus acquisition and growth under lime induced salinity stress. *Plants* 9(900). doi: 10.3390/plants9070900.

Ahmad S, Kamran M, Ding R, Meng X, Wang H, Ahmad I, Fahad S, Han Q (2019) Exogenous melatonin confers drought stress by promoting plant growth, photosynthetic capacity and antioxidant defense system of maize seedlings. *PeerJ* 7:e7793 http://doi.org/10.7717/peerj.7793.

Akram R, Turan V, Hammad HM, Ahmad S, Hussain S, Hasnain A, Maqbool MM, Rehmani MIA, Rasool A, Masood N, Mahmood F, Mubeen M, Sultana SR, Fahad S, Amanet K, Saleem M, Abbas Y, Akhtar HM, Waseem F, Murtaza R, Amin A, Zahoor SA, ul Din MS, Nasim W (2018a) Fate of organic and inorganic pollutants in paddy soils. In: Hashmi, MZ and Varma, A (eds.) *Environmental pollution of paddy soils*. Springer International Publishing, Cham, Switzerland, pp. 197–214.

Akram R, Turan V, Wahid A, Ijaz M, Shahid MA, Kaleem S, Hafeez A, Maqbool MM, Chaudhary HJ, Munis, MFH, Mubeen M, Sadiq N, Murtaza R, Kazmi DH, Ali S, Khan N, Sultana SR, Fahad S, Amin A, Nasim W (2018b) Paddy land pollutants and their role in climate change. In: Hashmi, MZ, Varma, A (eds.)

Environmental pollution of paddy soils. Springer International Publishing, Cham, Switzerland, pp. 113–124.

Alharby AF, Fahad S (2020) Melatonin application enhances biochar efficiency for drought tolerance in maize varieties: Modifications in physio-biochemical machinery. *Agron J*: 1–22.

Anderegg WRL, Prall JW, Harold J, Schneider SH (2010) Expert credibility in climate change. *PNAS* 107:12107–12109.

Barrett B, Charles JW, Temte JL (2015) Climate change, human health, and epidemiological transition. *Prev Med* 70:69–75.

Cannon T (2014) World disasters report 2014 – Focus on culture and risk.

Council NR (2001). *Climate change science: An analysis of some key questions*. The National Academies Press, Washington, DC.

Dudley B (2012) BP statistical review of world energy. London, UK.

Fahad S, Bano A (2012) Effect of salicylic acid on physiological and biochemical characterization of maize grown in saline area. *Pak J Bot* 44:1433–1438.

Fahad S, Chen Y, Saud S,Wang K, Xiong D, Chen C,Wu C, Shah F, Nie L, Huang J (2013) Ultraviolet radiation effect on photosynthetic pigments, biochemical attributes, antioxidant enzyme activity and hormonal contents of wheat. *J Food Agric Environ* 11(3&4):1635–1641.

Fahad S, Hussain S, Bano A, Saud S, Hassan S, Shan D, Khan FA, Khan F, Chen Y, Wu C, Tabassum MA, Chun MX, Afzal M, Jan A, Jan MT, Huang J (2014a) Potential role of phytohormones and plant growth-promoting rhizobacteria in abiotic stresses: Consequences for changing environment. *Environ Sci Pollut Res* 22(7):4907– 4921. https://doi.org/10.1007/s11356-014-3754-2.

Fahad S, Hussain S, Matloob A, Khan FA, Khaliq A, Saud S, Hassan S, Shan D, Khan F, Ullah N, Faiq M, Khan MR, Tareen AK, Khan A, Ullah A, Ullah N, Huang J (2014b) Phytohormones and plant responses to salinity stress: A review. *Plant Growth Regul* 75(2):391–404. https://doi.org/10.1007/s10725-014-0013-y.

Fahad S, Hussain S, Saud S, Tanveer M, Bajwa AA, Hassan S, Shah AN, Ullah A,Wu C, Khan FA, Shah F, Ullah S, Chen Y, Huang J (2015a) A biochar application protects rice pollen from high-temperature stress. *Plant Physiol Biochem* 96:281–287.

Fahad S, Nie L, Chen Y, Wu C, Xiong D, Saud S, Hongyan L, Cui K, Huang J (2015b) Crop plant hormones and environmental stress. *Sustain Agric Rev* 15:371–400.

Fahad S, Hussain S, Saud S, Hassan S, Chauhan BS, Khan F et al. (2016a) Responses of rapid viscoanalyzer profile and other rice grain qualities to exogenously applied plant growth regulators under high day and high night temperatures. *PLoS One* 11(7):e0159590. https://doi.org/10.1371/journal.pone.0159590.

Fahad S, Hussain S, Saud S, Hassan S, Ihsan Z, Shah AN,Wu C, Yousaf M, Nasim W, Alharby H, Alghabari F, Huang J (2016b) Exogenously applied plant growth regulators enhance the morphophysiological growth and yield of rice under high temperature. *Front Plant Sci* 7:1250. https://doi.org/10.3389/fpls.2016.01250.

Fahad S, Hussain S, Saud S, Hassan S, Tanveer M, Ihsan MZ, Shah AN, Ullah A, Nasrullah KF, Ullah S, Alharby HNW, Wu C, Huang J (2016c) A combined application of biochar and phosphorus alleviates heat-induced adversities on physiological, agronomical and quality attributes of rice. *Plant Physiol Biochem* 103:191–198.

Fahad S, Hussain S, Saud S, Khan F, Hassan S, Jr A, Nasim W, Arif M, Wang F, Huang J (2016d) Exogenously applied plant growth regulators affect heat-stressed rice pollens. *J Agron Crop Sci* 202:139–150.

Fahad S, Bajwa AA, Nazir U, Anjum SA, Farooq A, Zohaib A, Sadia S, NasimW, Adkins S, Saud S, Ihsan MZ, Alharby H,Wu C,Wang D, Huang J (2017) Crop production under drought and heat stress: Plant responses and Management Options. *Front Plant Sci* 8:1147. https://doi.org/10.3389/fpls.2017.01147.

Fahad S, Ishan MZ, Khaliq AK, Daur I, Saud S, Alzamanan S, Nasim W, Abdullah MA, Khan IA, Wu C, Wang D, Huang J (2018) Consequences of high temperature under changing climate optima for rice pollen characteristics-concepts and perspectives, *Archives Agron Soil Sci*. DOI: 10.1080/03650340.2018.1443213.

Fahad S, Adnan M, Hassan S, Saud S, Hussain S, Wu C, Wang D, Hakeem KR, Alharby HF, Turan V, Khan MA, Huang J. (2019a). Rice responses and tolerance to high temperature, in: Hasanuzzaman M, Fujita M, Nahar K, Biswas JK (eds.) *Advances in rice research for abiotic stress tolerance*. Woodhead, Cambridge, pp. 201–224.

Fahad S, Rehman A, Shahzad B., Tanveer M, Saud S, Kamran M, Ihtisham M, Khan, SU, Turan, V, Rahman MHU (2019b) Rice responses and tolerance to metal/metalloid toxicity, in: Hasanuzzaman M, Fujita M, Nahar K, Biswas JK (eds.) *Advances in rice research for abiotic stress tolerance*. Woodhead, Cambridge, pp. 299–312.

Farhat A, Hafiz MH, Wajid I, Aitazaz AF, Hafiz FB, Zahida Z, Fahad S, Wajid F, Artemi C (2020) A review of soil carbon dynamics resulting from agricultural practices. *J Environ Manag* 268:110319.

Gregory J, Stouffer RJ, Molina M, Chidthaisong A, Solomon S, Raga G, Friedlingstein P, Bindoff NL, Le Treut H, Rusticucci M (2007) Climate change 2007: The physical science basis.

Gul F, Ahmed I, Ashfaq M, Jan D, Shah F, Li X, Wang D, Fahad M, Fayyaz M, Shah AS (2020) Use of crop growth model to simulate the impact of climate change on yield of various wheat cultivars under different agro-environmental conditions in Khyber Pakhtunkhwa, Pakistan. *Arabian J Geosci* 13:112. https://doi.org/10.1007/s12517-020-5118-1.

Habib ur Rahman M, Ahmad A, Wajid A, Hussain M, Rasul F, Ishaque W, Islam MA, Shiela V, Awais M, Ullah A, Wahid A, Sultana SR, Saud S, Khan S, Shah F, Hussain M, Hussain S, Nasim W (2017) Application of CSM-CROPGRO-Cotton model for cultivars and optimum planting dates: Evaluation in changing semi-arid climate. *Field Crops Res*. http://dx.doi.org/10.1016/j.fcr.2017.07.007.

Hammad HM, Farhad W, Abbas F, Fahad S, Saeed S, Nasim W, Bakhat HF (2016) Maize plant nitrogen uptake dynamics at limited irrigation water and nitrogen. *Environ Sci Pollut Res* 24(3):2549–2557. https://doi.org/10.1007/s11356-016-8031-0.

Hammad HM, Abbas F, Saeed S, Shah F, Cerda A, Farhad W, Bernado CC, Wajid N, Mubeen M, Bakhat HF (2018) Offsetting land degradation through nitrogen and water management during maize cultivation under arid conditions. *Land Degrad Dev* 1–10. DOI: 10.1002/ldr.2933.

Hammad HM, Ashraf M, Abbas F, Bakhat HF, Qaisrani SA, Mubeen M, Shah F, Awais M (2019) Environmental factors affecting the frequency of road traffic accidents: a case study of sub-urban area of Pakistan. *Environ Sci Pollut Res*. https://doi.org/10.1007/s11356-019-04752-8.

Hammad HM, Abbas F, Ahmad A, Bakhat HF, Farhad W, Wilkerson CJ W, Shah F, Hoogenboom G (2020a) Predicting kernel growth of maize under controlled water and nitrogen applications. *Int J Plant Prod*. https://doi.org/10.1007/s42106-020-00110-8.

Hammad HM, Khaliq A, Abbas F, Farhad W, Shah F, Aslam M, Shah GM, Nasim W, Mubeen M, Bakhat HF (2020b) Comparative effects of organic and inorganic fertilizers on soil organic carbon and wheat productivity under arid region. *Communications in Soil Science and Plant Analysis*. DOI: 10.1080/00103624.2020.1763385.

Hussain MA, Fahad S, Rahat S, Muhammad FJ, Muhammad M, Qasid A, Ali A, Husain A, Nooral A, Babatope SA, Changbao S, Liya G, Ibrar A, Zhanmei J, Juncai H (2020) Multifunctional role of brassinosteroid and its analogues in plants. *Plant Growth Regul* https://doi.org/10.1007/s10725-020-00647-8.

Ilyas M, Mohammad N, Nadeem K, Ali H, Aamir HK, Kashif H, Fahad S, Aziz K, Abid U, (2020) Drought tolerance strategies in plants: A mechanistic approach. *J Plant Growth Regulation*. https://doi.org/10.1007/s00344-020-10174-5.

Jan M, Anwar-ul-Haq M, Shah AN, Yousaf M, Iqbal J, Li X, Wang D, Shah F (2019) Modulation in growth, gas exchange, and antioxidant activities of salt-stressed rice (Oryza sativa L.) genotypes by zinc fertilization. *Arabian J Geosci*. 12:775 https://doi.org/10.1007/s12517-019-4939-2.

Kamarn M, Wenwen C, Irshad A, Xiangping M, Xudong Z, Wennan S, Junzhi C, Shakeel A, Fahad S, Qingfang H, Tiening L (2017) Effect of paclobutrazol, a potential growth regulator on stalk mechanical strength, lignin accumulation and its relation with lodging resistance of maize. *Plant Growth Regul* 84:317–332. https://doi.org/10.1007/s10725-017-0342-8.

Khan A, Tan DKY, Munsif F, Afridi MZ, Shah F, Wei F, Fahad S, Zhou R (2017a) Nitrogen nutrition in cotton and control strategies for greenhouse gas emissions: A review. *Environ Sci Pollut Res* 24:23471–23487. https://doi.org/10.1007/s11356-017-0131-y.

Khan A, Kean DKY, Afridi MZ, Luo H, Tung SA, Ajab M, Fahad S (2017b) Nitrogen fertility and abiotic stresses management in cotton crop: A review. *Environ Sci Pollut Res* 24:14551–14566. https://doi.org/10.1007/s11356-017-8920-x.

Khatib H (2012) IEA world energy outlook 2011 – A comment. *Energy Policy* 48:737–743.

Levitus S, Antonov JI, Boyer TP, Stephens C (2000) Warming of the world ocean. *Science* 287:2225–2229.

Lindsay S, Martens W (1998) Malaria in the African highlands: past, present and future. *Bulletin of the World Health Organization* 76:33.

McCarthy JJ, Canziani OF, Leary NA, Dokken DJ, White KS (2001). *Climate Change 2001: Impacts, Adaptation, and Vulnerability: Contribution of Working Group II to the Third Assessment Report of the Intergovernmental Panel on Climate Change.* Cambridge: Cambridge University Press.

McMichael AJ, Campbell-Lendrum DH, Corvalán CF, Ebi KL, Githeko A, Scheraga JD, Woodward A (2003). *Climate Change and Human Health: Risks and Responses.* Geneva: World Health Organization.

Meier MF, Dyurgerov MB, Rick UK, O'Neel S, Pfeffer WT, Anderson RS, Anderson SP, Glazovsky AF (2007) Glaciers dominate eustatic sea-level rise in the 21st century. *Science* 317:1064–1067.

Mubeen M, Ashfaq A, Hafiz MH, Muhammad A, Hafiz UF, Mazhar S, Muhammad Sami ul Din, Asad A, Amjed A, Fahad S, Wajid N (2020) Evaluating the climate change impact on water use efficiency of cotton-wheat in semi-arid conditions using DSSAT model. *J Water Climate Change.* doi/10.2166/wcc.2019.179/622035/jwc2019179.pdf.

Muhammad B, Adnan M, Munsif F, Fahad S, Saeed M, Wahid F, Arif M, Jr. Amanullah, Wang D, Saud S, Noor M, Zamin M, Subhan F, Saeed B, Raza MA, Mian IA (2019) Substituting urea by organic wastes for improving maize yield in alkaline soil. *J Plant Nutrition.* doi.org/10.1080/01904167.2019.1659344.

Nasim W, Ahmad A, Amin A, Tariq M, Awais M, Saqib M, Jabran K, Shah GM, Sultana SR, Hammad HM, Rehmani MIA, Hashmi MZ, Habib Ur Rahman M, Turan V, Fahad S, Suad S, Khan A, Ali S (2017) Radiation efficiency and nitrogen fertilizer impacts on sunflower crop in contrasting environments of Punjab. *Pakistan Environ Sci Pollut Res* 25:1822–1836. https://doi.org/10.1007/s11356-017-0592-z.

Pachauri RK, Allen MR, Barros VR, Broome J, Cramer W, Christ R, Church JA, Clarke L, Dahe Q, Dasgupta P (2014). Climate change 2014: Synthesis Report. Contribution of Working Groups I, II and III to the fifth assessment report of the Intergovernmental Panel on Climate Change, *IPCC.*

Rehman M, Fahad S, Saleem MH, Hafeez M, Muhammad Habib ur Rahman, Liu F, Deng G (2020) Red light optimized physiological traits and enhanced the growth of ramie (Boehmeria nivea L.). *Photosynthetica* 58 (4):922–931.

Sajjad H, Muhammad M, Ashfaq A, Waseem A, Hafiz MH, Mazhar A, Nasir M, Asad A, Hafiz UF, Syeda RS, Fahad S, Depeng W, Wajid N (2019) Using GIS tools to detect the land use/land cover changes during forty years in Lodhran district of Pakistan. *Environ Sci Pollut Res* https://doi.org/10.1007/s11356-019-06072-3.

Saleem MH, Fahad S, Adnan M, Mohsin A, Muhammad SR, Muhammad K, Qurban A, Inas AH, Parashuram B, Mubassir A, Reem MH (2020a) Foliar application of gibberellic acid endorsed phytoextraction of copper and alleviates oxidative stress in jute (Corchorus capsularis L.) plant grown in highly copper-contaminated soil of China. *Environ Sci Pollution Res* https://doi.org/10.1007/s11356-020-09764-3.

Saleem MH, Fahad S, Shahid UK, Mairaj D, Abid U, Ayman ELS, Akbar H, Analía L, Lijun L (2020b) Copper-induced oxidative stress, initiation of antioxidants and phytoremediation potential of flax (Linum usitatissimum L.) seedlings grown under the mixing of two different soils of China. *Environ Sci Poll Res* https://doi.org/10.1007/s11356-019-07264-7.

Saleem MH, Rehman M, Fahad S, Tung SA, Iqbal N, Hassan A, Ayub A, Wahid MA, Shaukat S, Liu L, Deng G (2020c) Leaf gas exchange, oxidative stress, and physiological attributes of rapeseed (Brassica napus L.) grown under different light-emitting diodes. *Photosynthetica* 58 (3):836–845.

Saud S, Chen Y, Long B, Fahad S, Sadiq A (2013) The different impact on the growth of cool season turf grass under the various conditions on salinity and drought stress. *Int J Agric Sci Res* 3:77–84.

Saud S, Li X, Chen Y, Zhang L, Fahad S, Hussain S, Sadiq A, Chen Y (2014) Silicon application increases drought tolerance of Kentucky bluegrass by improving plant water relations and morph physiological functions. *SciWorld J* 2014:1–10. https://doi.org/10.1155/2014/368694.

Saud S, Chen Y, Fahad S, Hussain S, Na L, Xin L, Alhussien SA (2016) Silicate application increases the photosynthesis and its associated metabolic activities in Kentucky bluegrass under drought stress and post-drought recovery. *Environ Sci Pollut Res* 23(17):17647–17655. https://doi.org/10.1007/s11356-016-6957-x.

Saud S, Fahad S, Yajun C, Ihsan MZ, Hammad HM, Nasim W, Amanullah Jr, Arif M and Alharby H (2017) Effects of nitrogen supply on water stress and recovery mechanisms in Kentucky bluegrass plants. *Front. Plant Sci.* 8:983. doi: 10.3389/fpls.2017.00983.

Saud S, Fahad S, Cui G, Chen Y, Anwar S (2020) Determining nitrogen isotopes discrimination under drought stress on enzymatic activities, nitrogen isotope abundance and water contents of Kentucky bluegrass. *Sci Rep* 10:6415. https://doi.org/10.1038/s41598-020-63548-w.

Shafi MI, Adnan M, Fahad S, Fazli W, Ahsan K, Zhen Y, Subhan D, Zafar-ul-Hye M, Brtnicky M, Datta R (2020) Application of single superphosphate with humic acid improves the growth, yield and phosphorus uptake of wheat (Triticum aestivum L.) in calcareous soil. *Agron* (10):1224. doi: 10.3390/agronomy10091224.

Shah F, Lixiao N, Kehui C, Tariq S, Wei W, Chang C, Liyang Z, Farhan A, Fahad S, Huang J (2013) Rice grain yield and component responses to near 2°C of warming. *Field Crop Res* 157:98–110.

Solomon S, Qin D, Manning M (2007) IPCC (2007) Summary for policymakers. Climate Change 2007: The Physical Science Basis. *Contribution of Working Group I to the Fourth Assessment Report of the Intergovernmental Panel on Climate Change*.

Subhan D, Zafar-ul-Hye M, Fahad S, Saud S, Brtnicky M, Hammerschmiedt T, Datta R (2020) Drought stress alleviation by ACC deaminase producing achromobacter xylosoxidans and enterobacter cloacae, with and without timber waste biochar in maize. *Sustain* 12(6286). doi: 10.3390/su12156286.

Thompson LG (2010) Climate change: The evidence and our options. *The Behavior Analyst* 33:153–170.

Wahid F, Fahad S, Subhan D, Adnan M, Zhen Y, Saud S, Manzer HS, Martin B, Tereza H, Rahul D (2020) Sustainable management with mycorrhizae and phosphate solubilizing bacteria for enhanced phosphorus uptake in calcareous soils. *Agri* 10 (334). doi:10.3390/agriculture10080334.

Wang B, Pan S-Y, Ke R-Y, Wang K, Wei Y-M (2014) An overview of climate change vulnerability: A bibliometric analysis based on Web of Science database. *Nat Haz* 74:1649–1666.

Wang D, Shah F, Shah S, Kamran M, Khan A, Khan MN, Hammad HM, Nasim W (2018) Morphological acclimation to agronomic manipulation in leaf dispersion and orientation to promote 'Ideotype' breeding: Evidence from 3D visual modeling of 'super' rice (Oryza sativa L.). *Plant Physiol Biochem*. https://doi.org/10.1016/j.plaphy.2018.11.010.

Wu C, Tang S, Li G, Wang S, Fahad S, Ding Y (2019) Roles of phytohormone changes in the grain yield of rice plants exposed to heat: A review. *PeerJ* 7:e7792 DOI: 10.7717/peerj.7792

Wu C, Kehui C, She T, Ganghua L, Shaohua W, Fahad S, Lixiao N,Jianliang H, Shaobing P, Yanfeng D (2020) Intensified pollination and fertilization ameliorate heat injury in rice (Oryza sativa L.) during the flowering stage. *Field Crops Res* 252:107795.

Yang Z, Zhang Z, Zhang T, Fahad S, Cui K, Nie L, Peng S, Huang J (2017) The effect of season-long temperature increases on rice cultivars grown in the central and southern regions of China. *Front Plant Sci* 8:1908. https://doi.org/10.3389/fpls.2017.01908.

Zafar-ul-Hye M, Naeem M, Danish S, Fahad S, Datta R, Abbas M, Rahi AA, Brtnicky M, Holatko, Jiri, Tarar ZH, Nasir M (2020a) Alleviation of cadmium adverse effects by improving nutrients uptake in bitter gourd through cadmium tolerant rhizobacteria. *Environ* 7(54). doi:10.3390/environments7080054

Zafar-ul-Hye M, Tahzeeb-ul-Hassan M, Abid M, Fahad S, Brtnicky M, Dokulilova T, Datta R D, Danish S (2020b) Potential role of compost mixed biochar with rhizobacteria in mitigating lead toxicity in spinach. *Scientific Rep* 10:12159; https://doi.org/10.1038/s41598-020-69183-9.

Zahida Z, Hafiz FB, Zulfiqar AS, Ghulam MS, Fahad S, Muhammad RA, Hafiz MH,Wajid N, Muhammad S (2017) Effect of water management and silicon on germination, growth, phosphorus and arsenic uptake in rice. *Ecotoxicol Environ Saf* 144:11–18.

Zaman QU, Aslam Z, Yaseen M, Ishan MZ, Khan A, Shah F, Basir S, Ramzani PMA, Naeem M (2017) Zinc biofortification in rice: leveraging agriculture to moderate hidden hunger in developing countries. *Arch Agron Soil Sci* 64:147–161. https://doi.org/10.1080/03650340.2017.1338343.

Zamin M, Khattak AM, Salim AM, Marcum KB, Shakur M, Shah S, Jan I, Fahad S (2019) Performance of Aeluropus lagopoides (mangrove grass) ecotypes, a potential turfgrass, under high saline conditions. *Environ Sci Pollut Res*. https://doi.org/10.1007/s11356-019-04838-3.

13

Plants, Environmental Constraints, and Climate Change

Waqar Islam[1], Tayeba Sanaullah[2], Noreen Khalid[3], Muhammad Aqeel[3], Sibgha Noreen[4], Muhammad Kashif Irshad[2], and Ali Noman[5]
[1]*Institute of Geography, Fujian Normal University, Fuzhou, P.R. China*
[2]*Institute of Pure and Applied Biology, Bahauddin Zakariya University, Multan, Pakistan*
[3]*Department of Botany, Govt. College Women University, Sialkot, Pakistan*
[4]*School of Life Sciences, Lanzhou University, Lanzhou, Gansu Province, P.R. China*
[5]*Department of Environmental Science, Govt. College University Faislabad, Pakistan*

13.1 Introduction

Adverse changes in the environment badly affect plant production, human health, and natural systems (Zuzelo, 2018). Sudden variations in the environment impose harsh effects on productivity of plants. Concentration of CO_2 increased from 280 μmol^{-1} to 400 μmol^{-1} in the atmosphere due to increased deforestation and greater use of fossil fuels (Myers et al., 2014). It is considered that at the end of this century, concentration of CO_2 will double to 800 μmol^{-1} (Koytsoumpa et al., 2018). The key factors responsible for average warm temperatures and greenhouse effects is the discharge of various gases, mainly CO_2 (Lin et al., 2014). The effect of changes in environment and climate variations are mostly determined by their effects on everyday life and harm to living organisms (Luo et al., 2007). Abiotic stresses badly affect plant growth and productivity. Drought, salinity, water logging, and heat and cold stress are the main environmental stresses to plants (Adnan et al., 2018, 2020; Ahanger et al., 2017; Ahmad et al., 2019; Akram et al., 2018a,b; Wang et al., 2018; Gul et al., 2020; Farhat et al., 2020; Habib ur Rahman et al., 2017; Hammad et al., 2016, 2018, 2019, 2020a,b; Hussain et al., 2019, 2020; Ilyas et al., 2020; Jan et al., 2019; Kamarn et al., 2017; Khan et al., 2017a,b; Mubeen et al., 2020; Muhammad et al., 2019; Nasim et al., 2017; Rehman et al., 2020; Saleem et al, 2020a,b,c; Saud et al., 2013, 2014, 2016, 2017, 2020; Shafi et al., 2020; Shah et al., 2013; Subhan et al., 2020; Wahid et al., 2020; Wu et al., 2019, 2020; Yang et al., 2017; Zafar-ul-Hye et al., 2020a,b; Zahida et al., 2017; Zaman et al., 2017; Zamin et al., 2019). However, UV-B light intensities, gas discharge, flooding, and chemical and physical factors which impose more stressful situations are also included in abiotic stresses (Cramer et al., 2011; Fahad et al., 2013). The average temperature of earth is predicted to rise from 2°C to 4.5°C in the 21st century according to Intergovernmental Panel on Climate Change (IPCC), 2014. The time period considered the most warming was between the 19th and 21st centuries (Fahad et al., 2013, 2014a,b, 2015a,b, 2016a,b,c,d, 2017, 2018, 2019a,b; Hesham and Fahad, 2020; Pachauri et al., 2014). Floods due to high rainfall cause destruction; on the other hand, shortage or total absence of rainfall causes drought conditions (Ali et al., 2017). Industrialization increases the temperature that is changing the environment continuously worldwide. Global warming is predicted to increase because of harsh weather events, which will finally destroy the ecosystems (Fahad and Bano, 2012; Zaval et al., 2014). Decrease in crop production due to high intensity of temperature variations, drought, salt stress, rainfall, and insect pest attack lead to greater risk of malnourishment (UNFCCC, 2007). This chapter discusses the effect of climate constraints and weather variations on physiology of plants, which includes the general idea of change in climate, stresses developed as a result of climate fluctuations, followed by recommendations of ways we can develop strategies to defeat climate hurdles.

13.2 Environmental Changes and Their Impact on Plant Physiology

13.2.1 Elevated CO_2 Levels

Concentration of CO_2 has increased because of the start of the industrialization period up to 100 ppm and they are predicted to reach 550 ppm in 2050 (Pachauri et al., 2014). The main physiological functions of plants that are influenced primarily by different levels of CO_2 are stomatal conductance and photosynthesis (Daszkowska-Golec and Szarejko, 2013). Under the highest level of CO_2, prominent reduction in rates of stomatal conductance has been studied via observing alterations in aperture of stomata or concentration of stomata (Xu et al., 2016). These variations are mostly attained by changes in rate of evapotranspiration and water use efficiency (Guerrieri et al., 2016; Hatfield and Dold, 2019).

According to earlier studies, current atmospheric levels of CO_2 reduced the substrates of C_3 photosynthesis. A further effect of imposing a presently CO_2 limited C_3 species to greater concentration of CO_2 is the reduction of photorespiration, obtained by the highest carboxylation and electron transfer rates (Hagemann and Bauwe, 2016). This is significantly vital for plants since they have the greatest prospective for improved photosynthesis. In perennial rye grass, Kentucky blue grass, and tall fescue, a significant improvement in photosynthetic rates were observed (Loka et al., 2019). Burgess and Huang (2014, 2019) experienced a substantial increase in total photosynthetic rate of creeping bent grass because of doubled (800 ppm) CO_2 concentration. These endorsed either the improved accessibility of CO_2 or high activation conditions of Rubisco. Casella and Soussana (2007) reported a significant enhancement in rates of leaf photosynthesis of perennial rye grass after a continuing disclosure to variable CO_2 contents. They noticed that increases were dependent on supply of nutrients, specifically uptake of nitrogen (N). Ainsworth and Rogers (2007) explored that under variable concentration of CO_2, acclimation of photosynthesis in perennial rye grass was due to prominent decrease in the sink potency. In perennial rye grass, variable contents of CO_2 had no influence on photochemical efficiency of photosystem II and light saturated photosynthesis (A_{sat}) (Ali et al., 2008; Farfan-Vignolo and Asard, 2012). We explore that this inconsistency was due to the shorter interval to CO_2 exposure.

Whereas the influence of enhanced CO_2 on plants is usually helpful to photosynthesis, the influence of variable contents of CO_2 on rate of respiration has been more feasible. Song et al. (2014a) showed that when Kentucky blue grass was exposed to higher CO_2 as compared to control, a significantly lower respiration rate was studied because of suppression of respiratory action. Root respiration is sensitive to atmospheric CO_2 contents (Maček et al., 2005). Exposure of variable levels of CO_2 to perennial rye grass enhanced the rate of root respiration (Casella and Soussana, 2007). Abiotic stresses influenced plants in different ways which depends greatly on species, the kinds of tissues studied, and the growth level at which stress takes place (Cramer et al., 2011). Nonetheless, evidently more analysis of respiration responses of roots and shoots to CO_2 levels in the environment is required in order to explain the direct influence on each species of plants.

13.2.2 Flooding

About 10% land area has been affected with flooding worldwide and future climate predictions have concerns about increase in intensity and frequency of high precipitation incidents (Pachauri et al., 2014). Water logging and submergence are included in flooding depending on the amount of water concerned. Only roots suffer from wet situation under waterlogging. On the other way in submergence, whole or partial immersion was noted. Highly water saturated soil shows a harmful effect on the ecosystem as a whole because all the oxygen reachable to soil is used up rapidly by microbial organisms of soil and by respiration of plant roots (Gougoulias et al., 2014). Plants under water logging condition still provide oxygen to the roots during shoot photosynthesis but the supply rate is dependant greatly on porosity of tissue, sink strength, and demand of root respiration (Ben-Noah and Friedman, 2018). Nonetheless, the less oxygen in soil environment forces roots to the mode of anaerobic to sustain production of energy and plant function. Root functions are affected with limited nutrients uptake and translocation. All this may lead to plant leaves senescence and eventual plant death (Ben-Noah and Friedman, 2018; Iqbal et al., 2017).

Excess water in the soil limits the physiological and metabolic factors and as a result, plant growth is reduced between 15% and 80% of the norm. In addition to plant type and growth period, various factors are involved in damaging the plant growth such as soil type, temperature, duration, and depth of flood (Bailey-Serres and Voesenek, 2008; Visser et al., 2003). Significant variability of flooding tolerance persists within plants (Sarkar and Bhattacharjee, 2011; Vergara et al., 2014). Therefore, there is high loss of yield because of flooding but there is also possibility of alleviating this stress through recognition and utilization of adaptive germplasm. In various regions of high rainfall that are liable to flooding, the effects of atmospheric change are likely to develop flooding events more harsh. Therefore, more study is required on morphology, metabolism, and physiology of plants in order to identify tolerant varieties for plant breeding in future. Flooding variably influences plant physiological functions, where the damage is greatly dependent on the plant sensitivity to prolonged excess water conditions.

Plant nutrient availability is greatly affected to water logging because of variations in solubility of elements (Akhtar and Nazir, 2013). Furthermore, changes occurring in ion selectivity by root cells under waterlogging environments can cause a decrease in uptake of nutrient, distribution, and translocation (Najeeb et al., 2015). Adams and Akhtar (1994) studied marked increase of manganese and iron content in the perennial rye grass leaves under waterlogging as compared to control. Uptake of nitrogen significantly improved because of increased denitrification in the soil. The substantial reductions noted in the rate of photosynthesis of waterlogged plants has been attributed to nutrient deficiencies (Wang et al., 2017). Therefore, the primary explanation is that gas diffusion in water is 10^{-3} times lesser as compared to air. So plant gas exchange functions are damaged seriously (Kozlowski and Pallardy, 2002). Reduction in stomatal conductance under water logged situations has been seen in various species of plants (Ashraf, 2012). There is remarkable decrease in stomatal conductance rates in perennial rye grass under waterlogged conditions with reduced photosynthetic (Huang et al., 1998). Photosynthetic rate was also got reduced with compromise in PSII photochemical efficiency in perennial rye grass under submergence situations (Yu et al., 2012). Alternatively, Ploschuk et al. (2017) in experiments with tall fescue and *phalaris aquatica* plants showed that waterlogging of 15 days had no influence on stomatal conductance, and net rate of photosynthesis remained unchanged. Moreover, in submerged *phalaris aquatica,* photosynthetic rates remained uninfluenced after submergence of 30 days (Vervuren et al., 1999).

Oxygen supply to the roots is seriously prohibited and leads to substantial reduction in root respiration under waterlogging situations. Plants decrease their root respiration rate in response to flooding (Herrera, 2013) so function of mitochondria is greatly dependant on supply of oxygen (Igamberdiev and Hill, 2009). In waterlogged Kentucky, blue grass prominent mitochondrial swelling was noted. Huang et al. (1998) studied noticeable increase in creeping bent grass leaf respiration rates under low aeration. Finally, more study and more comprehensive experiments are required in order to compensate the influence of excess water on plant' photosynthetic apparatus.

13.2.3 Elevated Ozone Concentrations

Ozone (O_3), a ground level pollutant, is produced when sunlight reacts with air containing certain pollutants (Liu et al., 2016). Recently 35 ppb ozone contents were compared to 10 ppb for the duration of pre-industrial epoch and are considered to reach 70 ppb by 2050, whereas regional spikes as greater as 200 ppb have become somewhat common (Ainsworth et al., 2012).

Ozone is determined as the most dangerous air pollutant because of its greatly unstable and reactive nature and imposes great consequences on morphology, metabolism, and plant physiology (Ghorani-Azam et al., 2016). Experiments with plant species and monocultures have explored that ozone can have remarkable influence on distribution of plant species and balance in an environment (Fuhrer and Booker, 2003; Hayes et al., 2009). The approach of damage depends on various attributes together with ozone contents, that is, canopy architecture, rate of entry into the plant, and growth level of the plants (Fuhrer, 2009). Therefore, studies have shown that in response to high ozone levels, an increase in leaf senescence, photosynthesis reduction, and assimilate accessibility as well as high oxidative damage were observed in plants (Fiscus et al., 2005; Fuhrer, 2009).

Exposure of high ozone contents reduces the photosynthesis in various plant species (Fiscus et al., 2005; Rai and Agrawal, 2012). Stomatal and non-stomatal limitations also result of such reductions (Ainsworth et al., 2012). Changes in ion channel function of guard cells and hormone regulation calcium homeostasis, and decrease in internal CO_2 level are accountable for stomatal limitations (Engineer et al., 2016; Jezek and Blatt, 2017). On the other way, non-stomatal mutations are connected to compromises in carboxylation and photochemical efficiency of PS II (Calatayud and Barreno, 2001). Hayes et al. (2009) showed that early exposure of ozone to perennial rye grass had no influence on photosynthetic ability and carboxylation efficiency.

Ozone is a major dilemma because after inflowing the apoplast through stomata, it is degraded quickly to develop reactive oxygen species, for example, superoxide anion, hydrogen peroxide, and singlet oxygen, that, in turn, damage cell wall components and disturb membrane stability (Choudhury et al., 2017). Exposure to high ozone level reduced the membrane stability due to increase in leakage of electrolytes and lipid peroxidation (Valenzuela et al., 2017). High ozone level also influenced plasmodesmata connections and translocation of assimilation rates (Liesche et al., 2011). Pasqualetti et al. (2015) worked on Italian rye grass (*Lolium multiforum*) and explored a substantial increase in hydrogen peroxide levels following exposure to high contents of ozone.

13.2.4 Drought Stress

In the recent era of climate change, adaptation to water shortage is one of the main challenges for biotechnologist and plant scientists. Scientists are enhancing their struggle to alleviate climate activated metabolic events at gene and cellular levels (Brooks et al., 2011; Noman et al., 2015, 2018a). Research experiments studying the genome of plant for tolerance against water stress and increased production carried out worldwide with the premier objective of more crops per drop (Hussain et al., 2018). There is emerging tendency to enhance the crops water use efficiency to promote the more proficient use of water table (Hensley et al., 2011).

The considerable method for attaining enhanced water stress tolerance is to devise difference between conventional and transgenic approaches. One important transgenic method is the genes engineering of vital defensive and metabolic pathways, for example, synthesizing pathways of osmoprotectants and resistance system of antioxidants (Khan et al., 2015). Microarray techniques are used to identify various stress inducible genes, but as up till now, their work within the molecular apparatus for crop stress reaction and tolerance still required to interpret. For example, the phyto hormone abscisic acid (ABA) production causes closure of stomata and also stimulates the stress-responsive genes expression (Defez et al., 2017). Hoth (2002) studied about 1,354 genes in *Arabidopsis thaliana* wild type and abi 11 mutant seedlings that were up or down, regulated under ABA treatment, which code for signal transduction.

Pinheiro and Chaves (2011) observed low stomatal conductance as well as prolonged irradiance in relation with high levels of CO_2 in intercellular spaces. This was correlated with the fact of lowering of light absorption in Calvin cycle. This photo inhibition mechanism reduces the photosynthetic rate which reflects a defensive strategy in plants by C_3 pathway (Öquist et al., 1992). Unregulated rate of light harvests in plants depicts high photosynthesis rates. Ashraf and Harris (2013) gave detailed description on the advancement made in last eras in introducing new transgenic lines of C_3 crops having elevated photosynthetic rates. These new transgenic lines can be introduced by ingress of genes encoding C_4 enzymes or overexpression of C_3 enzymes or modification in transcription factors (TF). C_4 and CAM plants widely adapted to arid conditions due to efficient photosynthesis and WUE in contrast of C_3 plants (Kumar et al., 2017). Kim et al. (2014) also examined the overexpression of drought stress responsive 6 (CaDSR6 *Capsicum annuum*) in *Arabidopsis* plants with elevated drought tolerance comparable to wild type. Saad et al. (2013) stated that under drought stress, wheat regulatory component of ABA and salinity-responsive NAC1 (*SNAC1*) gene regulate sucrose phosphate synthase type 2C protein phosphatases, 1-phosphatidylinositol-3-phosphate-5-kinase. Variety of drought-tolerance genes is characterized in *Arabidopsis* and few have been introduced in domestic crops either in controlled conditions or in field trials.

13.2.5 Salinity Stress

Salt stress is the most devastating environmental constraint which negatively affects plant growth and development due to toxicity of ions that cause disturbance in water uptake, fluctuation in hormonal activity, and oxidative stress (Acosta-Motos et al., 2017; Ahmad et al., 2013). Plants triggered multiple tolerance responses to cope deleterious effects of abiotic stresses. When exposed to salinity stress, plants compartmentalize and exclude toxic ions (Na^+ and Cl^-) from their vital tissues like photosynthetic tissues where plants absorbed more K^+ to antiport sodium ions into apoplastic sites or the vacuole (Tavakkoli et al., 2010).

Root apoplastic sites are one of the best salinity tolerance strategies to maintain elevated levels of K^+ and retention of toxic ions or solutes. Through this strategy, plants can maintain high K^+/Na^+ in sensitive parts like mesophyll tissues (Assaha et al., 2017). Pandolfi et al. (2012) proposed that if glycophytes are exposed to mild salt stress for short intervals then they can cope with high levels of salinity for long intervals due to the presence of a set of physiological mechanisms like regulation of uptake of ions and toxic ions compartmentalization in xylem tissues (Pandolfi et al., 2012).

Halophytes express high tolerance against salt stress because they show elevated expressions of ion transporter genes (Volkov, 2015). *Aeluropus littoralis* show high salt tolerance due to overexpression of Na^+/H^+ antiporter (AlNHX) as this helps in regulation of Na^+ and K^+/Na^+ ratio (Zhang et al., 2008). *Spartina alterniflora* express high ratio of cell expansion and elevated rate of photosynthesis as they show overexpression of vacuolar ATPase subunit c1 (*SaVHAc1*) gene linked to increase K^+/Na^+ ratio which helps to maintain enhanced relative water content (RWC) in rice (Baisakh et al., 2012). Wheat show high tolerance to salinity and enhanced K^+/Na^+ ratio as wheat transporter gene *TaNHX2* exhibit increased overexpression (Bulle et al., 2016). All data reveal that halophyte plants show high salt tolerance as they have salt-tolerant genes so salt-sensitive cash crops should be engineered to cope with salinity stress.

13.2.6 Cold Stress

Plants have adapted several physiological and molecular mechanisms on exposure to low temperature stress (Hasanuzzaman et al., 2013; Noman et al., 2017). Plants growing on extreme latitudes are more likely to be confounded by photoperiods. Deacclimation and reacclimation to low temperature affects the WUE, growth and development, energy metabolism, and response to photoperiod (Miura and Furumoto, 2013). Kalberer et al. (2006) stated that plants have tolerance against low temperature stress as they can accumulate compatible solutes, antioxidants, protective membranes, and cold regulatory genes. Under chilling conditions expression of cold responsive genes alters their expression and led to code for synthesis of enzymes and proteins involved in metabolism and respiration of photosynthates as well as production of phenylpropanoids, lipids, antioxidants, and those coding for chaperones and antifreeze proteins. An array of genes show their expression for regulation of freezing-induced dehydration (Hannah et al., 2005). Plant survival and yield under cold stress is dependent on synthesis of freezing protective proteins via the process of altered gene expression (Barrero-Sicilia et al., 2017).

Under low temperature, methionine metabolism pathways played a pivotal role in the synthesis of compatible metabolic solutes like polyols and polyamines that helped in cold stress tolerance, but their actual role in cold tolerance is not yet clearly known (Obata and Fernie, 2012). Overexpression of methionine sulphoxide reductase A (MsrA) involved in the regulation of methionine metabolism appeared to boost competence to oxidative damage under cold conditions. Mutation in methionine sulphoxide reductase B3 (MsrB3) observed in *Arabidopsis* plants have showed more sensitivity to low temperature than their respective wild-type and MsrB3 transgenic plants. MsrB3 plays a ubiquitous role in scavenging reactive oxygen species (ROS) and methionine sulphoxide (MetO) accumulating in the endoplasmic reticulum during cold stress (Kwon et al., 2007).

Photosystems and photosynthetic pigments are believed to be highly damaged due to alteration in photosynthetic gene expression under cold stress (Hajihashemi et al., 2018). Han et al. (2010) isolated the violaxanthin de-epoxidase gene (*LeVDE*), a gene expressed according to photoperiods, from *Lycopersicon esculentum*. Overexpression of this gene elevated non-photochemical quenching, F_v/F_m and quantum yield, oxidizable P700, and alteration in the activity of the xanthophyll cycle and alleviated PSI and PSII photoinhibition under cold spells. Recently, in transgenic tobacco, comparative observations have been accounted by *LeLUT1* (carotenoid epsln-ring hydrolase gene present in tomato) introgression, as a result production of ROS is reduced and consequently membrane integrity is maintained (Miller et al., 2010). Over expression of this stress-responsive gene enables plants to gain advantages from their important roles in improvement of photo-oxidation and photo-inhibition, which as a result decline the cold sensitivity of photosynthetic apparatus of plant. The gene *AtICEI*, which under cold stress stimulates the expression of CBF/DREB in *Arabidopsis*, enhanced cold stress tolerance when introduced in the rice (Xiang et al., 2008). Overexpression of *CcCDR* (an effective drought and cold regulatory protein gene) in transgenic *A. thaliana* by improving many physio-biochemical characteristics, for example, accumulation of osmolytes, enhanced antioxidant activity, and enhanced tolerance to salinity, cold, and low temperatures (Tamirisa et al., 2014).

Guo et al. (2013) discovered the up- or down-regulated genes by utilizing suppression subtractive hybridization (SSH) from pepper seedlings which were pre-treated with ABA and incubated at 6°C for 2 days. It has been reported that 49.32% of unigenes demonstrated fewer functions or unknown functions while 50.68% demonstrated resemblance to genes with known functions. Under chilling stress, in seedlings which were pre-treated with ABA, level of expression of 10 genes was at least twofold higher than in plants which were not treated with ABA, which suggested that under cold stress, ABA positively or negatively regulates the genes in pepper plants.

Cold stimulates the activity of galactosyl synthase and accumulation of oligosaccharides. Synthesis of galactinol is mediated by galactinol synthase, which act as a donor of galactosyl during oligosaccharides synthesis, of raffinose family (Salvi et al., 2016). A galactinol synthase gene *AmGSI* from *Ammopiptanthus mongolicus* which is cold tolerant tree to *Photiniaserrulata* provides enhanced cold tolerant by over expression of this gene (Song et al., 2013).

Raffinose and galactinol are dynamic scavengers of hydroxyl redicals. The role of galactinol synthase is well recognized in salinity and drought; however, there are only few reports related to its potential role in cold tolerance. One of few studies is by Zhou et al. (2013) who observed the increased cold tolerance by improving the formation of raffinose, galactinol, and stachyose by introgression and over expression of *MfGolSl* in Tobacco. Explanation of mechanism of cold stress tolerance at molecular and biochemical levels in plants which are tolerant to cold stress can be incredibly useful in enhancing our understanding of reputed cold responsive genes and their successive introgression for improving tolerance to cold stress of economic crops.

13.2.7 High Temperature

A number of physiological and growth processes are badly affected by high temperature including reproductive processes, seed germination, photosynthesis, and subsequent development (Fahad et al., 2014a,b, 2015a,b, 2016a,b,c,d, 2017, 2018, 2019a,b; Noman et al., 2018b, 2019; Hesham and Fahad, 2020). For instance, pollen grain swelling which leads to indehiscence of anther and dispersal of perturbed pollen is the result of impaired reproductive growth due to high temperature, which eventually badly affects production of seeds (Harsant et al., 2013; Fahad et al., 2014a,b, 2015a,b, 2016a,b,c,d, 2017, 2018, 2019a,b; Hesham and Fahad, 2020).

It is essential to understand the mechanisms of tolerance to high temperature at molecular, biochemical, and physiological levels in the light of global warming for more successful efforts in development of crop cultivars tolerant to high temperature. Molecular and genetic mechanisms, for escaping deleterious changes induced by high temperature, play an important role in survival of plants under such conditions. Sensing of stress of high temperature and rising tolerance is very difficult, which involves networks operating in different compartments of cell. Different presumed sensors, for example, protein sensors located in a plasma membrane channel instigating inward calcium flux and cytoplasm and endoplasmic reticulum, and histone sensors in the nucleus, mediate activation of heat-stress responsive genes engaged in thermotolerance (Song et al., 2012).

The essence behind the effective acclimation of plants to high temperature relies upon the massive accumulation of transcripts which code HSPs and ROS detoxifying enzymes, for example, ascorbate peroxidase (APX). For instance, *Arabidopsis* and *Zea mays* mutants for HSP100 demonstrated hindered growth and adaptation to high temperature (Hong and Vierling, 2002; Nieto-Sotelo et al., 2002). In the same way, in tomato, tolerance to heat stress is reduced by silencing of expression of HSP100/ClpB protein gene present in chloroplast (Yang et al., 2006). Reports relating to the sensitivity of crop plants to high temperature because of silencing/mutation of HSPs (Ashraf et al., 2018) help our understanding of how vital these HSPs are for plants in activating expression of genes which response to heat.

By directing the detoxification, pathways of reactive oxygen species tolerance to temperature can be achieved in better ways, for example, Shi et al. (2001) cloned the APX-encoding gene *HvAPX1*, present in peroxisome and its introgression in the *Arabidopsis* enhanced tolerance to heat stress by increasing activity of APX, thus exhibiting low peroxidation of lipid. Gene LePHGPx, which encodes the phospholipid hydroperoxide glutathione peroxidase, protects the yeast cells from lethal effects. Though, its introgression and overexpression in tomato plants protects them from lethal salinity and temperature levels by reduction of apoptosis levels (Chen et al., 2004).

Up to this point, genes which have been discovered or introgressed in various transgenic plants generally relate to the regulation of the oxidative defense system. Thus, metabolic systems and activities of many other plants are influenced by changes in temperature, and can possibly be handled in genetically modified plants. Likewise, a rise in surrounding temperature which can be observed from past 10 years is an ongoing challenge for crop productivity, arising need to develop stress tolerant plants with tolerance to heat.

13.2.8 Nutrient Stress

Nutrient uptake and their assimilation in plants is directly influenced by the changes in the environmental conditions (Tardieu, 2013). Nutrient deficiencies which are commonly reported include those of zinc, calcium, and iron deficiencies while other nutrients deficiency disorders are believed to be uncommon (Bradley and Hosier, 1999; Hermans et al., 2006). Biomass might be improved by the use of many chemical fertilizers, which are enriched in desired nutrients, but their role in improvement of nutritional value for consumption is reduced either through volatilization, surface run-off, leaching, or microbial consumption. Enhancing nutrient use efficiency (NUE) among crops by efficient means is essential to prevent mineral losses. During the past few decades, conventional breeding extensively contributed in improvement of NUE in plants (Zafar et al., 2019), but those accomplishments through complex molecular techniques have not been enough. For attaining enhanced nutrient uptake, it is essential to involve proficient working of transporters and enzymes in nutrient assimilation, which directly influenced crop yield status. For instance, increased nitrogen accumulation in grains and shoot in wheat plants is due to over expression of glutamine synthetase gene (*GS1*) (Habash et al., 2001), though (30%) kernal number in maize is enhanced by overexpression of *GS1–3*.

Data reveals that a transporter's function in improving NUE development yield under both ideal and suboptimal nitrogen circumstances (Ranathunge et al., 2014). NRT1.1 works as a nitrate sensor and can improve the capacity of NRT1.1 to intervene effective nitrate transport in the protein kinase CIPK23-dependent phosphorylation and intracellular threoninde phosphorylation (Parker and Newstead, 2014). In addition, nodule initiation (NIN)-like protein (NLP), TFs are the principal regulators of nitrate reaction and activate nitrate-responsive transcription when binding with the nitrate-responsive cis-component, which is additionally modulated by post-translation nitrate signalling. Repression of NLP function results in hindered expression of multiple genes induced by nitrate (Konishi and Yanagisawa, 2011). Castaings et al. (2009) noted that nlp7 mutants demonstrate hindered transduction of nitrate signals and that their pattern of expression and function in nitrogen sensing are strongly related to each other. Kuo and Chiou (2011) proposed that post-transcription, micro-RNAs have a putative role in controlling the genes of nutrient starvation.

From the current germplasm or by genetic manipulation, it is possible to selectively develop nutrient-rich cultivars. To ameliorate nutrient concentrations, customizing crop genetic makeup is picking up interest as a way to lessen malnutrition. Significant outcomes for studying alterations in gene expression in reaction to nutritional stress can be produced by microarray and sequence-based transcription profiling technology (Newell-McGloughlin, 2008). In nutrient-starved crops, transient modifications in gene expression are well achieved. Differential gene expression was described by (Bi et al., 2007) in Arabidopsis under moderate nitrogen stress that was acting as putative controllers of nitrogen stress reaction. Due to the lack of a main regulatory gene NLA, a faulty Arabidopsis mutant has shown modified transcriptional reactions to nitrogen restriction in developing adequate reactions to nitrogen stress (Liu et al., 2017). A system method is being implemented in rice and maize, primarily aimed at profiling genes in reaction to individual or combined nutrient stress at transcriptional concentrations.

Genetic engineering approaches for NUE enhancement range from improving mineral nutrient solubility and remobilization inside the plant to transporting and accumulating within storage organs. Transgenic soybean have been recently created by (Zhou et al., 2014) in which constitutive overexpression of GmEXPB2 (β-expansin) has expanded leaf expansion and enhanced phosphate effectiveness. Excessive expression of genes of expansion, for example, HvEXPB1 in barley (Kwasniewski and Szarejko, 2006) and OsEXPA17 in rice (Yu et al., 2011) enhanced phosphate uptake by inducing better development of root hair even under phosphate deficiencies. A significant strategy for effective nutrient remobilization is manipulation of transporter genes as remobilization of nutrients inside a plant is unfavourable for their survival. For instance, GmPT1 transporter gene overexpression improved phosphate removal, yield, and associated characteristics such as phosphorus use soybean effectiveness and quantum yield (Song et al., 2014b). A comprehensive knowledge of the variables regulating effective mineral uptake and remobilization inside the plant is needed to improve NUE through genetic manipulations. Pre-requisites are the use of techniques such as transcription profiling, faulty mutant analysis in their reaction to mineral deficiency and plant research demonstrating ordinary development under nutrient stress.

Over the past few decades, biofortification of major seed plants to optimize micronutrient collection has been the topic of extensive studies (Garg et al., 2018). For instance, Saeed Akram et al. (2009) found during a sunflowers study that a potassium sulphate foliage spray fundamentally improved K^+ content of the shoot and leaf, whereas no change was seen in Mg^{2+}, Ca^{2+}, or N content of the leaf and root under non-stress and saline situations. Likewise, foliar application of Zn, Fe, and K had an important impact on crop nutrients and nutrient concentrations in wheat grain and straw (Zafar et al., 2016).

For biofortification of significant crops, the transgenic approach has been favoured over the traditional one since it is an appropriate and time-saving approach to overcoming issues with nutrient deficiencies (Garg et al., 2018). It has been noted that rice plants overexpressing iron storage protein ferritin have enhanced content of seed iron (Lucca et al., 2001). GM crop adoption is a tedious procedure as several policies and laws required to be adopted before such cultivars or varieties can be commercially released. Issues of human health can occur due to up and down regulation of a cascade of genes that might be required by a cultivar or variety that is transgenic to one of the nutrients. Various government organizations are working to survey these issues mostly in developed countries. In fundamental plant science studies, the transgenic method is considered as an incredibly helpful instrument, but knowing the gene networks and molecular physiology of crop reactions to nutrient deficiencies or excess is a precondition.

13.2.9 Heavy Metals

In those countries where economy is significantly heavily industrially dependent, heavy metal pollution and contamination is a noteworthy determinant of global plant species dispersion as well as agricultural productivity. When a concentration of metal exceeds a particular amount of limit numerous crops demonstrate side effects of heavy metal toxicity. Overall, the first noticeable signs of heavy metal exposure are necrosis and stunted shoot development (Emamverdian et al., 2015).

Heavy metals interfere with several aspects of plant growth and physiology. The amount of mitochondrial cristae prompting disabled oxidative phosphorylation is reduced by heavy metals such as lead (Pb) and cadmium (Cd). Heavy metals encourage chromatin aggregation and condensation as well as impaired replication and transcription when binding to nucleic acids (Emamverdian et al., 2015). Pb's and Cd's affinity to enzyme groups of sulphohydryl contributes to their inactivation. Moreover, heavy metal stress disturbs the ordinary metabolism as it contributes to increased ROS production and accumulation and therefore oxidative stress (Valko et al., 2005). As a way to remedy ecological contaminants, for example, heavy metals, phytoremediation is widely promoted (Placek et al., 2016). In addition, rhizoremediation involving crops and their natural or introduced rhizospheric microbes also helps to degrade or reduce contaminant concentrations and encourage ordinary plant development (Marques et al., 2009; Kala, 2014; Noman et al., 2019).

Most heavy metal salts are hard to distinguish by physical separation techniques because in wastewater, they are hydrophilic and readily soluble. Physicochemical techniques can be ineffectual or expensive at low concentrations of heavy metals. Alternative techniques for removing heavy metals incorporate biosorption or bioaccumulation. An efficient strategy for overcoming or minimizing heavy metal emissions is utilization of microorganisms and crops for remediation (Das and Dash, 2014). Regardless of the capability of these methodologies to contribute to the reclaiming of contaminated soils, thorough data on the fundamental processes is not accessible in the literature and attempts are extremely difficult to convert these strategies from effective laboratory or greenhouse studies to field sites. Two variables that make these approaches not very effective are (i) the various stress variables accessible in the field are not utilized in laboratory and greenhouse research and (ii) there is a lack of effective and sufficient methodologies and methods that can be used to determine whether or not distinct contaminant levels are diminishing (Das and Dash, 2014).

Phytoremediation is currently being utilized widely to clean overwhelming metals from the earth or render them innocuous as it is an environmentally friendly, cost-effective, and non-obtrusive approach. Various processes are used to detoxify harmful heavy metals and metalloids, such as chelation, trafficking, and compartimentation (Flora and Pachauri, 2010). In addition, detoxification through binding poisonous metals and metalloids is mediated by the development of high affinity ligands such as phytochelatins (PCs) and metallothioneins (MTs) and cysteine-rich thiol-reactive peptides (Gasic and Korban, 2007; Guo et al., 2008; Pal and Rai, 2010). It is necessary for heavy metal resistance to form PSc-metal or MT-metal complexes and their successive sequestration into the vacuole. PC synthesis using GSH or π-glutamyl cysteine as a substratum is mediated by phytochelatin synthase (PCS) enzyme (Cobbett and Goldsbrough, 2002). Viable research on genes encoding PCs has been carried out, with various genes cloned to date, for example, OsPCS1, TaPCS1, AtPCS1, and CePCS1 from rice, corn, and Arabidopsis (Gasic and Korban, 2007; Ha et al., 2007; Vatamaniuk et al., 2002) and BjPCS1 from metal-tolerant crops *Brassica juncea* and *Allium sativum*, respectively (Heiss et al., 2003; Zhang et al., 2005).

Guo et al. (2008) explained that simultaneous overexpression of *AsPCS1* and *GSH1* (from *A. sativum* and *S. cerevisiae*) induced tolerance against heavy metals and metalloids in *A.* thaliana. It was examined that single-gene transgenic lines appeared to be highly tolerant and accumulated more Cd and As than wild-type plants rather than dual gene transgenic lines which showed more tolerance and accumulated (twofold) more Cd and As compared with single gene transformed lines. The enhanced synthesis of GSH and PCs led to more accumulation and tolerance to Cd and As (Li et al., 2004). Phytoextraction is done in those plants having inherent property for accumulation of metalloids in plant total biomass (Pajević et al., 2016). Intensive studies are on board for genetic manipulations to develop specific morphological attributes and special anatomical characters to accumulate excess metalloids (Kotrba et al., 2009). A set of genes involved in plant tolerance for more uptake of metalloids and induce high metal binding capacity at intracellular sites, which efficiently sequestrate into the vacuole for deposition and detoxification (Viehweger, 2014). For example, yeast protein YCF1 intervenes the sequestration of Pb and Cd into the vacuole. Genetically modified *A. thaliana* plants showed overexpressing YCF1 to tolerate Pb and Cd (Song et al., 2003). Furthermore, high translocation of metals to aerial region of plants through apoplast or symplast, and their subsequent exclusion to trichomes, contributes to increased metal tolerance and remediation (Clemens et al., 2002).

Extensive studies have been done to document the role of PCs and MTs in heavy metal detoxification. Gonzalez-Mendoza et al. (2007) reported that enhanced expression of *AvPCS* and *AvMt2* in *Avicenniagerminans* under Cd and Cu stress depicts that the PCs and MTs pave the ways to detoxification response applied for exclusion of non-essential metals. Scientists are struggling hard to manipulate the genes expressed for PCs and MTs for enhancing the detoxifying potential of crop plants, for example, *Nicotianatabacum* (Wojas et al., 2010) and *B. juncea* (Gasic and Korban, 2007) overexpressing *AtPCS1*(PCS) resulted improved tolerance to Cd by maintaining higher levels of PCs in the cytosol and vacuole. *Arabidopsis thaliana* overexpressing *PCs1* appeared to be high tolerant to Cd and arsenic (As) and accumulated more biomass (Verbruggen et al., 2009). Furthermore, when Cd and As levels declined, thiol peptide enhanced in shoot biomass (Li et al., 2004). In addition to boosting tolerance, transgenic plants acclimatize less metal content in the aerial parts of plants (Gasic and Korban, 2007). Overexpression of *AtPCS1* in *N. tabacum* harbouring the *Agrobacterium rhizogenesrolB* oncogene increased their Cd tolerance; tolerance was more increased with GSH supplementation of medium (Pomponi et al., 2006). Genetically modified plants have increased expression of *O*-acetylserine (thiol) lyase (OASTL), improved catalysis of cysteine formation (from sulphide and O-acetylserine) which is a vital limiting factor in the accumulation of GSH for heavy metal tolerance (Ning et al., 2010). The advances in the use of transgenic techniques found to be very successful way towards toxic metalloids extraction (especially Se and As) and metals (particularly Cu, Pb and Cd) with the involvement of metal transporters present in the aboveground plant parts upgrades metal uptake activity, higher production of enzymes enhanced production of chelators involved in metal detoxification such as PCs and MTs. Some events such as genetic basis of resistance or tolerance, hyperaccumulation of metals and metalloids, pathways of translocation and other factors that influence phytoremediation can be better studied with the help of advanced mechanistic basis of transgenic approaches.

13.2.10 Increased UV-B Radiation Fluxes

In the 1970s, the threat of the ozone layer for stratospheric layer depletion was pointed out. Ozone layer is the major zone for absorption of harmful ultraviolet rays especially UV-B radiation. These cause very strong damages to almost all living organisms and even a small change in ozone layer causes large implications of UV-B negative effects on living beings. Measurements and roles during 1980s and 1990s of continuously increasing UV-B radiation enthused many research projects in this field. Montreal Protocol has been found very effective in the control of rate of release of UV destructive compounds present in the atmosphere and lead to the stabilized stratospheric ozone levels and also recovered to levels of mid-twentieth century by the end of the century (Fahad et al., 2013; McKenzie et al., 2011). Nevertheless, studies on this global change driver are continuously under steady study. At the beginning, studies were performed under controlled conditions and UV-B radiation under greenhouse conditions assured significant effects on plant responses. Later on field work experiments largely refused plants sensitivity toward UV-B light, thus earlier work gained relatively limited height. However, natural environmental levels of UV-B radiation show very significant effects on plant growth that can be easily obtained under controlled or unnatural experimental conditions. Under the field higher plants are naturally protected from significant damage done by UV-B radiations (Day and Neale, 2002). Increased UV-B fluxes, when checked experimentally have shown abnormalities in plant morphology and growth form, changed timing of flowering, enhanced number UV absorbing pigments (importantly phenolics), changes in DNA (Deoxyribonucleic acid), and improved antioxidants levels. When experimental data was analysed by formal data analysis, significant effects of increased UV-B can be seen only in biomass height and leaf area of plant. When plant species are grown in a mixture, it allows major effects to play a role in respect to those plants that are grown in isolation thus increased radiations can shift in the relative accomplishment of some species (Zlatev et al., 2012). Two plant species that are competing and accumulated over a six year time period studied in a field experiment where in favour of wheat altered competitive balance was recorded for wheat and wild oats results shown harmful impacts on health and leaf area growth rate of these competing plant species. Wheat plant showed changes that allowed it to compete more effectively for some light (Evans, 2013). According to data obtained from data analysis from a huge number of experimental material that have been designed with appropriate techniques are available and according to a rough estimate, under simulation of 20% ozone loss, we have to bear 7–15% drop in biomass production. Amusingly woody plants showed adverse effects than herbaceous plant species and even a large number of woody species that were even simulated by increased UV-B radiation showed non-significant results (Fahad et al., 2013; Singh et al., 2006). Under the experiments of adversaries of natural UV-B by the use of different types of filters a range of experiments has been performed in which plants are sparked that removed different wavelengths of UV-B spectrum. In both native and agricultural crops, UV-B radiation causes severe reduction in insect herbivore by affecting various properties of plant host tissues and this one is the most important observations that UV-B radiations impart on plants (Rozema et al., 2009). These mechanisms are likely to be initiated by alterations in plants secondary chemistry or changes in plant nitrogen or plant sugar content but the net change in the rate of herbivore is unsure (Fahad et al., 2013; Sullivan and Teramura, 1992). Studies at the genetic level show that there are a number of genes that are regulated by UV-B radiations that facilitated DNA repair, improved antioxidant synthesis, and enhanced concentration of pigments that absorbed UV-B radiation. Changes in plant tissue chemicals induced by UV-B also increase concentrations of UV-B screening pigments and some other compounds that appear to have additional effects, for example, defense against plant pathogens, and even insect herbivory-studies at genetic level also expose that UV-B radiation can stimulate group of genes influence regulation of radiation destruction and wounding responses of insects. Recent research works using non-damaging procedures that also allow improved sampling techniques have begun to illuminate leaf epidermis in protecting the machinery of photosynthesis from UV-B radiations. At the same time allows the movement of photosynthetically active UV-B radiations (Björn, 1996).

13.3 Biotic Stresses

A huge number of biotic stresses affect plant growth and morphology may include insect attachment, weed growth, and attack of different diseases with the use of externally applied nutrients (Pandey et al., 2017). Mostly weeds grow and appear in the soil where nutrients are depleted and by the application of adequate nutrients to the root environment weed development can be controlled with lessened side effects on plant growth. With improved nutrient use efficiency, plant vigour can be achieved and plants will be able to cope with the nutrients deficiency caused by weeds (Bindraban et al., 2014, 2015).

Moreover mineral nutrients have complex relations with plant diseases such as some pathogens live phytium and phytophtera require higher levels of humidity and extra nitrogen contents for their proper growth and development on plants (Lamichhane et al., 2017). Grasses also overcome their diseases easily when Mn and Si elements are externally added to their solution, for example, turf grasses. Insects' injury more severely affects plants with nutrient deficiencies and also less time is required for their recovery. In an experiment on turf grasses growth, addition of Al and Si in growth medium protected plants from insect's injury and also helped in production of some unpleasant chemical that repelled insects from damaging plants (Datnoff, 2008).

13.4 Conclusion

Plants exhibit various internal protective mechanisms to multiple abiotic stresses. Exploring the regulatory systems participating in starting these tolerance events has prevailed under rigorous investigation for decades. Testing of viability of these suppositions has given new ways regarding the good interpretation and illumination of stress-induced variation in plants. Physiologist and biochemist of plants have prevailed the vital players in exploring the basics of these events that are now being investigated greatly at the molecular and genetic level using many genomic, molecular, and biotechnological applications. The assortment and recognition of main stress accessible genes and their frequent introgression for producing tolerant crop varieties in the course of conventional breeding protocols are time-consuming. Plant biotechnology is very efficient technique than conventional breeding in spite of expensive. Various stress approachable genes have been studied and established effectively into other crops to develop transgenic crops with increased stress resistance. Therefore, it is imperative to reveal here that at some stage in the development of transgenic crop cultivars, concern is required to develop genes that result in increased resistance to various abiotic stresses, markedly at the whole plant level. This requires the improvement of set of markers to increase stress resistance. The benefits of biotechnology in transgenic plant development for effective crop cultivars are certainly vast. In addition hazard evaluation of transgenic crops is one of the main points obligatory before the release or makes use of transgenic plants. In calculation, the danger to the environment from the transgenic plants must be evaluated with various ground tests proceeding to commercialization, assessment on plant varieties, and sufficient organization implementations in place to handle inherent danger. There are many equipped governmental system in various countries for the security evaluation of GM crops. Additionally, there is some few international consensus that manage the farming and commercialization of transgenic plants and their imitative. The main objective of these systems and danger estimation approach is concentrated on defending the environment and human health all over the world. The development of transgenic plants totally depends on the appraisal of the harm or benefits, time period, regulatory agreement cost, economic rank, as well as commercialization, requirement, and standards of various countries.

REFERENCES

Acosta-Motos J, Ortuño M, Bernal-Vicente A et al. (2017) Plant Responses to Salt Stress: Adaptive Mechanisms. *Agronomy*. doi: 10.3390/agronomy7010018.

Adams WA, Akhtar N (1994) The possible consequences for herbage growth of waterlogging compacted pasture soils. *Plant Soil*. doi: 10.1007/BF01416085.

Adnan M, Shah Z, Fahad S, Arif M, Alam M, Khan IA, Mian IA, Basir A, Ullah H, Arshad M, Rahman I-U, Saud S, Ihsan MZ, Jamal Y, Amanullah, Hammad HM, Nasim W (2018) Phosphate-solubilizing bacteria nullify the antagonistic effect of soil calcification on bioavailability of phosphorus in alkaline soils. *Sci Rep* 8:4339. https://doi.org/10.1038/s41598-018-22653-7.

Adnan M, Fahad S, Zamin M, Shah S, Mian IA, Danish S, Zafar-ul-Hye M, Battaglia ML, Naz RMM, Saeed B, Saud S, Ahmad I, Yue Z, Brtnicky M, Holatko J, Datta R (2020) Coupling Phosphate-solubilizing bacteria with phosphorus supplements improve maize phosphorus acquisition and growth under lime induced salinity stress. *Plants* 9(900). doi: 10.3390/plants9070900.

Ahanger MA, Akram NA, Ashraf M et al. (2017) Plant responses to environmental stresses-From gene to biotechnology. *AoB Plants* 27 (4):plx025.

Ahmad P, Azooz MM, Prasad MNV (2013) *Ecophysiology and responses of plants under salt stress*. doi: 10.1007/978-1-4614-4747-4.

Ahmad S, Kamran M, Ding R, Meng X, Wang H, Ahmad I, Fahad S, Han Q (2019) Exogenous melatonin confers drought stress by promoting plant growth, photosynthetic capacity and antioxidant defense system of maize seedlings. *PeerJ* 7:e7793 http://doi.org/10.7717/peerj.7793.

Ainsworth EA, Rogers A (2007) The response of photosynthesis and stomatal conductance to rising CO_2: Mechanisms and environmental interactions. *Plant, Cell Environ* 30 (3):258–270.

Ainsworth EA, Yendrek CR, Sitch S et al. (2012) The effects of tropospheric ozone on net primary productivity and implications for climate change. *Annu Rev Plant Biol*. doi: 10.1146/annurev-arplant-042110-103829.

Akhtar I, Nazir N (2013) Effect of waterlogging and drought stress in plants. *Int J Water Resour Environ Sci*. doi: 10.5829/idosi.ijwres.2013.2.2.11125.

Akram R, Turan V, Hammad HM, Ahmad S, Hussain S, Hasnain A, Maqbool MM, Rehmani MIA, Rasool A, Masood N, Mahmood F, Mubeen M, Sultana SR, Fahad S, Amanet K, Saleem M, Abbas Y, Akhtar HM, Waseem F, Murtaza R, Amin A, Zahoor SA, ul Din MS, Nasim W (2018a) Fate of organic and inorganic pollutants in paddy soils In: Hashmi, MZ and Varma, A (eds.) *Environmental pollution of paddy soils*. Springer International Publishing, Cham, Switzerland, pp. 197–214.

Akram R, Turan V, Wahid A, Ijaz M, Shahid MA, Kaleem S, Hafeez A, Maqbool MM, Chaudhary HJ, Munis, MFH, Mubeen M, Sadiq N, Murtaza R, Kazmi DH, Ali S, Khan N, Sultana SR, Fahad S, Amin A, Nasim W (2018b) Paddy land pollutants and their role in climate change. In: Hashmi, MZ, Varma, A (eds.) *Environmental pollution of paddy soils*. Springer International Publishing, Cham, Switzerland, pp. 113–124.

Ali Q, Athar HUR, Ashraf M (2008) Modulation of growth, photosynthetic capacity and water relations in salt stressed wheat plants by exogenously applied 24-epibrassinolide. *Plant Growth Regul*. doi: 10.1007/s10725-008-9290-7.

Ali S, Liu Y, Ishaq M et al. (2017) Climate change and its impact on the yield of major food crops: Evidence from Pakistan. *Foods*. doi: 10.3390/foods6060039.

Ashraf M, Harris PJC (2013) Photosynthesis under stressful environments: An overview. *Photosynthetica* 51:163–190.

Ashraf MF, Yang S, Wu R et al. (2018) Capsicum annuum HsfB2a positively regulates the response to Ralstonia solanacearum infection or high temperature and high humidity forming transcriptional cascade with CaWRKY6 and CaWRKY40. *Plant Cell Physiol*. doi: 10.1093/pcp/pcy181.

Ashraf MA (2012) Waterlogging stress in plants: A review. *African J Agric Res*. doi: 10.5897/ajarx11.084.

Assaha DVM, Ueda A, Saneoka H et al. (2017) The role of Na^+ and K^+ transporters in salt stress adaptation in glycophytes. *Front Physiol* 8:509. doi: 10.3389/fphys.2017.00509.

Bailey-Serres J, Voesenek LACJ (2008) Flooding stress: Acclimations and genetic diversity. *Annu Rev Plant Biol*. doi: 10.1146/annurev.arplant.59.032607.092752.

Baisakh N, Ramanarao MV, Rajasekaran K et al. (2012) Enhanced salt stress tolerance of rice plants expressing a vacuolar H + -ATPase subunit c1 (SaVHAc1) gene from the halophyte grass Spartina alterniflora Löisel. *Plant Biotechnol J*. doi: 10.1111/j.1467-7652.2012.00678.x.

Barrero-Sicilia C, Silvestre S, Haslam RP, Michaelson LV (2017) Lipid remodelling: Unravelling the response to cold stress in Arabidopsis and its extremophile relative Eutrema salsugineum. *Plant Sci*.

Ben-Noah I, Friedman SP (2018) Review and evaluation of root respiration and of natural and agricultural processes of soil aeration. *Vadose Zo J*. doi: 10.2136/vzj2017.06.0119.

Bi YM, Wang RL, Zhu T, Rothstein SJ (2007) Global transcription profiling reveals differential responses to chronic nitrogen stress and putative nitrogen regulatory components in Arabidopsis. *BMC Genom*. doi: 10.1186/1471-2164-8-281.

Bindraban PS, Dimkpa CO, Nagarajan L et al. (2014) Towards fertilisers for improved uptake by plants. *Proc – Int Fertil Soc*.

Bindraban PS, Dimkpa C, Nagarajan L et al. (2015) Revisiting fertilisers and fertilisation strategies for improved nutrient uptake by plants. *Biol Fertil Soils*.

Björn LO (1996) Effects of ozone depletion and increased UV-B on terrestrial ecosystems. *Int J Environ Stud*. doi: 10.1080/00207239608711082.

Bradley L, Hosier S (1999) Guide to Symptoms of Plant Nutrient Deficiencies. *Univ Arizona Coop Ext*.

Brooks AN, Turkarslan S, Beer KD et al. (2011) Adaptation of cells to new environments. Wiley Interdiscip. *Rev Syst Biol Med*.

Bulle M, Yarra R, Abbagani S (2016) Enhanced salinity stress tolerance in transgenic chilli pepper (Capsicum annuum L.) plants overexpressing the wheat antiporter (TaNHX2) gene. *Mol Breed*. doi: 10.1007/s11032-016-0451-5.

Burgess P, Huang B (2014) Root protein metabolism in association with improved root growth and drought tolerance by elevated carbon dioxide in creeping bentgrass. *F Crop Res*. doi: 10.1016/j.fcr.2014.05.003.

Burgess P, Huang B (2019) Leaf protein abundance associated with improved drought tolerance by elevated carbon dioxide in creeping bentgrass. *J Am Soc Hortic Sci*. doi: 10.21273/jashs.141.1.85.

Calatayud A, Barreno E (2001) Chlorophyll a fluorescence, antioxidant enzymes and lipid peroxidation in tomato in response to ozone and benomyl. *Environ Pollut*. doi: 10.1016/S0269-7491(01)00101-4.

Casella E, Soussana J-F (2007) Long-term effects of CO_2 enrichment and temperature increase on the carbon balance of a temperate grass sward. *J Exp Bot*. doi: 10.1093/jxb/48.6.1309.

Castaings L, Camargo A, Pocholle D et al. (2009) The nodule inception-like protein 7 modulates nitrate sensing and metabolism in Arabidopsis. *Plant J*. doi: 10.1111/j.1365-313X.2008.03695.x.

Chen S, Vaghchhipawala Z, Li W et al. (2004) Tomato phospholipid hydroperoxide glutathione peroxidase inhibits cell death induced by bax and oxidative stresses in yeast and plants. *Plant Physiol.* doi: 10.1104/pp.103.038091.

Choudhury FK, Rivero RM, Blumwald E, Mittler R (2017) Reactive oxygen species, abiotic stress and stress combination. *Plant J.* doi: 10.1111/tpj.13299.

Clemens S, Palmgren MG, Krämer U (2002) A long way ahead: Understanding and engineering plant metal accumulation. *Trends Plant Sci.*

Cobbett C, Goldsbrough P (2002) Phytochelatins and metallotioneins: roles in heavy metal detoxification and homeostasis. *Annu Rev Plant Biol.* doi: doi:10.1146/annurev.arplant.53.100301.135154.

Cramer GR, Urano K, Delrot S et al. (2011) Effects of abiotic stress on plants: A systems biology perspective. *BMC Plant Biol* 11 (163).

Das S, Dash HR (2014) Microbial bioremediation: A potential tool for restoration of contaminated areas. In: *Microbial biodegradation and bioremediation.*

Daszkowska-Golec A, Szarejko I (2013) Open or close the gate – Stomata action under the control of phytohormones in drought stress conditions. *Front Plant Sci.* doi: 10.3389/fpls.2013.00138.

Datnoff LE (2008) *Silicon in the life and performance of turfgrass. ats.* doi: 10.1094/ats-2005-0914-01-rv.

Day TA, Neale PJ (2002) Effects of UV-B radiation on terrestrial and aquatic primary producers. *Annu Rev Ecol Syst.* doi: 10.1146/annurev.ecolsys.33.010802.150434.

Defez R, Andreozzi A, Dickinson M et al. (2017) Improved drought stress response in alfalfa plants nodulated by an IAA over-producing Rhizobium Strain. *Front Microbiol.* doi: 10.3389/fmicb.2017.02466.

Emamverdian A, Ding Y, Mokhberdoran F, Xie Y (2015) Heavy metal stress and some mechanisms of plant defense response. *Sci World J.*

Engineer CB, Hashimoto-Sugimoto M, Negi J et al. (2016) CO_2 sensing and CO_2 regulation of stomatal conductance: Advances and open questions. *Trends Plant Sci.*

Evans JR (2013) Improving Photosynthesis. *Plant Physiol.* doi: 10.1104/pp.113.219006.

Fahad S, Bano A (2012) Effect of salicylic acid on physiological and biochemical characterization of maize grown in saline area. *Pak J Bot* 44:1433–1438.

Fahad S, Chen Y, Saud S, Wang K, Xiong D, Chen C, Wu C, Shah F, Nie L, Huang J (2013) Ultraviolet radiation effect on photosynthetic pigments, biochemical attributes, antioxidant enzyme activity and hormonal contents of wheat. *J Food Agric Environ* 11(3&4):1635–1641.

Fahad S, Hussain S, Bano A, Saud S, Hassan S, Shan D, Khan FA, Khan F, Chen Y, Wu C, Tabassum MA, Chun MX, Afzal M, Jan A, Jan MT, Huang J (2014a) Potential role of phytohormones and plant growth-promoting rhizobacteria in abiotic stresses: Consequences for changing environment. *Environ Sci Pollut Res* 22(7):4907–4921. https://doi.org/10.1007/s11356-014-3754-2.

Fahad S, Hussain S, Matloob A, Khan FA, Khaliq A, Saud S, Hassan S, Shan D, Khan F, Ullah N, Faiq M, Khan MR, Tareen AK, Khan A, Ullah A, Ullah N, Huang J (2014b) Phytohormones and plant responses to salinity stress: A review. *Plant Growth Regul* 75(2):391–404. https://doi.org/10.1007/s10725-014-0013-y.

Fahad S, Hussain S, Saud S, Tanveer M, Bajwa AA, Hassan S, Shah AN, Ullah A, Wu C, Khan FA, Shah F, Ullah S, Chen Y, Huang J (2015a) A biochar application protects rice pollen from high-temperature stress. *Plant Physiol Biochem* 96:281–287.

Fahad S, Nie L, Chen Y, Wu C, Xiong D, Saud S, Hongyan L, Cui K, Huang J (2015b) Crop plant hormones and environmental stress. *Sustain Agric Rev* 15:371–400.

Fahad S, Hussain S, Saud S, Hassan S, Chauhan BS, Khan F et al. (2016a) Responses of rapid viscoanalyzer profile and other rice grain qualities to exogenously applied plant growth regulators under high day and high night temperatures. *PLoS One* 11(7):e0159590. https://doi. org/10.1371/journal.pone.0159590.

Fahad S, Hussain S, Saud S, Hassan S, Ihsan Z, Shah AN,Wu C, Yousaf M, Nasim W, Alharby H, Alghabari F, Huang J (2016b) Exogenously applied plant growth regulators enhance the morphophysiological growth and yield of rice under high temperature. *Front Plant Sci* 7:1250. https://doi.org/10.3389/fpls.2016.01250.

Fahad S, Hussain S, Saud S, Hassan S, Tanveer M, Ihsan MZ, Shah AN, Ullah A, Nasrullah KF, Ullah S, AlharbyH NW, Wu C, Huang J (2016c) A combined application of biochar and phosphorus alleviates heat-induced adversities on physiological, agronomical and quality attributes of rice. *Plant Physiol Biochem* 103:191–198.

Fahad S, Hussain S, Saud S, Khan F, Hassan S, Jr A, Nasim W, Arif M, Wang F, Huang J (2016d) Exogenously applied plant growth regulators affect heat-stressed rice pollens. *J Agron Crop Sci* 202:139–150.

Fahad S, Bajwa AA, Nazir U, Anjum SA, Farooq A, Zohaib A, Sadia S, NasimW, Adkins S, Saud S, Ihsan MZ, Alharby H,Wu C,Wang D, Huang J (2017) Crop production under drought and heat stress: Plant responses and management options. *Front Plant Sci* 8:1147. https://doi.org/10.3389/fpls.2017.01147.

Fahad S, Ishan MZ, Khaliq AK, Daur I, Saud S, Alzamanan S, Nasim W, Abdullah MA, Khan IA, Wu C, Wang D, Huang J (2018) Consequences of high temperature under changing climate optima for rice pollen characteristics-concepts and perspectives. *Archives Agron Soil Sci* DOI: 10.1080/03650340.2018.1443213.

Fahad, S, Adnan, M, Hassan, S, Saud, S, Hussain, S, Wu, C, Wang, D, Hakeem, KR, Alharby, HF, Turan, V, Khan, MA, Huang, J, (2019a). Rice responses and tolerance to high temperature. In: Hasanuzzaman M, Fujita M, Nahar K, Biswas JK (eds.), *Advances in Rice Research for Abiotic Stress Tolerance.* Cambridge: Woodhead, pp. 201–224.

Fahad, S, Rehman, A, Shahzad, B, Tanveer, M, Saud, S, Kamran, M, Ihtisham, M, Khan, SU, Turan, V, Rahman, MHU, (2019b). Rice responses and tolerance to metal/metalloid toxicity, in Hasanuzzaman M, Fujita M, Nahar K Biswas, JK (eds.), *Advances in Rice Research for Abiotic Stress Tolerance.* Cambridge: Woodhead, pp. 299–312.

Farfan-Vignolo ER, Asard H (2012) Effect of elevated CO_2 and temperature on the oxidative stress response to drought in Lolium perenne L. and Medicago sativa L. *Plant Physiol Biochem.* doi: 10.1016/j.plaphy.2012.06.014.

Gul F, Ahmed I, Ashfaq M, Jan D, Shah F, Li X, Wang D, Fahad M, Fayyaz M, Shah AS (2020) Use of crop growth model to simulate the impact of climate change on yield of various wheat cultivars under different agro-environmental conditions in Khyber Pakhtunkhwa, Pakistan. *Arabian J Geosci* 13:112. https://doi.org/10.1007/s12517-020-5118-1.

Farhat A, Hafiz MH, Wajid I, Aitazaz AF, Hafiz FB, Zahida Z, Fahad S, Wajid F, Artemi C (2020) A review of soil carbon dynamics resulting from agricultural practices. *J Environ Manag* 268:110319.

Fiscus EL, Booker FL, Burkey KO (2005) Crop responses to ozone: Uptake, modes of action, carbon assimilation and partitioning. *Plant Cell Environ* 28 (8):997–1011.

Flora SJS, Pachauri V (2010) Chelation in metal intoxication. *Int J Environ Res Public Health.*

Fuhrer J (2009) Ozone risk for crops and pastures in present and future climates. *Naturwissenschaften.*

Fuhrer J, Booker F (2003) Ecological issues related to ozone: Agricultural issues. *Environ Int.*

Garg M, Sharma N, Sharma S et al. (2018) Biofortified crops generated by breeding, agronomy, and transgenic approaches are improving lives of millions of people around the world. *Front Nutr.* doi: 10.3389/fnut.2018.00012.

Gasic K, Korban SS (2007) Transgenic Indian mustard (Brassica juncea) plants expressing an Arabidopsis phytochelatin synthase (AtPCS1) exhibit enhanced As and Cd tolerance. *Plant Mol Biol.* doi: 10.1007/s11103-007-9158-7.

Ghorani-Azam A, Riahi-Zanjani B, Balali-Mood M (2016) Effects of air pollution on human health and practical measures for prevention in Iran. *J Res Med Sci.*

Gonzalez-Mendoza D, Moreno AQ, Zapata-Perez O (2007) Coordinated responses of phytochelatin synthase and metallothionein genes in black mangrove, Avicennia germinans, exposed to cadmium and copper. *Aquat Toxicol.* doi: 10.1016/j.aquatox.2007.05.005.

Gougoulias C, Clark JM, Shaw LJ (2014) The role of soil microbes in the global carbon cycle: Tracking the below-ground microbial processing of plant-derived carbon for manipulating carbon dynamics in agricultural systems. *J Sci Food Agric.* 94:2362–2371.

Guerrieri R, Lepine L, Asbjornsen H et al. (2016) Evapotranspiration and water use efficiency in relation to climate and canopy nitrogen in U.S. forests. *J Geophys Res Biogeosci.* doi: 10.1002/2016JG003415.

Guo J, Dai X, Xu W, Ma M (2008) Overexpressing GSH1 and AsPCS1 simultaneously increases the tolerance and accumulation of cadmium and arsenic in Arabidopsis thaliana. *Chemosphere.* doi: 10.1016/j.chemosphere.2008.04.018.

Guo WL, Chen RG, Gong ZH et al. (2013) Suppression Subtractive Hybridization Analysis of Genes Regulated by Application of Exogenous Abscisic Acid in Pepper Plant (Capsicum annuum L.) Leaves under Chilling tress. *PLoS One.* doi: 10.1371/journal.pone.0066667.

Ha S-B, Smith AP, Howden R et al. (2007) Phytochelatin Synthase Genes from Arabidopsis and the Yeast Schizosaccharomyces pombe. *Plant Cell*. doi: 10.2307/3870806.

Habash DZ, Massiah AJ, Rong HL et al. (2001) The role of cytosolic glutamine synthetase in wheat. In: *Annals of applied biology*.

Habib ur Rahman M, Ahmad A, Wajid A, Hussain M, Rasul F, Ishaque W, Islam MA, Shiela V, Awais M, Ullah A, Wahid A, Sultana SR, Saud S, Khan S, Shah F, Hussain M, Hussain S, Nasim W (2017) Application of CSM-CROPGRO-Cotton model for cultivars and optimum planting dates: Evaluation in changing semi-arid climate. *Field Crops Res* http://dx.doi.org/10.1016/j.fcr.2017.07.007.

Hammad HM, Farhad W, Abbas F, Fahad S, Saeed S, Nasim W, Bakhat HF (2016) Maize plant nitrogen uptake dynamics at limited irrigation water and nitrogen. *Environ Sci Pollut Res* 24(3):2549–2557. https://doi.org/10.1007/s11356-016-8031-0.

Hammad HM, Abbas F, Saeed S, Shah F, Cerda A, Farhad W, Bernado CC, Wajid N, Mubeen M, Bakhat HF (2018) Offsetting land degradation through nitrogen and water management during maize cultivation under arid conditions. *Land Degrad Dev.* 1–10. DOI: 10.1002/ldr.2933.

Hammad HM, Ashraf M, Abbas F, Bakhat HF, Qaisrani SA, Mubeen M, Shah F, Awais M (2019) Environmental factors affecting the frequency of road traffic accidents: A case study of sub-urban area of Pakistan. *Environ Sci Pollut Res.* https://doi.org/10.1007/s11356-019-04752-8.

Hammad HM, Abbas F, Ahmad A, Bakhat HF, Farhad W, Wilkerson CJ W, Shah F, Hoogenboom G (2020a) Predicting kernel growth of maize under controlled water and nitrogen applications. *Int J Plant Prod.* https://doi.org/10.1007/s42106-020-00110-8.

Hammad HM, Khaliq A, Abbas F, Farhad W, Shah F, Aslam M, Shah GM, Nasim W, Mubeen M, Bakhat HF (2020b) Comparative effects of organic and inorganic fertilizers on soil organic carbon and wheat productivity under arid region. *Communications in Soil Science and Plant Analysis.* DOI: 10.1080/00103624.2020.1763385.

Hagemann M, Bauwe H (2016) Photorespiration. In: *Encyclopedia of applied plant sciences*.

Hajihashemi S, Noedoost F, Geuns JMC et al. (2018) Effect of cold stress on photosynthetic traits, carbohydrates, morphology, and anatomy in nine cultivars of Stevia rebaudiana. *Front Plant Sci.* doi: 10.3389/fpls.2018.01430.

Han H, Gao S, Li B et al. (2010) Overexpression of violaxanthin de-epoxidase gene alleviates photoinhibition of PSII and PSI in tomato during high light and chilling stress. *J Plant Physiol.* doi: 10.1016/j.jplph.2009.08.009.

Hannah MA, Heyer AG, Hincha DK (2005) A global survey of gene regulation during cold acclimation in Arabidopsis thaliana. *PLoS Genet.* doi: 10.1371/journal.pgen.0010026.

Harsant J, Pavlovic L, Chiu G et al. (2013) High temperature stress and its effect on pollen development and morphological components of harvest index in the C_3 model grass Brachypodium distachyon. *J Exp Bot.* doi: 10.1093/jxb/ert142.

Hasanuzzaman M, Nahar K, Alam MM, Roychowdhury R, Fujita M (2013) Physiological, biochemical, and molecular mechanisms of heat stress tolerance in plants. *Int J Mol Sci.*

Hatfield JL, Dold C (2019) Water-use efficiency: Advances and challenges in a changing climate. *Front Plant Sci.* doi: 10.3389/fpls.2019.00103.

Hayes F, Mills G, Ashmore M (2009) Effects of ozone on inter- and intra-species competition and photosynthesis in mesocosms of Lolium perenne and Trifolium repens. *Environ Pollut.* doi: 10.1016/j.envpol.2008.07.002.

Heiss S, Wachter A, Bogs J, Cobbett C, Rausch T (2003) Phytochelatin synthase (PCS) protein is induced in Brassica juncea leaves after prolonged Cd exposure. *J Exp Bot.* doi: 10.1093/jxb/erg205.

Hensley M, Bennie ATP, van Rensburg LD, Botha JJ (2011) Review of 'plant available water' aspects of water use efficiency under irrigated and dryland conditions. *Water SA*.

Hermans C, Hammond JP, White PJ, Verbruggen N (2006) How do plants respond to nutrient shortage by biomass allocation? *Trends Plant Sci.*

Herrera A (2013) Responses to flooding of plant water relations and leaf gas exchange in tropical tolerant trees of a black-water wetland. *Front Plant Sci.* doi: 10.3389/fpls.2013.00106.

Hesham FA, Fahad S (2020) Melatonin application enhances biochar efficiency for drought tolerance in maize varieties:Modifications in physio-biochemical machinery. *Agronomy Journal* 112 (4):2826–2847.

Hong S-W, Vierling E (2002) Mutants of Arabidopsis thaliana defective in the acquisition of tolerance to high temperature stress. *Proc Natl Acad Sci.* doi: 10.1073/pnas.97.8.4392.
Hoth S (2002) Genome-wide gene expression profiling in Arabidopsis thaliana reveals new targets of abscisic acid and largely impaired gene regulation in the abi1-1 mutant. *J Cell Sci.* doi: 10.1242/jcs.00175.
Huang B, Liu X, Fry JD (1998) Shoot physiological responses of two bentgrass cultivars to high temperature and poor soil aeration. *Crop Sci.* doi: 10.2135/cropsci1998.0011183X003800050018x.
Hussain A, Li X, Weng Y et al. (2018) CaWRKY22 acts as a positive regulator in pepper response to ralstonia solanacearum by constituting networks with caWRKY6, caWRKY27, caWRKY40, and caWRKY58. *Int J Mol Sci.* doi: 10.3390/ijms19051426.
Hussain S, Mubeen M, Ahmad A, Akram W, Hammad HM, Ali M, Masood N, Amin A, Farid HU, Sultana SR, Shah F, Wang D, Nasim W (2019) Using GIS tools to detect the land use/land cover changes during forty years in Lodhran district of Pakistan. *Environ Sci Pollut Res.* https://doi.org/10.1007/s11356-019-06072-3.
Hussain MA, Fahad S, Sharif R, Jan MF, Mujtaba M, Ali Q, Ahmad A, Ahmad H, Amin N, Ajayo BS, Sun C, Gu L, Ahmad I, Jiang Z, Hou J (2020) Multifunctional role of brassinosteroid and its analogues in plants. *Plant Growth Regul* https://doi.org/10.1007/s10725-020-00647-8.
Igamberdiev AU, Hill RD (2009) Plant mitochondrial function during anaerobiosis. *Ann. Bot.*
Ilyas M, Mohammad N, Nadeem K, Ali H, Aamir HK, Kashif H, Fahad S, Aziz K, Abid U, (2020) Drought tolerance strategies in plants: A mechanistic approach. *J Plant Growth Regul.* https://doi.org/10.1007/s00344-020-10174-5.
Iqbal N, Khan NA, Ferrante A et al. (2017) Ethylene role in plant growth, development and senescence: Interaction with other phytohormones. *Front Plant Sci.* doi: 10.3389/fpls.2017.00475.
Jan M, Anwar-ul-Haq M, Shah AN, Yousaf M, Iqbal J, Li X, Wang D, Shah F (2019) Modulation in growth, gas exchange, and antioxidant activities of salt-stressed rice (Oryza sativa L.) genotypes by zinc fertilization. *Arabian J Geosci* 12:775 https://doi.org/10.1007/s12517-019-4939-2.
Jezek M, Blatt MR (2017) The membrane transport system of the guard cell and its integration for stomatal dynamics. *Plant Physiol.* doi: 10.1104/pp.16.01949.
Kala DS (2014) Rhizoremediation: A promising rhizosphere technology. *IOSR J Environ Sci Toxicol Food Technol.* doi: 10.9790/2402-08822327.
Kalberer SR, Wisniewski M, Arora R (2006) Deacclimation and reacclimation of cold-hardy plants: Current understanding and emerging concepts. *Plant Sci.*
Kamarn M, Wenwen C, Irshad A, Xiangping M, Xudong Z, Wennan S, Junzhi C, Shakeel A, Fahad S, Qingfang H, Tiening L (2017) Effect of paclobutrazol, a potential growth regulator on stalk mechanical strength, lignin accumulation and its relation with lodging resistance of maize. *Plant Growth Regul* 84:317–332. https://doi.org/10.1007/ s10725-017-0342-8.
Khan MS, Ahmad D, Khan MA (2015) Utilization of genes encoding osmoprotectants in transgenic plants for enhanced abiotic stress tolerance. *Electron J Biotechnol.*
Khan A, Tan DKY, Munsif F, Afridi MZ, Shah F, Wei F, Fahad S, Zhou R (2017a) Nitrogen nutrition in cotton and control strategies for greenhouse gas emissions: A review. *Environ Sci Pollut Res* 24:23471–23487. https://doi.org/10.1007/s11356-017-0131-y.
Khan A, Kean DKY, Afridi MZ, Luo H, Tung SA, Ajab M, Fahad S (2017b) Nitrogen fertility and abiotic stresses management in cotton crop: A review. *Environ Sci Pollut Res* 24:14551–14566. https://doi.org/10.1007/s11356-017-8920-x.
Kim EY, Seo YS, Park KY et al. (2014) Overexpression of CaDSR6 increases tolerance to drought and salt stresses in transgenic Arabidopsis plants. *Gene.* doi: 10.1016/j.gene.2014.09.028.
Konishi M, Yanagisawa S (2011) The regulatory region controlling the nitrate-responsive expression of a nitrate reductase gene, NIA1, in Arabidopsis. *Plant Cell Physiol.* doi: 10.1093/pcp/pcr033.
Kotrba P, Najmanova J, Macek T et al. (2009) Genetically modified plants in phytoremediation of heavy metal and metalloid soil and sediment pollution. *Biotechnol. Adv.*
Koytsoumpa EI, Bergins C, Kakaras E (2018) The CO_2 economy: Review of CO_2 capture and reuse technologies. *J Supercrit Fluids.*
Kozlowski ATT, Pallardy SG (2002) Acclimation and adaptive responses of woody plants to environmental stresses acclimation and adaptive responses of woody plants to environmental stresses. *BioOne.* doi: 10. 1663/0006-8101(2002)068.

Kumar V, Sharma A, Kumar Soni J et al. (2017) Physiological response of C_3, C_4 and CAM plants in changeable climate. *Pharma Innov* 6 (9):70–79.
Kuo H-F, Chiou T-J (2011) The role of microRNAs in phosphorus deficiency signaling. *Plant Physiol*. doi: 10.1104/pp.111.175265.
Kwasniewski M, Szarejko I (2006) Molecular cloning and characterization of β-Expansin gene related to root hair formation in Barley. *Plant Physiol*. doi: 10.1104/pp.106.078626.
Kwon SJ, Kwon S Il, Bae MS et al. (2007) Role of the methionine sulfoxide reductase MsrB3 in cold acclimation in Arabidopsis. *Plant Cell Physiol*. doi: 10.1093/pcp/pcm143.
Lamichhane JR, Dürr C, Schwanck AA et al. (2017) Integrated management of damping-off diseases. A review. *Agron Sustain Dev*.
Li Y, Dhankher OP, Carreira L et al. (2004) Overexpression of phytochelatin synthase in Arabidopsis leads to enhanced arsenic tolerance and cadmium hypersensitivity. *Plant Cell Physiol*. doi: 10.1093/pcp/pch202.
Liesche J, Martens HJ, Schulz A (2011) Symplasmic transport and phloem loading in gymnosperm leaves. *Protoplasma*.
Lin MT, Occhialini A, Andralojc PJ et al. (2014) A faster Rubisco with potential to increase photosynthesis in crops. *Nature*. doi: 10.1038/nature13776.
Liu SK, Cai S, Chen Y et al. (2016) The effect of pollutional haze on pulmonary function. *J Thorac Dis*.
Liu W, Sun Q, Wang K et al. (2017) Nitrogen Limitation Adaptation (NLA) is involved in source-to-sink remobilization of nitrate by mediating the degradation of NRT1.7 in Arabidopsis. *New Phytol*. doi: 10. 1111/nph.14396.
Loka D, Harper J, Humphreys M et al. (2019) Impacts of abiotic stresses on the physiology and metabolism of cool-season grasses: A review. *Food Energy Secur*.
Lucca P, Hurrell R, Potrykus I (2001) Genetic engineering approaches to improve the bioavailability and the level of iron in rice grains. *Theor Appl Genet*. doi: 10.1007/s001220051659.
Luo Z, Sun OJ, Ge Q et al. (2007) Phenological responses of plants to climate change in an urban environment. *Ecol Res*. doi: 10.1007/s11284-006-0044-6.
Marques APGC, Rangel AOSS, Castro PML (2009) Remediation of heavy metal contaminated soils: Phytoremediation as a potentially promising clean-Up technology. *Crit Rev Environ Sci Technol*.
Maček I, Pfanz H, Francetič V et al. (2005) Root respiration response to high CO_2 concentrations in plants from natural CO_2 springs. *Environ Exp Bot*. doi: 10.1016/j.envexpbot.2004.06.003.
McKenzie RL, Aucamp PJ, Bais AF et al. (2011) Ozone depletion and climate change: Impacts on UV radiation. *Photochem Photobiol Sci*.
Miller G, Suzuki N, Ciftci-Yilmaz S Mittler R (2010) Reactive oxygen species homeostasis and signalling during drought and salinity stresses. *Plant, Cell Environ*. doi: 10.1111/j.1365-3040.2009.02041.x.
Miura K, Furumoto T (2013) Cold signaling and cold response in plants. *Int J Mol Sci*.
Mubeen M, Ashfaq A, Hafiz MH, Muhammad A, Hafiz UF, Mazhar S, Muhammad Sami ul Din, Asad A, Amjed A, Fahad S, Wajid N (2020) Evaluating the climate change impact on water use efficiency of cotton-wheat in semi-arid conditions using DSSAT model. *J Water Climate Change*. doi/10.2166/wcc. 2019.179/622035/jwc2019179.pdf.
Muhammad B, Adnan M, Munsif F, Fahad S, Saeed M, Wahid F, Arif M, Jr. Amanullah, Wang D, Saud S, Noor M, Zamin M, Subhan F, Saeed B, Raza MA, Mian IA (2019) Substituting urea by organic wastes for improving maize yield in alkaline soil. *J Plant Nutr*. doi.org/10.1080/01904167.2019.1659344.
Myers SS, Zanobetti A, Kloog I et al. (2014) Increasing CO_2 threatens human nutrition. *Nature*. doi: 10.1038/ nature13179.
Najeeb U, Bange MP, Tan DKY, Atwell BJ (2015) Consequences of waterlogging in cotton and opportunities for mitigation of yield losses. *AoB Plants*. doi: 10.1093/aobpla/plv080.
Nasim W, Ahmad A, Amin A, Tariq M, Awais M, Saqib M, Jabran K, Shah GM, Sultana SR, Hammad HM, Rehmani MIA, Hashmi MZ, Habib Ur Rahman M, Turan V, Fahad S, Suad S, Khan A, Ali S (2017) Radiation efficiency and nitrogen fertilizer impacts on sunflower crop in contrasting environments of Punjab. *Pak Environ Sci Pollut Res* 25:1822–1836. https://doi.org/10.1007/s11356-017-0592-z.
Newell-McGloughlin M (2008) Nutritionally Improved Agricultural Crops. *Plant Physiol*. doi: 10.1104/pp. 108.121947.
Nieto-Sotelo J, Martínez LM, Ponce G et al. (2002) Maize HSP101 Plays Important Roles in Both Induced and Basal Thermotolerance and Primary Root Growth. *Plant Cell*. doi: 10.1105/tpc.010487.

Ning H, Zhang C, Yao Y, Yu D (2010) Overexpression of a soybean O-acetylserine (thiol) lyase-encoding gene GmOASTL4 in tobacco increases cysteine levels and enhances tolerance to cadmium stress. *Biotechnol Lett*. doi: 10.1007/s10529-009-0178-z.

Noman A, Ali S, Naheed F et al. (2015) Foliar application of ascorbate enhances the physiological and biochemical attributes of maize (Zea mays L.) cultivars under drought stress. *Arch Agron Soil Sci*. doi: 10.1080/03650340.2015.1028379.

Noman A, Kanwal H, Khalid N et al. (2017) Perspective Research Progress in Cold Responses of Capsella bursa-pastoris. *Front Plant Sci*. doi: 10.3389/fpls.2017.01388.

Noman A, Ali Q, Maqsood J et al. (2018a) Deciphering physio-biochemical, yield, and nutritional quality attributes of water-stressed radish (Raphanus sativus L.) plants grown from Zn-Lys primed seeds. *Chemosphere* 195:175–189. doi: 10.1016/j.chemosphere.2017.12.059.

Noman A, Ali Q, Naseem J et al. (2018b) Sugar beet extract acts as a natural bio-stimulant for physio-biochemical attributes in water stressed wheat (Triticum aestivum L.). *Acta Physiol Plant*. doi: 10.1007/s11738-018-2681-0.

Noman A, Hussain A, Ashraf MF et al. (2019) CabZIP53 is targeted by CaWRKY40 and act as positive regulator in pepper defense against Ralstonia solanacearum and thermotolerance. *Environ Exp Bot*. doi: 10.1016/j.envexpbot.2018.12.017.

Obata T, Fernie AR (2012) The use of metabolomics to dissect plant responses to abiotic stresses. *Cell. Mol. Life Sci*.

Öquist G, Chow WS, Anderson JM (1992) Photoinhibition of photosynthesis represents a mechanism for the long-term regulation of photosystem II. *Planta*. doi: 10.1007/BF00195327.

Pachauri RK, Allen MR, Barros VR et al. (2014) IPCC, 2014. *Clim Chang 2014 Synth Rep Summ Policymakers*. doi: citeulike-article-id:2297298.

Pajević S, Borišev M, Nikolić N et al. (2016) Phytoextraction of heavy metals by fast-growing trees: A review. In: *Phytoremediation: Management of environmental contaminants*, volume 3.

Pal R, Rai JPN (2010) Phytochelatins: Peptides involved in heavy metal detoxification. *Appl Biochem Biotechnol*.

Pandey P, Irulappan V, Bagavathiannan MV, Senthil-Kumar M (2017) Impact of Combined Abiotic and Biotic Stresses on Plant Growth and Avenues for Crop Improvement by Exploiting Physio-morphological Traits. *Front Plant Sci*. doi: 10.3389/fpls.2017.00537.

Pandolfi C, Mancuso S, Shabala S (2012) Physiology of acclimation to salinity stress in pea (Pisum sativum). *Environ Exp Bot*. doi: 10.1016/j.envexpbot.2012.04.015.

Parker JL, Newstead S (2014) Molecular basis of nitrate uptake by the plant nitrate transporter NRT1.1. *Nature*. doi: 10.1038/nature13116.

Pasqualetti CB, Sandrin CZ, Pedroso ANV et al. (2015) Fructans, ascorbate peroxidase, and hydrogen peroxide in ryegrass exposed to ozone under contrasting meteorological conditions. *Environ Sci Pollut Res*. doi: 10.1007/s11356-014-3965-6.

Pinheiro C, Chaves MM (2011) Photosynthesis and drought: Can we make metabolic connections from available data? *J Exp Bot*.

Placek A, Grobelak A, Kacprzak M (2016) Improving the phytoremediation of heavy metals contaminated soil by use of sewage sludge. *Int J Phytoremediation*. doi: 10.1080/15226514.2015.1086308.

Ploschuk RA, Grimoldi AA, Ploschuk EL, Striker GG (2017) Growth during recovery evidences the waterlogging tolerance of forage grasses. *Crop Pasture Sci*. doi: 10.1071/CP17137.

Pomponi M, Censi V, Di Girolamo V et al. (2006) Overexpression of Arabidopsis phytochelatin synthase in tobacco plants enhances Cd2+ tolerance and accumulation but not translocation to the shoot. *Planta*. doi: 10.1007/s00425-005-0073-3.

Rai R, Agrawal M (2012) Impact of tropospheric ozone on crop plants. *Proc Natl Acad Sci. India Sect B - Biol Sci*.

Ranathunge K, El-Kereamy A, Gidda S et al. (2014) AMT1;1 transgenic rice plants with enhanced NH^{4+} permeability show superior growth and higher yield under optimal and suboptimal NH^{4+} conditions. *J Exp Bot*. doi: 10.1093/jxb/ert458.

Rehman M, Fahad S, Saleem MH, Hafeez M, Habib ur Rahman M, Liu F, Deng G (2020) Red light optimized physiological traits and enhanced the growth of ramie (Boehmeria nivea L.). *Photosynthetica* 58 (4):922–931.

Rozema J, Blokker P, Mayoral Fuertes MA, Broekman R (2009) UV-B absorbing compounds in present-day and fossil pollen, spores, cuticles, seed coats and wood: Evaluation of a proxy for solar UV radiation. *Photochem Photobiol Sci*.

Saad ASI, Li X, Li HP et al. (2013) A rice stress-responsive NAC gene enhances tolerance of transgenic wheat to drought and salt stresses. *Plant Sci*. doi: 10.1016/j.plantsci.2012.12.016.

Saeed Akram M, Ashraf M, Aisha Akram N (2009) Effectiveness of potassium sulfate in mitigating salt-induced adverse effects on different physio-biochemical attributes in sunflower (Helianthus annuus L.). *Flora Morphol Distrib Funct Ecol Plants*. doi: 10.1016/j.flora.2008.05.008.

Saleem MH, Fahad S, Adnan M, Mohsin A, Muhammad SR, Muhammad K, Qurban A, Inas AH, Parashuram B, Mubassir A, Reem MH (2020a) Foliar application of gibberellic acid endorsed phytoextraction of copper and alleviates oxidative stress in jute (Corchorus capsularis L.) plant grown in highly copper-contaminated soil of China. *Environ Sci Pollution Res* https://doi.org/10.1007/s11356-020-09764-3.

Saleem MH, Fahad S, Shahid UK, Mairaj D, Abid U, Ayman ELS, Akbar H, Analía L, Lijun L (2020b) Copper-induced oxidative stress, initiation of antioxidants and phytoremediation potential of flax (Linum usitatissimum L.) seedlings grown under the mixing of two different soils of China. *Environ Sci Poll Res* https://doi.org/10.1007/s11356-019-07264-7.

Saleem MH, Rehman M, Fahad S, Tung SA, Iqbal N, Hassan A, Ayub A, Wahid MA, Shaukat S, Liu L, Deng G (2020c) Leaf gas exchange, oxidative stress, and physiological attributes of rapeseed (Brassica napus L.) grown under different light-emitting diodes. *Photosynthetica* 58 (3):836–845.

Salvi P, Saxena SC, Petla BP et al. (2016) Differentially expressed galactinol synthase(s) in chickpea are implicated in seed vigor and longevity by limiting the age induced ROS accumulation. *Sci Rep*. doi: 10. 1038/srep35088.

Sarkar RK, Bhattacharjee B (2011) Rice genotypes with SUB1 QTL differ in submergence tolerance, elongation ability during submergence, and re-generation growth at re-emergence. *Rice*. doi: 10.1007/ s12284-011-9065-z.

Saud S, Chen Y, Long B, Fahad S, Sadiq A (2013) The different impact on the growth of cool season turf grass under the various conditions on salinity and drought stress. *Int J Agric Sci Res* 3:77–84.

Saud S, Li X, Chen Y, Zhang L, Fahad S, Hussain S, Sadiq A, Chen Y (2014) Silicon application increases drought tolerance of Kentucky bluegrass by improving plant water relations and morph physiological functions. *SciWorld J* 2014:1–10. https://doi.org/10.1155/2014/ 368694.

Saud S, Chen Y, Fahad S, Hussain S, Na L, Xin L, Alhussien SA (2016) Silicate application increases the photosynthesis and its associated metabolic activities in Kentucky bluegrass under drought stress and post-drought recovery. *Environ Sci Pollut Res* 23(17):17647–17655. https://doi.org/10.1007/s11356-016-6957-x.

Saud S, Fahad S, Yajun C, Ihsan MZ, Hammad HM, Nasim W, Amanullah Jr, Arif M, Alharby H (2017) Effects of Nitrogen Supply on Water Stress and Recovery Mechanisms in Kentucky Bluegrass Plants. *Front Plant Sci* 8:983. doi: 10.3389/fpls.2017.00983.

Saud S, Fahad S, Cui G, Chen Y, Anwar S (2020) Determining nitrogen isotopes discrimination under drought stress on enzymatic activities, nitrogen isotope abundance and water contents of Kentucky bluegrass. *Sci Rep* 10:6415. https://doi.org/10.1038/s41598-020-63548-w.

Shafi MI, Adnan M, Fahad S, Fazli W, Ahsan K, Zhen Y, Subhan D, Zafar-ul-Hye M, Brtnicky M, Datta R (2020) Application of single superphosphate with humic acid improves the growth, *yield and phosphorus uptake of wheat (Triticum aestivum L.) in calcareous soil. Agronomy Journal* (10):1224. doi:10. 3390/agronomy10091224.

Shah F, Lixiao N, Kehui C, Tariq S, Wei W, Chang C, Liyang Z, Farhan A, Fahad S, Huang J (2013) Rice grain yield and component responses to near 2°C of warming. *Field Crop Res* 157:98–110.

Shi WM, Muramoto Y, Ueda A, Takabe T (2001) Cloning of peroxisomal ascorbate peroxidase gene from barley and enhanced thermotolerance by overexpressing in Arabidopsis thaliana. *Gene*. doi: 10.1016/ S0378-1119(01)00566-2.

Singh SS, Kumar P, Rai AK (2006) Ultraviolet radiation stress: Molecular and physiological adaptations in trees. In: *Abiotic Stress Tolerance in Plants*.

Song J, Liu J, Weng M et al. (2013) Cloning of galactinol synthase gene from Ammopiptanthus mongolicus and its expression in transgenic Photinia serrulata plants. *Gene*. doi: 10.1016/j.gene.2012.10.058.

Song L, Jiang Y, Zhao H, Hou M (2012) Acquired thermotolerance in plants. *Plant Cell Tissue Organ Cult*.

Song WY, Sohn EJ, Martinoia E et al. (2003) Engineering tolerance and accumulation of lead and cadmium in transgenic plants. *Nat Biotechnol*. doi: 10.1038/nbt850.

Song Y, Yu J, Huang B (2014a) Elevated CO_2-mitigation of high temperature stress associated with maintenance of positive carbon balance and carbohydrate accumulation in Kentucky bluegrass. *PLoS One*. doi: 10.1371/journal.pone.0089725.

Song YY, Ye M, Li C et al. (2014b) Hijacking common mycorrhizal networks for herbivore-induced defence signal transfer between tomato plants. *Sci Rep*. doi: 10.1038/srep03915.

Subhan D, Zafar-ul-Hye M, Fahad S, Saud S, Brtnicky M, Hammerschmiedt T, Datta R (2020) Drought stress alleviation by ACC deaminase producing Achromobacter xylosoxidans and Enterobacter cloacae, with and without timber waste biochar in maize. *Sustainability* 12(6286):doi:10.3390/su12156286.

Sullivan JH, Teramura AH (1992) The effects of ultraviolet-B radiation on loblolly pine - 2. Growth of field-grown seedlings. *Trees*. doi: 10.1007/BF00202426.

Tamirisa S, Vudem DR, Khareedu VR (2014) Overexpression of pigeonpea stress-induced cold and drought regulatory gene (CcCDR) confers drought, salt, and cold tolerance in Arabidopsis. *J Exp Bot*. doi: 10.1093/jxb/eru224.

Tardieu F (2013) Plant response to environmental conditions: Assessing potential production, water demand, and negative effects of water deficit. *Front Physiol*.

Tavakkoli E, Rengasamy P, McDonald GK (2010) High concentrations of Na^+ and Cl^- ions in soil solution have simultaneous detrimental effects on growth of faba bean under salinity stress. *J Exp Bot*. doi: 10.1093/jxb/erq251.

Valenzuela JL, Manzano S, Palma F et al. (2017) Oxidative stress associated with chilling injury in immature fruit: Postharvest technological and biotechnological solutions. *Int J Mol Sci*.

UNFCCC (2007) Climate change: Impacts, vulnerabilities and adaptation in developing countries. *United Nations Framew Conv Clim Chang*. doi: 10.1029/2005JD006289.

Valko M, Morris H, Cronin M (2005) Metals, toxicity and oxidative stress. *Curr Med Chem*. doi: 10.2174/0929867053764635.

Vatamaniuk OK, Mari S, Lu Y-P, Rea PA (2002) AtPCS1, a phytochelatin synthase from Arabidopsis: Isolation and in vitro reconstitution. *Proc Natl Acad Sci*. doi: 10.1073/pnas.96.12.7110.

Verbruggen N, Hermans C, Schat H (2009) Mechanisms to cope with arsenic or cadmium excess in plants. *Curr Opin Plant Biol*.

Vergara GV, Nugraha Y, Esguerra MQ et al. (2014) Variation in tolerance of rice to long-term stagnant flooding that submerges most of the shoot will aid in breeding tolerant cultivars. *AoB Plants*. doi: 10.1093/aobpla/plu055.

Vervuren PJA, Beurskens SMJH, Blom CWPM (1999) Light acclimation, CO_2 response and long-term capacity of underwater photosynthesis in three terrestrial plant species. *Plant, Cell Environ*. doi: 10.1046/j.1365-3040.1999.00461.x.

Viehweger K (2014) How plants cope with heavy metals. *Bot Stud*.

Visser EJW, Voesenek LACJ, Vartapetian BB, Jackson MB (2003) Flooding and plant growth. *Ann Bot*.

Volkov V (2015) Salinity tolerance in plants. Quantitative approach to ion transport starting from halophytes and stepping to genetic and protein engineering for manipulating ion fluxes. *Front Plant Sci*. doi: 10.3389/fpls.2015.00873.

Wahid F, Fahad S, Subhan D, Adnan M, Zhen Y, Saud S, Manzer HS, Brtnicky M, Hammerschmiedt T, Datta R (2020) Sustainable Management with Mycorrhizae and Phosphate Solubilizing Bacteria for Enhanced Phosphorus Uptake in Calcareous Soils. *Agriculture* 10 (334). doi:10.3390/agriculture10080334.

Wang AF, Roitto M, Lehto T et al. (2017) Photosynthesis, nutrient accumulation and growth of two Betula species exposed to waterlogging in late dormancy and in the early growing season. *Tree Physiol*. doi: 10.1093/treephys/tpx021.

Wang D, Shah F, Shah S, Kamran M, Khan A, Khan MN, Hammad HM, Nasim W (2018) Morphological acclimation to agronomic manipulation in leaf dispersion and orientation to promote 'Ideotype' breeding: Evidence from 3D visual modeling of 'super' rice (Oryza sativa L.). *Plant Physiol Biochem*. https://doi.org/10.1016/j.plaphy.2018.11.010.

Wojas S, Ruszczyńska A, Bulska E et al. (2010) The role of subcellular distribution of cadmium and phytochelatins in the generation of distinct phenotypes of AtPCS1- and CePCS3-expressing tobacco. *J Plant Physiol*. doi: 10.1016/j.jplph.2010.02.010.

Wu C, Tang S, Li G, Wang S, Fahad S, Ding Y (2019) Roles of phytohormone changes in the grain yield of rice plants exposed to heat: A review. *PeerJ* 7:e7792 DOI 10.7717/peerj.7792.

Wu C, Kehui C, She T, Ganghua L, Shaohua W, Fahad S, Lixiao N, Jianliang H, Shaobing P, Yanfeng D (2020) Intensified pollination and fertilization ameliorate heat injury in rice (Oryza sativa L.) during the flowering stage. *Field Crops Res* 252:107795.

Xiang D jun, Xiang- Yang Hu, Zhang Y, Yin K-de (2008) Over-expression of ICE1 gene in transgenic rice improves cold tolerance. *Rice Sci.* doi: 10.1016/S1672-6308(08)60039-6.

Xu Z, Jiang Y, Jia B, Zhou G (2016) Elevated-CO_2 response of stomata and its dependence on environmental factors. *Front Plant Sci.* doi: 10.3389/fpls.2016.00657.

Yang JY, Sun Y, Sun AQ et al. (2006) The involvement of chloroplast HSP100/ClpB in the acquired thermotolerance in tomato. *Plant Mol Biol.* doi: 10.1007/s11103-006-9027-9.

Yang Z, Zhang Z, Zhang T, Fahad S, Cui K, Nie L, Peng S, Huang J (2017) The effect of season-long temperature increases on rice cultivars grown in the central and southern regions of China. *Front Plant Sci* 8:1908. https://doi.org/10.3389/fpls.2017.01908.

Yu Y, Streubel J, Balzergue S et al. (2011) Colonization of Rice Leaf Blades by an African Strain of Xanthomonas oryzae pv. oryzae Depends on a New TAL Effector That Induces the Rice Nodulin-3 Os11N3 Gene. *Mol Plant-Microbe Interact.* doi: 10.1094/mpmi-11-10-0254.

Yu J, Chen L, Xu M, Huang B (2012) Effects of elevated CO_2 on physiological responses of tall fescue to elevated temperature, drought stress, and the combined stresses. *Crop Sci.* doi: 10.2135/cropsci2012.01.0030.

Zafar S, Ashraf MY, Anwar S et al. (2016) Yield enhancement in wheat by soil and foliar fertilization of K and Zn under saline environment. *Soil Environ.*

Zafar S, Hasnain Z, Anwar S et al. (2019) Influence of melatonin on antioxidant defense system and yield of wheat (Triticum aestivum L.) genotypes under saline condition. *Pak J Bot.* doi: 10.30848/pjb2019-6(5).

Zafar-ul-Hye M, Naeem M, Danish S, Fahad S, Datta R, Abbas M, Rahi AA, Brtnicky M, Holatko, Jiri, Tarar ZH, Nasir M (2020a) Alleviation of cadmium adverse effects by improving nutrients uptake in bitter gourd through cadmium tolerant rhizobacteria. *Environments* 7(54). doi:10.3390/environments7080054.

Zafar-ul-Hye M, Tahzeeb-ul-Hassan M, Abid M, Fahad S, Brtnicky M, Dokulilova T, Datta R D, Danish S (2020b) Potential role of compost mixed biochar with rhizobacteria in mitigating lead toxicity in spinach. *Sci Rep* 10:12159; https://doi.org/10.1038/s41598-020-69183-9.

Zahida Z, Hafiz FB, Zulfiqar AS, Ghulam MS, Fahad S, Muhammad RA, Hafiz MH, Wajid N, Muhammad S (2017) Effect of water management and silicon on germination, growth, phosphorus and arsenic uptake in rice. *Ecotoxicol Environ Saf* 144:11–18

Zaval L, Keenan EA, Johnson EJ, Weber EU (2014) How warm days increase belief in global warming. *Nat Clim Chang.* doi: 10.1038/nclimate2093.

Zaman QU, Aslam Z, Yaseen M, Ishan MZ, Khan A, Shah F, Basir S, Ramzani PMA, Naeem M (2017) Zinc biofortification in rice: Leveraging agriculture to moderate hidden hunger in developing countries. *Arch Agron Soil Sci* 64:147–161. https://doi.org/10.1080/03650340.2017.1338343.

Zamin M, Khattak AM, Salim AM, Marcum KB, Shakur M, Shah S, Jan I, Fahad S (2019) Performance of Aeluropus lagopoides (mangrove grass) ecotypes, a potential turfgrass, under high saline conditions. *Environ Sci Pollut Res* https://doi.org/10.1007/s11356-019-04838-3.

Zhang H, Xu W, Guo J et al. (2005) Coordinated responses of phytochelatins and metallothioneins to heavy metals in garlic seedlings. *Plant Sci.* doi: 10.1016/j.plantsci.2005.07.010.

Zhang GH, Su Q, An LJ, Wu S (2008) Characterization and expression of a vacuolar Na^+/H^+ antiporter gene from the monocot halophyte Aeluropus littoralis. *Plant Physiol Biochem.* doi: 10.1016/j.plaphy.2007.10.022.

Zhou B, Deng YS, Kong FY et al. (2013) Overexpression of a tomato carotenoid e{open}-hydroxylase gene alleviates sensitivity to chilling stress in transgenic tobacco. *Plant Physiol Biochem.* doi: 10.1016/j.plaphy.2013.05.035.

Zhou J, Wang J, Yu J-Q, Chen Z (2014) Role and regulation of autophagy in heat stress responses of tomato plants. *Front Plant Sci.* doi: 10.3389/fpls.2014.00174.

Zlatev ZS, Lidon FJC, Kaimakanova M (2012) Plant physiological responses to UV-B radiation. *Emirates J Food Agric.* doi: 10.9755/ejfa.v24i6.14669.

Zuzelo PR (2018) The interconnectedness of climate, weather, and living organisms' health. *Holist Nurs Pract.* doi: 10.1097/HNP.0000000000000248.

14

Traditional Ecological Knowledge and Medicinal Systems from Gilgit-Baltistan, Pakistan: An Ethnoecological Perspective

Muhammad Asad Salim, Hafiz Muhammad Wariss, Muhammad Abbas Qazi, and Tika Khan
Kunming Institute of Botany, Chinese Academy of Sciences, Kunming, China

14.1 Introduction

14.1.1 Ethnoecology, Traditional Ecological Knowledge (TEK), and Climate Change

The mountains are home to biological diversity and a lot of these biodiversity agents are ecological indicators for mountainous communities (Khan et al., 2014a). These ecosystems are providing services to around 12% of the global masses by utilizing their traditional knowledge of natural resources (Khan et al., 2013a, 2014a). Aboriginal communities residing in these mountains rely on natural resources for their subsistence. Predominantly agrarian in nature, the rough terrain of the mountains is susceptible to harsh climatic conditions. Limited landholding, soil fragility especially vulnerable to erosion and unavailability of irrigation water adds further constraints to their efforts against poverty alleviation. Regardless, these communities have learned their way around these hardships by learning and understanding seasonal signals and adjusting their agricultural and transhumance activities. Living close to nature, aboriginal communities develop good understanding of the biodiversity and ecosystem (Yang et al., 2019). Therefore, ethnoecology is modestly the ecological knowledge of local communities that help them to adjust with their surrounding environment and adapt to it through cultural and social linkages (Gomez-Baggethum et al., 2013). Besides, it can also be referred to as human strategy for adjusting and transforming local environment and resources (Sujakhu, 2015). It is a multi-disciplinary approach that pursues valid and reasonable understanding of complicated relationships between indigenous people and how such relationships have continued over time (Virginia Dimasuay Nazarea, 1999; Salick and Byg, 2007). Ethnoecology governs the means for different communities and ethnic groups in order to establish and improve their traditional ecological knowledge (TEK). TEK is based on personal and collective knowledge and experience (Gomez-Baggethum et al., 2013; Nalau et al., 2017, 2018) and refers to the evolving knowledge acquired by indigenous and local peoples over hundreds or thousands of years through direct contact with the environment. It is also referred to as the communal capacity of consistently generating, adapting, and applying knowledge to contribute toward the resilience of a particular ecosystem (Nesheim et al., 2006; Ingty and Bawa, 2012; Gomez-Baggethum et al., 2013; Nalau et al., 2018). TEK is transferred to the younger generations through traditional songs, stories of the ancestors, and belief systems (Kimmerer, 2006; Franco, 2015). The role of TEK in explaining ecological, environmental, and social systems has been recognized at global level yet it is rapidly depleting as the socioeconomic changes lay a huge impact on the indigenous communities (Nesheim et al., 2006; Ingty and Bawa, 2012).

Humans have discovered with the passage of time different uses of plants such as food, medicine, and shelter (Winslow and Kroll, 1998; Akhtar et al., 2016b). Most folk medicine and treatments have been derived and evolved from plants through different human civilizations (Akhtar et al., 2016b). Traditional means of medication and the use of medicinal herbs and plants have been one of the major strategies toward good health and lifestyle in developing countries (UNESCO, 1996; Ateş and Turgay, 2003; Muthu et al., 2006; Makkar et al., 2007; Tushar et al., 2010). TEK has a very strong influence on the capacities of indigenous communities as it progresses and grows with changes to the ecological and socioeconomic systems, yet the recent industrial advancements, modernization of education, loss of primitive languages, domination of certain religions, and globalization have all inflicted negative impacts on the bearers of TEK (Colding et al., 2002; Berkes and Davidson-Hunt, 2006; Gomez-Baggethum et al., 2013). TEK still exists across the world especially with indigenous communities and smallholder agrarian societies (Gomez-Baggethum et al., 2013). Studying human and plant interaction and relation with ecosystem can best be described through the TEK of local communities as these communities rely on seasonal products from plant diversity including food, medicine, construction materials, and without proper interaction and understanding of ecosystem and biodiversity, it is not possible to collect such resources at proper time (Nesheim et al., 2006).

TEK of indigenous communities is rapidly depleting throughout the world and the story is no different in Pakistan (Shedayi and Gulshan, 2012; Khalil et al., 2014; Kodirekkala, 2017). For instance, even though the communities in Gilgit-Baltistan still rely on medicinal plants (Salim et al., 2019a), their utilization has decreased from 100% to only 20% over the past 50 years and is swiftly depleting further (Shedayi and Gulshan, 2012; Abbasi et al., 2013; Bano et al., 2014; Khan et al., 2014d; Shedayi et al., 2014; Abbas et al., 2016; Akhtar et al., 2016a,b; Ali et al., 2017). Although TEK still exists in the communities of Gilgit-Baltistan, the lack of transfer to the next generation is the main constraint as only 20–25% of the community members have been reported to have acquired and retained this knowledge (Shedayi and Gulshan, 2012; Bano et al., 2014; Shedayi et al., 2014; Kayani et al., 2015). It is also worth noting that the male community has retained more of this knowledge compared to the female as they are the ones mainly involved in farming and collection of plant resources (Shedayi et al., 2014).

14.1.2 TEK and Medicinal Plants – Global Perspective

The Chinese communities have a strong faith and belief in traditional Chinese medicine (TCM) and therefore they strongly rely on it (Salim et al., 2019a). TCM is one of the oldest and incredibly documented literature with more than 5,000 varieties of medicinal herbs being catalogued in different remedies (Gorman, 1992; UNESCO, 1996; Akhtar et al., 2016b; Salim et al., 2019a). Ayurveda in India has a history of over 5,000 years (Morgan, 2002; Akhtar et al., 2016b). There are more than 500 species of flowering plants in Pakistan utilized for medicinal purposes and are a vital part of the cultural, traditional, and spiritual setup within a community (Khan Shinwari et al., 2005). A study done by Hocking (1958) indicates up to 84% of the Pakistani population used to rely on traditional medication back in the 1950s.

The increased reliability on Western/scientific knowledge and rapidly fading TEK has led to a debate of the integration of both knowledge systems (Lynch et al., 2010; Bohensky and Maru, 2011; Evering, 2012; Gratani et al., 2014; Ludwig, 2016). Studies involving indigenous communities and their ethnoecological and ethnomedicinal knowledge are receiving substantial attention globally as TEK is based on experiences through interactions with different biodiversity agents like plants, animals, birds, and the associated social, ethical, and spiritual affiliations (Heinrich, 2000; Kassam et al., 2010; Ayyanar and Ignacimuthu, 2011; Amjad et al., 2017). It is about time that efforts are put into hybridization of knowledge by scientific confirmation and documentation of the indigenous knowledge before it is too late (Heinrich, 2000; Ayyanar and Ignacimuthu, 2011; Bibi et al., 2014; Ludwig, 2016; Hussain et al., 2017).

14.1.3 Regional Profile of Gilgit-Baltistan

Gilgit-Baltistan is located in the far north of Pakistan, with Afghanistan to the north and west, China to the north and east, and India to the south. The ten districts of Gilgit-Baltistan (Figure 14.1) cover an area of 72,496 square kilometres in 34.50 to 37.00 °N and 72.00 to 77.80 °E with an estimated population of ~1.8 million (Barcha, 2013). The region is rugged and mountainous, located among three of the highest mountain ranges – the Himalayas, the Karakoram, and the Hindukush (HKH), and home to the largest number of glaciers outside the polar region (Wazir et al., 2004; Khan et al., 2014e; Akhtar et al., 2016a). Basically an agrarian society, the region relies on irrigation water from streams and rivers fed by the glaciers and snowmelt (AKRSP and JICA, 2010). Gilgit-Baltistan has a temperate climate with annual temperatures ranging within −30°C to 47°C while the average winter and summer temperatures are around ±4°C and ±25.19°C (AKRSP and JICA, 2010; Abbas et al., 2018).

14.1.4 Ethnographic Profile of the Region

Archaeological studies throughout Gilgit-Baltistan reveal that the region remained under influence of cultural incursions from the Indian subcontinent, China, Scythia, Transoxiana, and ancient Greek, amongst others (Hauptmann, 2007; Barcha, 2013). The passes created by the Indus River system in Hunza, Shigar, Shyjok, Ghizer, Gilgit, and Astore valleys served as the main travel routes for these invasions and exchanges as Gilgit got its famous name 'gate to India' (Hauptmann, 2007). The Karakoram Highway (KKH) and the recent China–Pakistan Economic Corridor (CPEC) are following the travel routes which were used as the Silk Routes in ancient times (Hauptmann, 2007; Mark, 2014). Dating back to pre-Islamic history, the residents followed Buddhism, Bon (religion of Tibet), and Hinduism which is reflected through sacred rocks and sites found throughout the region especially Ghizer, Gilgit, Hunza, Diamer, Skardu, and Astore (Hauptmann, 2007; Kassam et al., 2010; Middleton, 2011; Barcha, 2013; Mark, 2014; Stevens et al., 2016). Before the region completely came under Dogra Raj of the Kashmir State, Hunza, Gilgit, Nagar, and Ghizer mainly remained under Chinese influence, while Skardu, Astore, Ghanche, and Diamer remained under Tibetan influence (Hauptmann, 2007; Barcha, 2013).

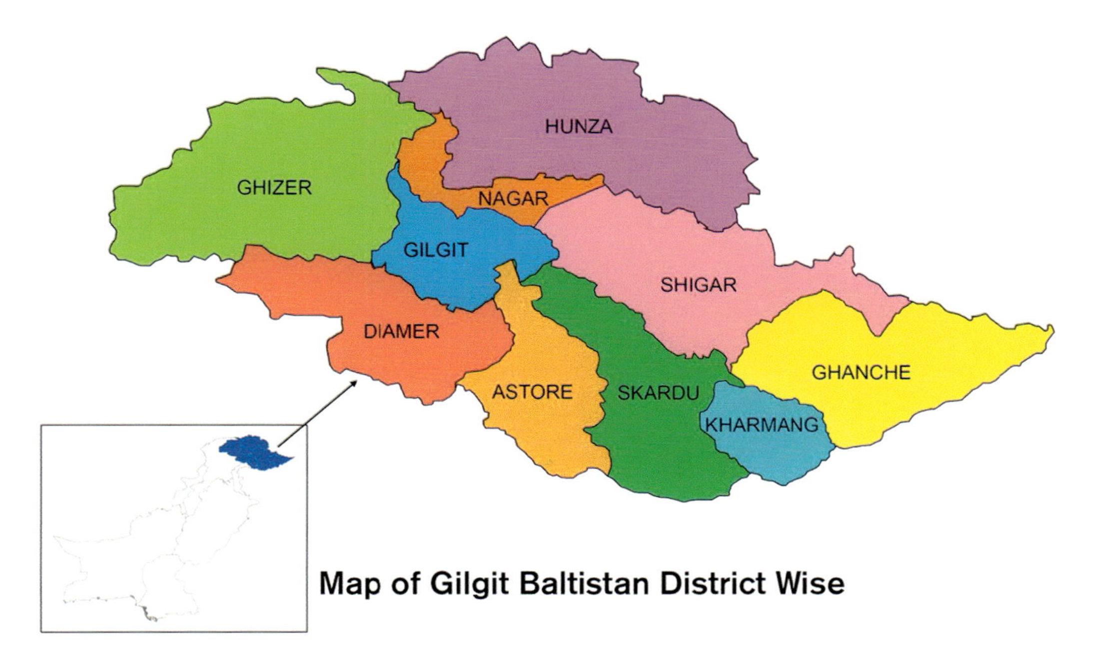

FIGURE 14.1 Map of Gilgit-Baltistan showing the districtwise breakup of the region. The map includes the newly designated districts of Nagar, Shigar, and Kharmang as well.

The residents of Gilgit-Baltistan are divided into sub-groups based on their origin as well as their ethno-linguistic clustering. Yashkun, Sheen/Shinaki, Wakhi (those who migrated from Wakhan), Burushos, Dom, Brokpa, and Balti are the main tribes of the area; some Kashmiris, Kohistani, Mongols, Mughals, Rajas, Pathans, Gujar, Soniwal, Mon, Hor, and Kashgari also reside here (Muhammad, 1905; Rhoades and Thompson, 1975; Kreutzmann, 1993; Khan et al., 2014c; IFAD, 2015; Abbas et al., 2016, 2017). The majority of Gilgit-Baltistan is sparsely populated with these tribes but Ghizer is dominated by Burushos; Gilgit, Hunza, and Nagar have Burushos, Sheen, and Yashkun; Diamer and Astore are majorly populated by Sheen, Yashkun, and Kohistani communities; and Skardu and Ghanche are predominantly Mon, Hor, and Mongols (Platon, 2012; IFAD, 2015). The languages spoken by Burushos, Sheen, and Yashkun are Shina, Burushaski, Wakhi, and Khowar (only Ghizer and parts of Hunza), while the Mongols, Mon, and Hor speak Balti (Bano et al., 2014; Khan et al., 2014b; IFAD, 2015; Ali et al., 2017).

14.1.5 Geographic Division of Gilgit-Baltistan

Previously called the Northern Areas of Pakistan, the Gilgit-Baltistan region of Pakistan is divided into two sub-regions: Gilgit and Baltistan. These sub-regions are further divided into ten districts where Gilgit, Ghizer, Hunza, and Nagar fall in the Gilgit region and Astore, Skardu, Shigar, Ghanche, Kharmang, and Diamer fall in the Baltistan region (Wikipedia, 2019). Further details of the districts are provided in Table 14.1.

14.1.6 Key Geographic Features of Gilgit-Baltistan

The three mightiest mountain ranges of the world, the Himalayas, Hindukush, and Karakoram, meet in this region, thus making Gilgit-Baltistan home to a few of the highest mountain peaks in the world including K-2, Nanga Parbat, Gasherbrum-1, Broad Peak, and Gasherbrum-2, which are all above 8,000 m in height (Barcha, 2013; Wikipedia, 2019). Popularly known as the third pole of the world, and famous for the Siachin, Biyafo, and Hispar glaciers, Gilgit-Baltistan is home to the largest glaciers out

TABLE 14.1

Districtwise details of Gilgit-Baltistan region of Pakistan showing the name of each district, the year of its establishment, its headquarter, population, area covered and languages spoken

Name	Year of Establishment	Headquarters	Population (2013)	Area (Sq km)	Languages
Astore	2004	Eidgah	71,666	5,092	Shina
Diamer	1972	Chilas	131,925	8,949	Shina
Skardu	1972	Skardu	305,000*	8,700	Balti
Ghizer	1974	Gahkuch	190,000	9,635	Khowar, Burushaski, Shina
Ghanche	1974	Khaplu	88,366	4,052	Balti
Gilgit	1972	Gilgit	222,000	14,672	Shina, Burushaski, Wakhi
Hunza	2015	Aliabad	70,000 (2015)	7,900	Shina, Burushaski, Wakhi, Domaki
Nagar	2015	Nagar	51,387 (1998)	5,000	Shina, Burushaski, Wakhi, Domaki
Kharmang	2015	Kharmang	--	5,500	Balti
Shigar	2015	Shigar	--	8,500	Balti

Source: Barcha, 2013; Wikipedia, 2019.

Notes

*Population of Skardu is a combined figure for Skardu, Shigar, and Kharmang districts as Shigar and Kharmang districts were previously part of Skardu district.

of the polar region (Steinbauer and Zeidler, 2008; Barcha, 2013; Bocchiola and Diolaiuti, 2013). Another attraction of the region are the rivers and lakes of Gilgit-Baltistan. All the rivers in the regions, including rivers from Gilgit, Shigar, Shiwak, and Zanskar, are tributaries to the Indus River system (Barcha, 2013). Some of the famous lakes in the region are the Bhorit Lake in Hunza, Paloga Lake in Kargha, Rama Lake in Astore, Satpara Lake in Skardu, Shandur Lake in Ghizer, Kachora Lakes in Skardu, Naltar Lake in Naltar, and the recent Glacial Lake Outburst Flood (GLOF)-caused Atta Abad Lake in Hunza. These lakes are famous for hosting the seasonal migratory birds, especially Siberian cranes (Barcha, 2013). There are five national parks in Gilgit-Baltistan which are famous for their natural beauty and play a major role in conservation of natural resources including flora and fauna.

14.1.7 Sources of Livelihoods for the Local Communities

Communities in Gilgit-Baltistan are dependent on agricultural resources. Land cover changes, lack of resource management and sustainable harvesting policies, and political interests at a massive scale in the Karakoram, Himalaya, Hindukush, and Pamir mountain ranges have severe and long lasting impacts on the region (Liniger and Schwilch, 2002; Gautam et al., 2002; Maikhuri et al., 2004; Khan et al., 2014f). Since the construction of the Karakoram Highway and strengthening of roads linking different valleys, the regional farmers established linkages with regional and national fruit and agriculture markets (AKRSP and JICA, 2010; Cook and Butz, 2013; Khan et al., 2014f). The horticulture sector has since been one of the major livelihood resources of the region as most of the fruit crops like apricots, apples, and cherries from the region gained a significant market share (AKRSP and JICA, 2010) injecting financial resources into the subsistence farming system of the region. However, it is clear that the potential of this sector is far from being realized.

Traditional knowledge of medicinal plants is often socially integrated through communal learning and intercultural exchange. Medicinal plants in mountainous terrain are known for their distribution in elevation corridors or endemism to a particular locality. It is therefore important to explore the richness of traditional knowledge of medicinal plants, their uses, distribution, and trade in the mountainous region of Gilgit-Baltistan. Documenting the knowledge exchange between the old and young generation as well as amongst different communities and localities of the region, and how medicinal systems like Chinese, Ayurveda, Tibetan, and Unani influenced the use of traditional medicine system in the region is vital for its sustainability and documentation. It is also important to explore the possible factors behind the general decline in knowledge of medicinal plants yet their continued use for treatment of different diseases and how the current markets and market players supplement this phenomenon.

14.2 Occurrence and Markets for Medicinal Plants in Gilgit-Baltistan

Interpolated results have revealed that the utilization as well as high occurrence of medicinal plants is mainly concentrated in two locations (Figure 14.2). Skardu, Gilgit, and Ghizer are reported for the highest number of medicinal plants and their uses (Salim et al., 2019a). There are no formal markets for medicinal plants in the region yet vegetable and fruit markets in Gilgit and Skardu serve this purpose in an informal manner (Abbas et al., 2016; Salim et al., 2019a). Most of the herbal medicine companies in Pakistan rely on supplies of raw materials from India (Salim et al., 2019a).

There are a number of species majorly used against wounds, skin infections, pain relief, kidney and uterus infections, diaphoretic, cardiac issues, stomach and intestine issues, respiratory disorders, inflammation, hepatitis, hypertension, liver disorders, gynaecological disorders, brain and nervous disorders, weight loss, eye disorders, diabetes, teeth and gums, blood purification, nausea/altitude sickness, livestock diseases, sexual diseases/stimulant, haemorrhoids/piles, maternal health, and ENT (ear, nose, and throat) challenges. *Thymus linearis*, *Delphinium brunonianum*, *Bergenia stracheyi*, *Saussurea heteromalla*, *Saussurea lappa*, *Carthamus tinctorius*, *Peganum harmala*, *Rheum emodi*, *Mentha longifolia*, *Mentha arvensis*, *Valeriana wallichii*, *Berberis lyceum*, and *Elaeagnus*

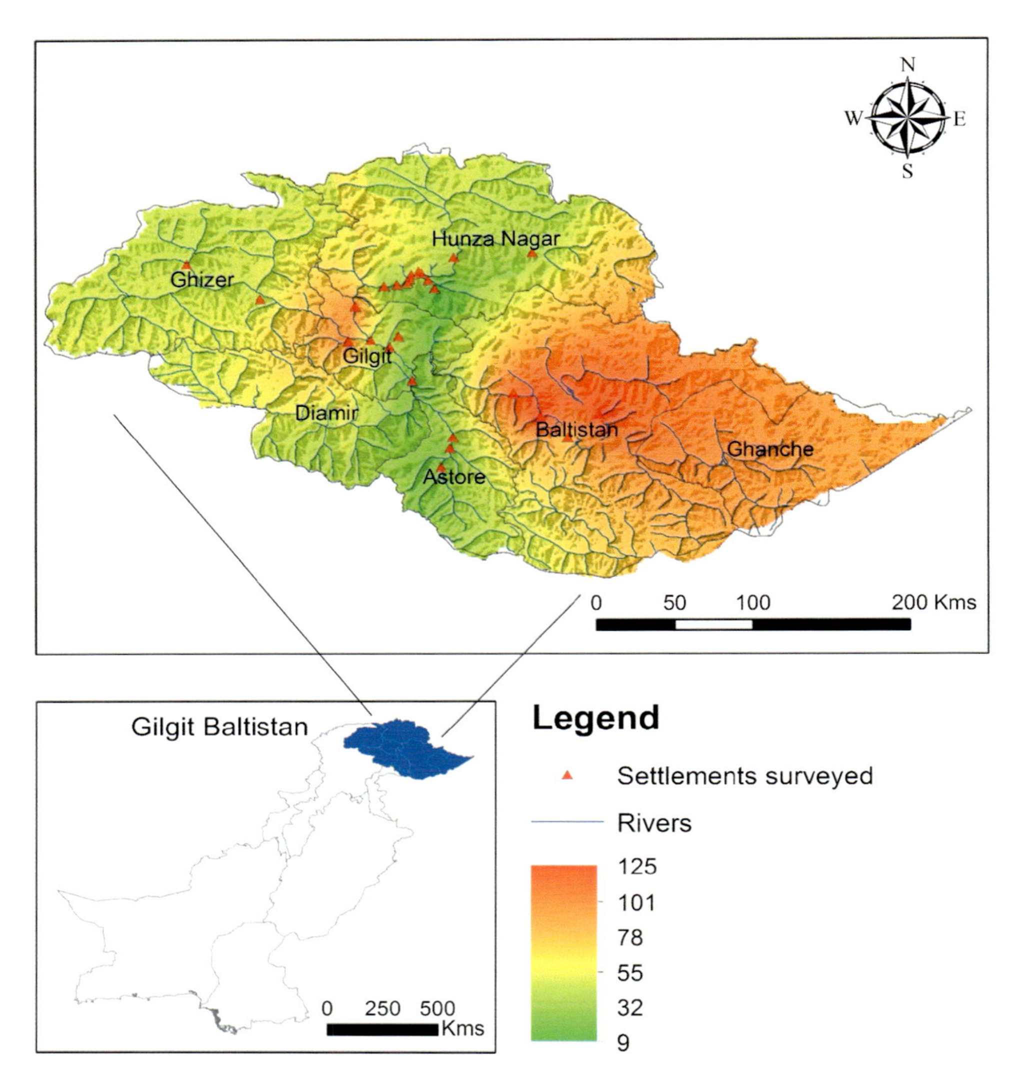

FIGURE 14.2 Map of Gilgit-Baltistan showing the occurrence of medicinal plants in the region. Red and yellow zones represent a high number of species reported.

rhamnoides account for the most demanded and utilized species from the region while a list of key medicinal plants species is provided in Table 14.2 (Qureshi et al., 2006; Khan and Khatoon, 2007; Khan et al., 2013b, 2014g; Shedayi et al., 2014; Salim et al., 2019a). Further details on these medicinal plants species and their uses can be found in Salim et al. (2019b). Although there is a large group of locals in Gilgit-Baltistan who rely on traditional medicine, these practices and the knowledge associated with them arerapidly depleting from the region due to a constantly growing gap between the old and young generations (Khan et al., 2013b; Shedayi et al., 2014; Salim et al., 2019a). The main source of knowledge transfer rests within the family followed by learning through other community members and traditional health practitioners and is mostly retained by the male community members above 40 years of age (Shedayi and Gulshan, 2012; Bano et al., 2014; Shedayi et al., 2014; Kayani et al., 2015; Salim et al., 2019a). This ever-increasing gap between the young and old generations is affecting the knowledge transfer mechanism, in combination with over-exploitation and lack of conservation strategies, and the impacts of climatic changes (Salim et al., 2019a). Such issues have also been observed in other studies (Shedayi and Gulshan, 2012; Bano et al., 2014; Shedayi et al.,

TABLE 14.2

Major species of medicinal plants commonly used in Gilgit-Baltistan in traditional medicinal systems, their vernacular names and number of diseases locals address with use of each species

Species	Family	Local Name	Number of Diseases Addressed
Allium carolinianum DC.	Amaryllidaceae	Chong	5
Allium sativum L.	Amaryllidaceae	Bukpa	5
Artemisia maritima L.	Compositae	Zoon	5
Astragalus frigidus (L.) A.Gray	Leguminosae	Shashal	3
Berberis brandisiana Ahrendt	Berberidaceae	Ishkeen	10
Berberis lycium Royle	Berberidaceae	Skyurboo	8
Berberis orthobotrys Bien. ex Aitch.	Berberidaceae	Ishkein	3
Berberis pseudumbellata R.Parker	Berberidaceae	Shokurum	3
Bergenia ciliata (Haw.) Sternb.	Saxifragaceae	Shafus	6
Capparis spinosa L.	Capparaceae	Kraba	4
Capsicum annuum L.	Solanaceae	Ma'rooch	2
Carthamus tinctorius L.	Compositae	Pock	8
Cerastium fontanum Baumg.	Caryophyllaceae	Bloghar	2
Datura stramonium L.	Solanaceae	Daturo	5
Delphinium brunonianum Royle	Ranunculaceae	Makhoting	5
Dracocephalum nuristanicum Rech.f. & Edelb.	Lamiaceae	Shamdun	5
Elaeagnus angustifolia L.	Elaeagnaceae	Shekarkuch	10
Elaeagnus rhamnoides (L.) A.Nelson	Elaeagnaceae	Buru	7
Ephedra gerardiana Wall. ex Stapf	Ephedraceae	Say	5
Ephedra intermedia Schrenk & C.A.Mey.	Ephedraceae	Shaay Soom	4
Fagopyrum esculentum Moench	Polygonaceae	Bro	3
Ficus carica L.	Moraceae	Faag	5
Glycyrrhiza glabra L.	Leguminosae	Shalako	4
Juglans regia L.	Juglandaceae	Achow	5
Juniperus communis L.	Cupressaceae	Oshuk	3
Linum usitatissimum L.	Linaceae	Human	10
Mentha longifolia (L.) L.	Lamiaceae	Fileel	10
Mentha royleana Wall. ex Benth.	Lamiaceae	Foling	4
Mentha spicata L.	Lamiaceae	Podina	6
Morus alba L.	Moraceae	Shae Maroch	4
Myricaria squamosa Desv.	Tamaricaceae	Targ	5
Nepeta floccosa Benth.	Lamiaceae	Buzlanj	5
Peganum harmala L.	Nitrariaceae	Spandur	3
Pimpinella diversifolia DC.	Apiaceae	Kohniod	3
Pistacia khinjuk Stocks	Anacardiaceae	Kakavomn	6
Plantago major L.	Plantaginaceae	Shiltive	5
Potentilla bifurca L.	Rosaceae	Tarqan	3
Potentilla salesoviana Steph.	Rosaceae	Karfo mindoq	3
Prunus armeniaca L.	Rosaceae	Chooli	8
Punica granatum L.	Lythraceae	Danooh	9
Salix acmophylla Boiss.	Salicaceae	Brawoon	5
Saxifraga hirculus L.	Saxifragaceae	Sitbark	4
Solanum nigrum L.	Solanaceae	Gabeeli	8
Sophora mollis (Royle) Baker	Leguminosae	Popshing	3
Thymus linearis Benth.	Lamiaceae	Tumuro	3

(*Continued*)

TABLE 14.2 (Continued)

Species	Family	Local Name	Number of Diseases Addressed
Thymus serphyllum L.	Lamiaceae	Tumuro	6
Trifolium pratense L.	Leguminosae	Chita-batta	2
Urtica dioica L.	Urticaceae	Kheshing	6

2014; Kayani et al., 2015). Traditional healers in the community are experts in recognizing and collecting medicinal plants while other community members are not fully aware of the exact timing for collection which lead to issues relating to conservation and malpractices in harvesting of these plant species (Angmo et al., 2012; Leto et al., 2013; Lulekal et al., 2013; Kayani et al., 2014; Soelberg and Jäger, 2016; Aziz et al., 2017; Salim et al., 2019a). The gradual expansion of trade and an increasing demand for medicinal plants in and outside the region has a positive impact on knowledge sharing (Salim et al., 2019a).

14.3 Medicinal Systems and Affiliations

Gilgit-Baltistan with its strategic position between China, Central Asian countries, and the Indian subcontinent make it home to traders who used the silk routes for trade and cultural exchanges (Hauptmann, 2007; Kassam et al., 2010; Middleton, 2011; Barcha, 2013; Attewell, 2014; Mark, 2014; Stevens et al., 2016; Salim et al., 2019a). It was through these exchanges that the local communities interacted with medicinal systems from China, Tibet, Central Asia, India, and Iran (Hauptmann, 2007; Barcha, 2013). These exchanges led to the transformation of TEK of medicinal plants. The major medicinal systems influencing different parts of the region include TCM, Ayurveda, and Unani medicine (Salim et al., 2019a). Most of the region is under a mix of these medicinal systems yet these systems can be seen to dominate different clusters and indigenous communities within the region. Astore, Skardu, and Ghanche, due to close proximity to the Indian subcontinent, are influenced by a mix of Ayurveda and Unani systems while Gilgit, Hunza, Nagar, and Ghizer are more inclined toward TCM (Salim et al., 2019a). THPs and wholesalers/retailers in the major markets are openly practising a mix of these medicinal systems in order to cope with patients from all over the region.

There is a need to study TEK of medicinal plants and their utilization from the region in detail with a clear goal of comparison with TCM, Ayurveda, and Unani medicinal systems in order to establish a clear line between the knowledge being practiced in the region and how it has been influenced by these systems.

14.4 Conclusion

Geographically, Gilgit-Baltistan is one of the hotspot areas for biodiversity and harbours a lot of fruit crops and medicinal plant species (Bano et al., 2014; Shedayi et al., 2014; Ishaq et al., 2015; Akhtar et al., 2016a). Traditional medicinal plant usage and trade play a major role in the livelihoods of local inhabitants of Gilgit-Baltistan (Sher et al., 2014; Abbas et al., 2016). Many local communities depend on the medicinal plant harvesting and trade but not all are familiar with the proper collection, parts to be used, preservation, and storage, which consequently lead to the over-exploitation of natural resources of the area (Shedayi and Gulshan, 2012; Bano et al., 2014; Shedayi et al., 2014; Kayani et al., 2015). Proper conservation strategies such as controlled grazing, reforestation, and rangeland management among many others may be adopted to promote the sustainable use of medicinal plants. This chapter clearly reveals the importance and contributions of THPs and retailers as well as the transfer of traditional knowledge from one generation to the next, where and when to acquire a particular species, and the utilization of medicinal plants. These factors point toward a high level of cooperation, collaboration, and openness to knowledge exchange amongst the ethnicities and tribes of Gilgit-Baltistan. Local public and private institutes can therefore play a vital role in clustering the knowledge and bridging the gaps by providing platforms for disseminating traditional knowledge and further formulation of natural product-based remedies. Hunza, Nagar, Ghizer, and Gilgit were influenced by the TCM system while the remaining part of the study area was dominated by a mix of Ayurveda and Unani systems which can be further evaluated to find the diverse uses of the same species. Medicinal plants indicating traditional uses against different diseases should be investigated phytochemically and pharmacologically to prove their efficacy. With the involvement of multiple stakeholders (the relevant local government departments, herbal medicine producing companies, THPs, and the interest of the national government), medicinal plants and associated traditional knowledge from Gilgit-Baltistan can make a substantial contribution to improve the livelihood resources of local communities. Practical steps should be taken immediately to ensure the inclusion of relevant flora within conservation designations for sustainable uses (Shinwari and Gilani, 2003). A re-evaluation by ethno-pharmacologists and other public health actors are needed for the authentication of this TEK. Future strategies for exploration, raising awareness, and conservation are recommended in order to use these natural resources and TEK in a more sustainable and effective way through the involvement from government institutions, research organizations, NGOs, donors, and the private sector.

REFERENCES

Abbas Z, Khan SM, Abbasi AM et al. (2016) Ethnobotany of the Balti community, Tormik valley, Karakorum range, Baltistan, Pakistan. *J Ethnobiol Ethnomed* 12:38. doi: 10.1186/s13002-016-0114-y.

Abbas A, Syed SA, Moosa A (2018) Emerging plant diseases of Gilgit-Baltistan (GB) Pakistan: A review. *Agric Res Technol* 18:001–007. doi: 10.19080/ARTOAJ.2018.18.556066.

Abbasi A, Khan M, Shah MH et al. (2013) Ethnobotanical appraisal and cultural values of medicinally important wild edible vegetables of Lesser Himalayas-Pakistan. *J Ethnobiol Ethnomed* 9:66. doi: 10.1186/1746-4269-9-66.

Akhtar N, Akhtar S, Kazim S, Khan T (2016a) Ethnomedicinal study of important medicinal plants used for gynecological issues among rural women folk in district Gilgit, Pakistan. *Nat Sci* 14:30–34. doi: 10.7537/marsnsj14091605.

Akhtar S, Akhtar N, Kazim S, Khan T (2016b) Study of ethno-gynecologically important medicinal and other plants used for women specific purposes in Murtazaabad, Hunza, Pakistan. *Nat Sci* 14:36–39. doi: 10.7537/marsnsj140616.07.

AKRSP, JICA (2010) Basic Study on Horticulture Sector in Gilgit-Baltistan.

Ali I, Hussain H, Batool H et al. (2017) Documentation of ethno veterinary practices in the CKNP region, gilgit-baltistan. *Int J Phytomedicine* 9:223–240.

Amjad MS, Qaeem MF, Ahmad I et al. (2017) Descriptive study of plant resources in the context of the ethnomedicinal relevance of indigenous flora: A case study from Toli Peer National Park, Azad Jammu and Kashmir, Pakistan. *PLoS One* 12:1–31. doi: 10.1371/journal.pone.0171896.

Angmo K, Adhikari BS, Rawat GS (2012) Changing aspects of Traditional Healthcare System in Western Ladakh, India. *J Ethnopharmacol* 143:621–630. doi: 10.1016/j.jep.2012.07.017.

Ateş AD, Turgay ERDO Ö (2003) Antimicrobial Activities of Various Medicinal and Commercial Plant Extracts. *Turk J Biol* 27:157–162.

Attewell G (2014) Compromised: Making institutions and indigenous medicine in Mysore State, circa 1908–1940. *Cult Med Psychiatry* 38:369–386. doi: 10.1007/s11013-014-9390-y.

Ayyanar M, Ignacimuthu S (2011) Ethnobotanical survey of medicinal plants commonly used by Kani tribals in Tirunelveli hills of Western Ghats, India. *J Ethnopharmacol* 134:851–864. doi: 10.1016/j.jep.2011.01.029.

Aziz MA, Khan AH, Adnan M, Izatullah I (2017) Traditional uses of medicinal plants reported by the indigenous communities and local herbal practitioners of Bajaur Agency, Federally Administrated Tribal Areas, Pakistan. *J Ethnopharmacol* 198:268–281. doi: 10.1016/j.jep.2017.01.024.

Bano A, Ahmad M, ben Hadda T et al. (2014) Quantitative ethnomedicinal study of plants used in the skardu valley at high altitude of Karakoram-Himalayan range, Pakistan. *J Ethnobiol Ethnomed* 10:43. doi: 10.1186/1746-4269-10-43.

Barcha SBAK (2013) Aks-e-Gilgit Baltistan. North Books Gilgit, Gilgit.

Berkes F, Davidson-Hunt IJ (2006) Biodiversity, traditional management systems, and cultural landscapes: examples from the boreal forest of Canada. *Int Soc Sci J* 58:35–47. doi: 10.1111/j.1468-2451.2006.00605.x.

Bibi S, Sultana J, Sultana H, Malik RN (2014) Ethnobotanical uses of medicinal plants in the highlands of Soan Valley, Salt Range, Pakistan. *J Ethnopharmacol* 155:352–361. doi: 10.1016/j.jep.2014.05.031.

Bocchiola D, Diolaiuti G (2013) Recent (1980–2009) evidence of climate change in the upper Karakoram, Pakistan. *Theor Appl Climatol* 113:611–641. doi: 10.1007/s00704-012-0803-y.

Bohensky EL, Maru Y (2011) Indigenous knowledge, science, and resilience: What have we learned from a decade of international literature on 'integration'? *Ecol Soc* 16:art6. doi: 10.5751/ES-04342-160406.

Colding J, Elmqvist T, Olsson P (2002) Living with disturbance: Building resilience in social–ecological systems. In: Folke C, Berkes F, Colding J (Eds.) *Navigating Social-Ecological Systems: Building Resilience for Complexity and Change*. Cambridge University Press, Cambridge, pp 163–186.

Cook N, Butz D (2013) The Atta Abad landslide and everyday mobility in Gojal, Northern Pakistan. *Mt Res Dev* 33:372–380. doi: 10.1659/mrd-journal-d-13-00013.1.

Evering B (2012) Relationships between knowledge (s): implications for 'knowledge integration'. *J Environ Stud Sci* 2 (4): 357–368.

Franco FM (2015) Calendars and ecosystem management: Some observations. *Hum Ecol*. doi: 10.1007/s10745-015-9740-6.

Gautam AP, Webb EL, Eiumnoh A (2002) GIS assessment of land use/land cover changes associated with community forestry implementation in the middle hills of Nepal. *Mt Res Dev* 22:63–69. doi: 10.1659/0276-4741(2002)022[0063:GAOLUL]2.0.CO;2.

Gomez-Baggethum E, Corbera E, Reyes-Garcia V (2013) Traditional ecological knowledge and global environmental change: Research findings and policy implications. *Ecol Soc* 18:72. doi: 10.5751/ES-06288-180472.

Gorman C (1992) The power of potions. *Time* 52–53.

Gratani M, Bohensky EL, Butler JRA et al. (2014) Experts' perspectives on the integration of indigenous knowledge and science in wet tropics natural resource management. *Aust Geogr* 45:167–184. doi: 10.1080/00049182.2014.899027.

Hauptmann H (2007) Pre-Islamic heritage in the Northern Areas of Pakistan. In: Bianca S (ed) *Karakoram: Hidden Treasures in the Northern Areas of Pakistan*, 2nd edn. Umberto Allemandi & Co., Turin, Italy, pp 21–40.

Heinrich M (2000) Ethnobotany and its role in drug development. *Phytother Res* 14:479–488. doi: 10.1002/1099-1573(200011)14:7<479::aid-ptr958>3.0.co;2-2.

Hocking GM (1958) Pakistan medicinal plants I. *Qual Plant Mater Veg* 5:145–153. doi: 10.1007/BF01099867.

Hussain Z, Thu HE, Shuid AN et al. (2017) Phytotherapeutic potential of natural herbal medicines for the treatment of mild-to-severe atopic dermatitis: A review of human clinical studies. *Biomed Pharmacother* 93:596–608. doi: 10.1016/j.biopha.2017.06.087.

IFAD (2015) Economic transformation initiative Gilgit-Baltistan.

Ingty T, Bawa KS (2012) Climate change and indigenous people. In: Arrawatia ML, Tambe S (eds.) *Climate change in Sikkim: Patterns, impacts and initiatives*. Information and Public Relations Department, Government of Sikkim, p. 424.

Ishaq S, Khan MZ, Begum F et al. (2015) Climate change impact on mountain biodiversity: A special reference to Gilgit-Baltistan of Pakistan. *J Mt Area Res* 1:53–63.

Kassam K, Karamkhudoeva M, Ruelle M, Baumflek M (2010) Medicinal plant use and health sovereignty: Findings from the Tajik and Afghan Pamirs. *Hum Ecol* 38:817–829. doi: 10.1007/s10745-010-9356-9.

Kayani S, Ahmad M, Zafar M et al. (2014) Ethnobotanical uses of medicinal plants for respiratory disorders among the inhabitants of Gallies – Abbottabad, Northern Pakistan. *J Ethnopharmacol* 156:47–60. doi: 10.1016/j.jep.2014.08.005.

Kayani S, Ahmad M, Sultana S et al. (2015) Ethnobotany of medicinal plants among the communities of Alpine and sub-Alpine regions of Pakistan. *J Ethnopharmacol* 164:186–202. doi: 10.1016/j.jep.2015.02.004.

Khalil AT, Shinwari ZK, Qaiser M, Marwat KB (2014) Phyto-therapeutic claims about Euphorbeaceous plants belonging to Pakistan; an ethnomedicinal review. *Pak J Bot* 46:1137–1144.

Khan Shinwari Z, Watanabe T, Ali M, Anwar R (2005) Medicinal plants in 21st century. In: *International Symposium Medicinal Plants: Linkages Beyond National Boundaries*. Kohat University of Science and Technology, Kohat, Pakistan, pp 12–17.

Khan SW, Khatoon S (2007) Ethnobotanical studies on useful trees and shrubs of Haramosh and Bugrote valleys In Gilgit Notheren areas of Pakistan. *Pak J Bot* 39:699–710.

Khan SM, Page S, Ahmad H et al. (2013a) Medicinal flora and ethnoecological knowledge in the Naran Valley, Western Himalaya, Pakistan. *J Ethnobiol Ethnomed* 9:4. doi: 10.1186/1746-4269-9-4.

Khan T, Khan IA, Rehman A et al. (2013b) Exploration of near-extinct folk wisdom on medicinally important plants from Shinaki Valley Hunza, Pakistan. *Int J Biosci* 3:180–186. doi: 10.12692/ijb/3.10.180-186.

Khan SM, Page S, Ahmad H, Harper D (2014a) Ethno-ecological importance of plant biodiversity in mountain ecosystems with special emphasis on indicator species of a Himalayan Valley in the northern Pakistan. *Ecol Indic* 37:175–185. doi: 10.1016/j.ecolind.2013.09.012.

Khan T, Khan IA, Rehman A et al. (2014b) Evaluation of human induced threats on berberis populations across cultural landscape of Karakorum mountain ranges. *J Biodivers Environ Sci* 5:333–342.

Khan T, Khan IA, Rehman A (2014c) Investigations into ecological consequences and threats from ethnoecological and ethnobotanical practices across Karakorum Mountain Ranges: A case study Berberis. *Curr World Environ* 9:713–720. doi: 10.12944/CWE.9.3.20.

Khan T, Khan IA, Rehman A et al. (2014d) Investigation into Ethnopharmacological perceptions and practices among traditional communities of Karakorum Mountain Ranges: special reference to Berberis species. *Asian J Biol Life Sci* 3:173–178.

Khan T, Khan IA, Rehman A (2014e) A review on Berberis species reported from Gilgit-Baltistan and Central Karakoram National Park, *Pak J Med plants Stud* 2:16–20.

Khan T, Khan IA, Rehman A et al. (2014f) Studies on population change dynamics of critically endanger Berberis species in Karakoram Mountain Ranges: An ethnoecological perspective. *Asian J Biol Life Sci* 3:206–211.

Khan T, Khan IA, Rehman A, Bibi Z (2014g) Determination of effectiveness of berberine, a characteristic phytochemical of Berberis species, against human proteome using in-silico analysis. *J Biodivers Environ Sci* 4:53–63.

Kimmerer RW (2006) Weaving traditional ecological knowledge into biological education: A call to action. *Bioscience* 52:432. doi: 10.1641/0006-3568(2002)052[0432:wtekib]2.0.co;2.

Kodirekkala KR (2017) Internal and external factors affecting loss of traditional knowledge: Evidence from a horticultural society in South India. *J Anthropol Res* 73:22–42. doi: 10.1086/690524.

Kreutzmann H (1993) Challenge and response in the Karakoram: Socioeconomic transformation in Hunza, Northern Areas, Pakistan. *Mt Res Dev* 13:19–39.

Leto C, Tuttolomondo T, La Bella S, Licata M (2013) Ethnobotanical study in the Madonie Regional Park (Central Sicily, Italy) - Medicinal use of wild shrub and herbaceous plant species. *J Ethnopharmacol* 146:90–112. doi: 10.1016/j.jep.2012.11.042.

Liniger H, Schwilch G (2002) Enhanced decision-making based on local knowledge. *Mt Res Dev* 22:14–18. doi: 10.1659/0276-4741(2002)022[0014:EDMBOL]2.0.CO;2.

Ludwig D (2016) Overlapping ontologies and Indigenous knowledge. From integration to ontological self-determination. *Stud Hist Philos Sci Part A* 59:36–45. doi: 10.1016/j.shpsa.2016.06.002.

Lulekal E, Asfaw Z, Kelbessa E, Van Damme P (2013) Ethnomedicinal study of plants used for human ailments in Ankober District, North Shewa Zone, Amhara Region, Ethiopia. *J Ethnobiol Ethnomed* 9:63. doi: 10.1186/1746-4269-9-63.

Lynch AJJ, Fell DG, McIntyre-Tamwoy S (2010) Incorporating indigenous values with 'Western' conservation values in sustainable biodiversity management. *Australas J Environ Manag* 17:244–255. doi: 10.1080/14486563.2010.9725272.

Maikhuri RK, Rao KS, Saxena KG (2004) Bioprospecting of wild edibles for rural development in the Central Himalayan Mountains of India. *Mt Res Dev* 24:110–113. doi: 10.1659/0276-4741(2004)024[0110:BOWEFR]2.0.CO;2.

Makkar HPS, Francis G, Becker K (2007) Bioactivity of phytochemicals in some lesser-known plants and their effects and potential applications in livestock and aquaculture production systems. *Animal* 1:1371–1391.

Mark JJ (2014) Silk road. https://www.ancient.eu/Silk_Road/. Accessed Mar 27, 2018.

Middleton R (2011) The great game – Myth or reality? *Tajikistan High Pamirs*.

Morgan C (2002) Medicine of the gods: Basic principles of ayurvedic medicine. *Mandrake*.
Muhammad G (1905) Festivals and folklore of Gilgit. *The Asiatic Society*, 57, Park Street, Culcutta.
Muthu C, Ayyanar M, Raja N, Ignacimuthu S (2006) Medicinal plants used by traditional healers in Kancheepuram district of Tamil Nadu, India. *J Ethnobiol Ethnomed* 2:43. doi: 10.1186/1746-4269-2-43.
Nalau J, Becken S, Noakes S, Mackey B (2017) Mapping tourism stakeholders' weather and climate information-seeking behavior in Fiji. *Weather Clim Soc* 9:377–391. doi: 10.1175/WCAS-D-16-0078.1.
Nalau J, Becken S, Schliephack J et al. (2018) The role of Indigenous and Traditional Knowledge in ecosystem-based adaptation: A review of the literature and case studies from the Pacific Islands. *Weather Clim Soc* 10:851–865. doi: 10.1175/WCAS-D-18-0032.1.
Nesheim I, Dhillion SS, Stølen KA (2006) What happens to traditional knowledge and use of natural resources when people migrate? *Hum Ecol* 34:99–131. doi: 10.1007/s10745-005-9004-y.
Platon A (2012) Pakistan ethnic map. https://commons.wikimedia.org/wiki/File:Pakistan_ethnic_map.svg. Accessed April 10, 2018.
Qureshi RA, Ghufran MA, Sultana KN et al. (2006) Ethnobotanical studies of medicinal plants of Gilgit District and surrounding areas. *Ethnobot Res Appl* 5:115–122.
Rhoades RE, Thompson SI (1975) Adaptive strategies in alpine environments: Beyond ecological particularism. *Am Ethnol* 2:535–551. doi: 10.2307/643726.
Salick J, Byg A (eds.) (2007) Indigenous peoples and climate change. *Indigenous peoples and climate change*. Tyndall Centre for Climate Change Research, Oxford, pp 1–32.
Salim MA, Ranjitkar S, Hart R et al. (2019a) Regional trade of medicinal plants has facilitated the retention of traditional knowledge: case study in Gilgit-Baltistan Pakistan. *J Ethnobiol Ethnomed* 15: doi: 10.1186/s13002-018-0281-0.
Salim MA, Ranjitkar S, Hart R et al. (2019b) Regional trade of medicinal plants has facilitated the retention of traditional knowledge: case study in Gilgit Baltistan Pakistan. *J Ethnobiol Ethnomed* 15:1–33.
Shedayi A, Gulshan B (2012) Ethnomedicinal uses of plant resources in Gilgit-Baltistan of Pakistan. *J Med Plants Res* 6:4540–4549. doi: 10.5897/JMPR12.719.
Shedayi AA, Xu M, Gulraiz B (2014) Traditional medicinal uses of plants in Gilgit-Baltistan, Pakistan. *J Med Plant Res* 8:992–1004. doi: 10.5897/JMPR2014.5461.
Sher H, Aldosari A, Ali A, de Boer HJ (2014) Economic benefits of high value medicinal plants to Pakistani communities: an analysis of current practice and potential. *J Ethnobiol Ethnomed* 10:1–16.
Shinwari ZK, Gilani SS (2003) Sustainable harvest of medicinal plants at Bulashbar Nullah, Astore. *J Ethnopharmacol* 84:289–298.
Soelberg J, Jäger AK (2016) Comparative ethnobotany of the Wakhi agropastoralist and the Kyrgyz nomads of Afghanistan. *J Ethnobiol Ethnomed* 12:24. doi: 10.1186/s13002-015-0063-x.
Steinbauer MJ, Zeidler J (2008) Climate change in the northern areas Pakistan impacts on glaciers, ecology and livelyhoods.
Stevens CJ, Murphy C, Roberts R et al. (2016) Between China and South Asia: A middle Asian corridor of crop dispersal and agricultural innovation in the Bronze Age. *The Holocene* 26:1541–1555. doi: 10.1177/0959683616650268.
Sujakhu NM (2015) *Community vulnerability and adaptive capacity to changing climate and related hazards in two sites in the Asian Highlands*. University of Chinese Academy of Sciences.
Tushar, Basak S, Sarma GC, Rangan L (2010) Ethnomedical uses of Zingiberaceous plants of Northeast India. *J Ethnopharmacol* 132:286–296. doi: 10.1016/j.jep.2010.08.032.
UNESCO (1996) Culture and health orientation texts on the 1996 theme.
Virginia Dimasuay Nazarea (ed.) (1999) *Ethnoecology: Situated knowledge/located lives*. University of Arizona Press.
Wazir SM, Dasti AA, Shah J (2004) Common medicinal plants of Chapursan Valley, Gojal II, Gilgit-Pakistan. *J Res (Science), Bahauddin Zakariya Univ Multan, Pakistan* 15:41–43.
Wikipedia (2019) Gilgit-Baltistan. https://en.wikipedia.org/wiki/Gilgit-Baltistan. Accessed February18, 2019
Winslow LC, Kroll DJ (1998) Herbs as medicines. *Arch Intern Med* 158:2192–2199. doi: 10.1001/archinte.158.20.2192.
Yang H, Ranjitkar S, Zhai D et al. (2019) Role of traditional ecological knowledge and seasonal calendars in the context of climate change: A case study from China. *Sustainability* 11:22. doi: 10.3390/su11123243.

Index